T0329587

Solid-State Sensors

Solid-State Sensors

Ambarish Paul
University of Glasgow
Glasgow, UK

Mitradip Bhattacharjee
Indian Institute of Science Education and Research Bhopal
Bhopal, Madhya Pradesh, India

Ravinder Dahiya
Northeastern University
Boston, MA, USA

IEEE Press Series on Sensors

IEEE PRESS

WILEY

Library of Congress Cataloging-in-Publication Data:

Names: Paul, Ambarish, author. | Bhattacharjee, Mitradip, author. | Dahiya, Ravinder S., author.
Title: Solid-state sensors / Ambarish Paul, Mitradip Bhattacharjee, Ravinder Dahiya.
Description: Hoboken, New Jersey : Wiley-IEEE Press, [2024] | Includes index.
Identifiers: LCCN 2023040077 (print) | LCCN 2023040078 (ebook) | ISBN 9781119473046 (cloth) | ISBN 9781119473053 (adobe pdf) | ISBN 9781119473077 (epub)
Subjects: LCSH: Detectors.
Classification: LCC TA165 .P37 2024 (print) | LCC TA165 (ebook) | DDC 621.3815/36–dc23/eng/20230905
LC record available at https://lccn.loc.gov/2023040077
LC ebook record available at https://lccn.loc.gov/2023040078

Cover Design: Wiley
Cover Image: © Andrew Berezovsky/Shutterstock

Set in 9.5/12.5pt STIXTwoText by Straive, Pondicherry, India

Contents

About the Authors

Ambarish Paul

Ambarish Paul obtained his doctoral degree from the Indian Institute of Technology Kharagpur, India, in 2015. He was awarded the Newton International Fellowship 2016 at the University of Glasgow, United Kingdom, where he worked on flexible and biodegradable sensors for medical applications. His research interest includes medical devices, nanomaterials, functionalized biomaterials, bendable electronics, sensors for environmental monitoring, intelligent systems, and robotics. He has authored nearly 17 research articles in international journals and conferences. He is the recipient of Marie Sklodowska-Curie Seal of Excellence Award by European Union (EU) in 2016 and 2017.

Mitradip Bhattacharjee

Mitradip Bhattacharjee (Senior Member, IEEE) received his PhD degree from the Indian Institute of Technology, Guwahati, India, in 2018. Subsequently, he joined the University of Glasgow, United Kingdom, as a postdoctoral fellow. He is currently an assistant professor with the Electrical Engineering and Computer Science Department, Indian Institute of Science Education and Research (IISER) Bhopal, India, where he is leading the i-Lab Research Group. His research interests include electronic sensors and systems, biomedical engineering, bioelectronics, flexible/printed and wearable electronics, wireless systems, IoT, micro/nanoelectronics, and reconfigurable sensing antenna. He has authored more than 50 research articles in reputed journals/conferences and filed more than 15 national/international patents. He also has authored several book chapters/books till date. He is recipient of several awards such Marie Sklodowska-Curie Seal of Excellence Award by the European Commission in 2019, Nanochallenge Award by PSG in 2018, Gandhiyan Young Technological Innovation Award at Rashtrapati Bhavan in 2016, among others. He served as the chair of IEEE sensors council young professionals in 2022. He is also an editorial team member of IEEE Sensors Alert.

Ravinder Dahiya

Ravinder Dahiya is a professor in the Department of Electrical and Computer Engineering at Northeastern University, Boston, United States. His group (Bendable Electronics and Sustainable Technologies [BEST]) conducts research in flexible printed electronics, electronic skin, and their applications in robotics, wearables, and interactive systems. He has authored or co-authored more than 500 publications, books, and submitted/ granted patents and disclosures. He has led or contributed to many international projects.

Prof. Dahiya is the president of IEEE Sensors Council, IEEE Division X Delegate-Elect/Director-Elect (2024) and editor-in-chief of *NPJ Flexible Electronics*. He launched the *IEEE Journal on Flexible Electronics* and also served as the founding editor-in-chief of this journal. He has been editorial boards of several other leading journals. He founded the IEEE International Conference on Flexible, Printed Sensors and Systems (IEEE FLEPS) and has served as General Chair or Technical Programme Chair of several international conferences such as IEEE Sensors and IEEE FLEPS. He is recipient of EPSRC Fellowship, Marie Curie Fellowship and Japanese Monbusho Fellowship and has received several awards, including Technical Achievement award from IEEE Sensors Council, Young Investigator Award from Elsevier, and 13 best journal/conference paper awards as author/co-author. He is Fellow of IEEE and the Royal Society of Edinburgh.

Preface

The field of solid-state sensors has broadened and expanded significantly in the last 20 years. The solid-state sensors are being used in the development of various devices in medicine, agriculture, industry, transport, environmental control, and other fields. The process of development of new sensors and technologies has been revolutionized with the emergence of printed electronics, flexible substrates, biodegradable materials, nanotechnology, smart materials, integrated devices through IoT, and advanced wireless data transmission systems. To meet the high demands of commercial market for solid-state sensors, researchers are working on the development of affordable sensing devices with low production cost. Further, to make the sensors suitable for wearable applications, it is necessary to develop sensors with small size and low power consumption.

In this book the readers will find detailed description of the design, development, and applications of capacitive, piezoelectric, chemical, optical, and magnetic solid-state sensors. This book takes the readers on a journey from the emergence of solid-state devices through their transition and miniaturization period to the present-day smart and integrated sensors with remote data access facilities. The aim of this book is to provide a detailed and updated assessment of the new trends in the field of solid-state sensors in terms of working principles, classifications, design architecture, fabrication techniques, active sensing materials, and new age applications. Since the microfabrication has reached its zenith of application with the advent of state-of-the-art tools and techniques and its versatility with different materials and processes, a chapter is dedicated for the discussion on these fabrication techniques. Further, different fabrication techniques in the form of solution processed routes and custom printed electronics are also discussed in a chapter. As circuit and system are very important aspects of sensor application, a chapter on interface circuits with different circuit components is included in this book. This chapter on interface circuit will help in understanding different circuit components and system aspect associated with the sensor system design. Each chapter presented in this book is written in lucid language and made interesting and attractive to students with the use of illustrations and schematic diagrams. Each chapter has a list of references to make the information accessible to any reader irrespective of their background.

This book will benefit students studying courses in the field of material sciences, chemical engineering, electrical and electronics, physics, environmental sciences, and relevant disciplines. Moreover, researchers and scientists working in the field of sensor design and fabrication, analytical control, and sensor systems will benefit from the contents of this book. Practicing engineers or project managers, who want to be acquainted with solid-state sensors, can also follow this book for required details and information. This book will also help masters' students and those in the early stage of research to design new devices and select best architecture for enhanced performance aim for specific applications. We believe that this book covers all aspects of solid-state sensors from fundamentals of operation to optimum materials and approaches for achieving better sensor performances.

Ambarish Paul, Mitradip Bhattacharjee, and Ravinder Dahiya

1

Introduction

1.1   Overview

A sensor is a device which detects or measures a physical, chemical, optical, electrical, and magnetic properties relating to temperature, pressure, strain, molecular finger prints, absorbance, and subsequently records, indicates, or responds to it. The term "solid state" became popular in the beginning of the semiconductor era in the 1960s to distinguish this new technology based on semiconducting materials such as silicon and its doped derivatives, in which the electronic action of such devices occurred within the material, which is in contrast with the vacuum tubes where electronic action occurred in a gaseous state. The solid-state sensor responds to external variations and induces change in properties of the semiconducting material, thereby producing variations in electric current through the semiconducting material. The measure of this electric current is a manifestation of the amount of external variation also called the measurand. However, nowadays with the advent of new materials, technologies, and smart sensor architectures and scopes, the term "solid state" is also used for devices in which the devices have no moving parts. Again, solid-state sensors must not be confused with transducers or actuators which react depending on the sensor response. A solid-state sensor is designed in such a way that the measurand, the physical property to be sensed, exploits a physical phenomenon within the sensor structure. This physical phenomenon if found in traces or is weak may not produce measurable amount of variation in electrical response that can lead to nondetection and poor sensor performance. Such electric signal must be detected by magnifying the weak signals with suitable electronics and signal processing stage. Thus, solid-state sensor with integrated electronics along with other intelligent computational processes is in high commercial demand.

However, the conventional "solid-state" semiconductor-based sensors have transformed largely over the years. With the discovery of new tools and

Solid-State Sensors, First Edition. Ambarish Paul, Mitradip Bhattacharjee, and Ravinder Dahiya.
© 2024 The Institute of Electrical and Electronics Engineers, Inc.
Published 2024 by John Wiley & Sons, Inc.

technologies and the exploration of new materials and synthesis techniques the solid-state sensors have evolved as one of the blooming areas of research because of its promise and the potential to redefine itself. Researchers have developed sensors using metal oxides, ceramics, nanomaterials, polymers, and biomaterials as the active sensing materials on disposable [1], bendable, and ultrathin transparent polymer substrates with no movable parts for different applications [1–8]. Such sensors with no movable parts and capable of generating electronics signals in response to the measurand are considered within the scope of this book. This book discusses the promise, benefits, fabrication techniques, sensing material commonly used, sensor architecture and its role on the performance, sensing strategies, important applications, and new trends for each type of sensors including capacitive, piezoelectric, piezoresistive, optical, chemical, and magnetic. The aim of the book is not only to educate the readers on the scopes and potential of each types of sensors, but also cultivate interest and encourage them to explore new dimensions of multivariant sensing.

1.1.1 Growth in Solid-State Sensor Market

The late twentieth century has witnessed a burst of technological advances in the field of solid-state sensors [9–11]. Industry report suggests that sensors especially solid-state sensors market is expanding its horizon every year [12]. In the present context, this expansion is such that the solid-state sensor market is comparable to leading markets like that of the computers and the communication market. This is because the solid-state sensors are being widely used in every aspect of our lives such as of smartphones, other electronic gadgets, automobiles, security systems, and even everyday objects like coffee makers, sanitizer dispensers, blood pressure monitors, and glucometers. In addition to consumer electronics, these sensors find application in Internet of things (IoT), medical, nuclear, defense, aviation, robotics, artificial intelligence (AI), environment, agriculture, and in geophysical and oceanographic explorations.

The requirement for the solid-state sensing devices have increased considerably over the years (Figure 1.1a) and is also expected to grow subsequently during the decade as predicted in Figure 1.1a, b. This is due to the increasing demand of solid-state sensors in wide cross-section of industrial applications. The demand for cost-efficient, reliable, and high-performance solid-state sensor market is also driving the market growth. There is increased use of solid-state sensors in multiutility devices such as watches and smartphones integrated with sensors, which in turn are assisting the development of solid-state sensor industry. The segmentation of the solid-state sensor market is performed based on classification and application. By classification, the market is divided into varied type of sensors and the respective market revenue per year is shown in Table 1.1. Based on application, the market is divided into automotive, oil and gas, consumer electronics, medical, health care, and others. In addition, the pressure and temperature sensor market were valued at $7.21 and $4.99 billion in 2016 and is projected to reach $12.07 and 6.86 billion by 2024, growing with a CAGR of 6.7% and 4.5%, respectively, during the forecast period of 2017–2024. The piezoresistive sensor market generated the highest revenue share in the global pressure sensor market. With the emergence of advance technologies for different gas sensing, the market is expected to be valued at $1.4 billion by 2024 at a CAGR of 6.86% in the forecasting period of 2017–2024. The rising demands of oxygen gas in the healthcare sector such as in the anesthesia machines, ventilators, oxygen monitors, and in automotive and transportation application are driving the oxygen gas sensor market. The consumer electronics market is expected to grow at the highest rate as the gas sensors are believed to be integrated with smartphones and other wearable devices to detect alcohols, carbon dioxide, carbon monoxide, and nitrogen dioxide. The optical sensing market is estimated to grow from $1.12 billion dollars in 2016 to $3.46 billion by 2023 at a CAGR of 15.47% between 2017 and 2023. The demand for optical sensors is attributed to its accuracy and the ability to withstand harsh environments in aerospace, defense sectors, and oil and gas industries. The increasing investments in the R&D activities on optical sensors drive the growth of the market. The magnetic sensors market is expected to reach $4.16 billion by 2022 at a CAGR of $4.16 between 2016 and 2022.

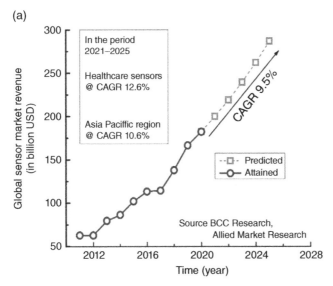

Figure 1.1 (a) Global sensor market revenue from 2010 to 2020 and predicted growth with CAGR of 9.5% for 2021–2025. (b) Global sensor market trends for different devices as predicted by IDTechEx. *Source:* (a) Adapted from BCC Research, Allied Market Research.

(b)

Sensor market trends in 2020–2030

Source:
IDTechEx Research Reports 2020–2030,

Global sensors market will be around $250 billion by 2041

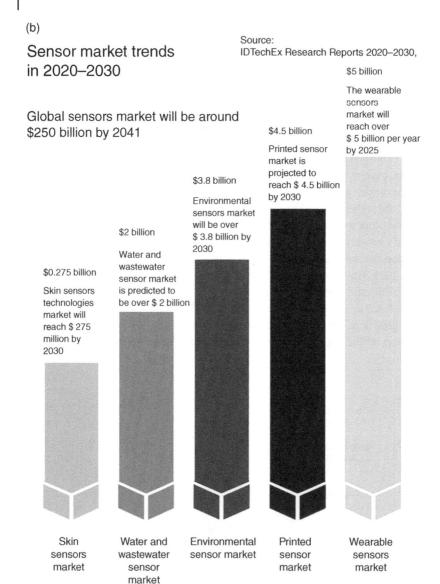

$0.275 billion

Skin sensors technologies market will reach $ 275 million by 2030

$2 billion

Water and wastewater sensor market is predicted to be over $ 2 billion

$3.8 billion

Environmental sensors market will be over $ 3.8 billion by 2030

$4.5 billion

Printed sensor market is projected to reach $ 4.5 billion by 2030

$5 billion

The wearable sensors market will reach over $ 5 billion per year by 2025

| Skin sensors market | Water and wastewater sensor market | Environmental sensor market | Printed sensor market | Wearable sensors market |

Figure 1.1 (Continued)

Table 1.1 Various sectors of solid-state sensor market and their respective predicted market revenue per year.

Materials/device/sensor market	Prediction until year	Predicted revenue per year (in billions)
Graphene market	2031	$0.7
Fluoropolymer/electronics market	2041	$14
Printed electronic materials market	2031	$6.9
3D printing materials	2030	$18.4
Wearable sensors market	2025	$5
CNT market	2030	$0.750
3D electronics	2030	$3
Advanced photovoltaics market beyond silicon	2040	$40
Sensors market	2041	$250
Printed sensor market	2030	$4.5
Skin sensors technologies	2030	$0.275
Wearable technology	2019 attained	$70
Environmental sensors market	2030	$3.8
Flexible electronics in health care	2030	$8.3
Flexible hybrid electronics	2030	$3
Biosafe polymer market	2030	New market
Harvesting for roads and sensing for IoT and healthcare market for piezo transducers	2030	$1
Piezoelectric-based sensing market in health care beyond imaging	2029	$1.04
RFID sensor tags and systems	2028	$0.904
Robotic sensing	2027	$16.1
Water and wastewater networks sensor market	2030	$2
Solid-state batteries	2030	$6
Wearable technology for animals market	2030	$22

1.1.2 Solid-State Sensors: A Recipe for Smart Sensing Systems

The rapid growth solid-state sensor market can be attributed to the development of computer-controlled processes and remote monitoring of industrial process in real time. With the advent of microfabrication and miniaturization of devices, system-on-chip (SOC) sensors have drawn significant attention from the researchers from academia and industries. Currently due to the growing demands of on-chip integrated devices, the class of solid-state sensors has surpassed the traditional sensing techniques such as the electrochemical and the chromatographic methods in terms of scopes, benefits, and reliability. The development of solid-state sensors not only promises integrated multiple sensors on a common chip, but also shows potential in closer interfacing with the computers for remote monitoring of various processes which aid computerized manufacturing and process control. There are many types of solid-state sensors, which are in everyday use or in the different stages of development. They encompass a wide realm of sensing technologies including the chemical sensors (gas and biosensors), physical sensors (e.g. strain, pressure, temperature), acoustic sensors (e.g. bulk and surface acoustic wave devices), optical sensors (optical waveguide, infrared detectors), thermochemistry (e.g. microcalorie and microenthalpy sensors), and magnetic sensors. The major advantages of the solid-state sensors lie in the simplicity, portability, microfabricability, and reliability. The simplicity in operation of the solid-state devices relates to the noninvolvement of complex equipment and skilled operators for sensing. The miniaturization compactness and feasibility for on-chip integration lends portability to solid-state sensors. The solid-state sensors are compliant to batch and planar fabrication techniques, which reduces the cost of fabrication. The solid-state sensors overcome the problems of inconsistent liquid-phase sensing processes, which make them reliable for long-term use.

1.2 Evolution of Solid-State Sensors

1.2.1 Origin and Early Developments in Detection Devices

The need for devices arises from its necessity in public welfare or from the demand in commercial market. Originated in necessity and demand, and conceptualized through suitable understanding, the evolution of sensors went through several scientific challenges and technological advancements, which sometimes spanned over many centuries. The ruler or the measuring rod can be considered as the first measuring tool which was reported to be used in the Indus Valley civilization in 2650 BCE [13]. However, an astronomical device and an inclinometer called the Astrolabe (Figure 1.2a), invented by Apollonius of Perga around 200 BCE, can be regarded as the first device to be invented in the history of mankind [14, 15]. The seismometer, invented by Zhang Heng of the Han dynasty in 132 CE [16], and the Mariner's compass, invented in Southern India in 400 CE [17], were a few of the early detection equipment that were developed in the Medieval era.

The evolution of transducer/sensor took place in diverse realms of fields, including those associated with health care originated in the same era but underwent

(a)   (b)   (c)

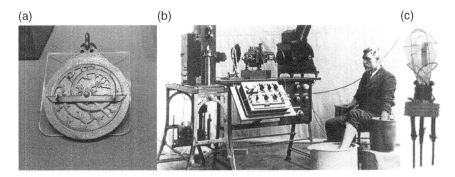

Figure 1.2 (a) The front face of the ancient Astrolabe. (b) Einthoven with his EKG machine. (c) The first vacuum tube invented in 1904 by John Ambrose Fleming. *Sources:* (a) Mustafa-trit20/Wikimedia Commons/CC BY-SA 4.0. (b) Informa UK Limited.

massive transformation through the modern era and continuing in the electronic era (post 1895). One such device which has been through the journey of evolution from its medieval form to its present form is the modern-day pulse meter. The modern-day pulse meter supported with wireless features and other physiological sensors and regulated through AI can be considered as a unique example of scientific, technological, and engineering device evolution. The authors have traced back the evolution of pulse meters since its inception when people relied on qualitative diagnosis to a more complex and detailed informative device in its present form. That pulse diagnosis directly related to human health was realized in 350 BCE by Chinese physician Bian Que (401–310 BCE). Later in 305 BCE, the Praxagoras of Cos (340–??[1] BCE) [18], the teacher of Herophilus of Chalcedon (335–280 BCE), became the first Greek physician to feel the pulse in order to observe the health of the ill. Having understood the importance of pulse diagnosis, the ancient researchers felt the need for a device that would help them to count the heart beat and investigate the pulse rate in unhealthy humans. In 290 BCE, Greek physician Herophilus of Chalcedon (335–280 BCE) devised a water clock to count the human pulse to analyze the rhythm and heartbeat, though it was not considered as a complete device [19]. Although the ancient inventions and breakthroughs are not well documented, but are reported in scriptures. A Polish Professor Joseph Struthius (1510–1568) was the first to present a graphic image of the pulse in 1555 and

---

1 There are no historical details about his personal life. It is believed that he suffered the ravages of the terrible fire at the Library of Alexandria or, in a speculative sense, he might have been persecuted for his probable forbidden practices of dissections on human cadavers. Therefore, what little we know about him comes from his colleagues and pupils, such as Galen, Herophilus, or Erasistratus.

introduced the concept of a device that could mechanically register the pulse [20]. It was after the Italian physician, mathematician, and philosopher Galilei (1564–1642) in 1601 who correlated his own pulse with the pendulum movements of a clock, an Italian Professor of Medicine Santorio Santorio (1561–1636) in the year 1602 successfully counted the pulse using this pendulum [21]. This was the first pulse meter. Incidentally, Galilei and Santorio invented the thermoscope and the clinical thermometer in 1593 and 1625, respectively [22]. In 1707, an English physician Sir John Floyer (1649–1734) invented the "one minute pulse watch" for a correct pulse measurement. Floyer was one of the first to measure the pulse in daily practice and used the device to obtain accurate pulse rate measurements [23]. The devices invented in the modern era were mostly mechanical devices with moving parts and was only considered as mechanical devices.

A more advanced electronic record of heartbeat was obtained in 1872 by Alexander Muirhead when he attached wires on the wrist of the patient. Augustus Waller invented an ECG machine in 1887, however the more practical and sensitive one was invented by Einthoven in 1901 [24] (Figure 1.2b). In 1924, Einthoven was awarded the Nobel Prize in Medicine for his pioneering work in developing the ECG. The electronic era commenced with the discovery of wireless transmission in 1895 by Russian scientists [25, 26] and riding on the advancements in electronics and miniaturization, wireless ECG-based heart rate monitor was invented in 1977 by Polar Electro as a training tool for the Finnish National Cross Country Ski Team [27]. In 2005, Textronics, Inc., introduced the first garment with integrated heart sensors in the form of a sports bra [28]. Special materials in the sports bra sense heart rate from the body and transmit it to a wrist receiver. Recently in 2020, researchers have developed smart healthcare system in IoT environment that can monitor a patient's basic health signs in real-time using integrated heart beat sensor, body temperature sensor, room temperature sensor, CO sensor, and CO_2 sensor in a sensor system [29].

Earlier the invention of thermionic emission in 1873 marked the beginning of a new era which revolutionized the scope of detection devices. Thomas Edison's discovery of electric current in 1883 was a huge leap in mankind and laid the foundation to the present-day electronics (Figure 1.3). Following the discovery of unilateral conduction across the junction of a semiconductor (Galena crystal) and a metal by KF Braun in 1974, JC Bose was the first to use this crystal semiconductor for detecting radiowaves in 1894 [30]. This concept of rectifying contact between metal semiconductor junctions was used by Braun and GW Pickard to develop radiowave detector (commonly called Cat whisker detector) in which they replaced Galena crystal with silicon [31]. The Cat whisker detector was the first solid-state electronic detection equipment that was patented in 1906. In 1916 and 1917, Paul Langevin and Chilowsky developed the first ultrasonic submarine detector using an electrostatic method (singing condenser) [32]. The amount of

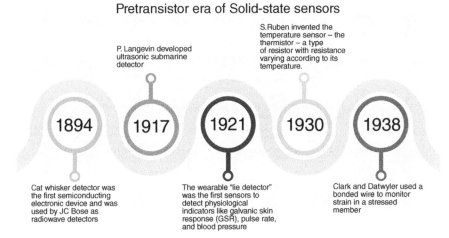

Figure 1.3 Timeline of key inventions of solid-state devices in the electronic era.

time taken by the signal to travel to the enemy submarine and echo back to the ship on which the device was mounted was used to calculate the distance under water. The first wearable solid-state detection equipment was the polygraph which was invented in 1921 by a medical student named John Augustus Larson and worked on the Galvanic response of the skin [33]. Subsequent inventions during the modern age continued in different field of research until the invention of the transistor in 1947 which boosted the field of solid-state electronics and sensing systems.

1.2.2 Solid-State Electronics: Post Transistor Era

Vacuum tubes, invented in 1904 by John Ambrose Fleming (Figure 1.2c), formed the basic electronic components for use in television, radar, telephone, and industrial process control applications [34]. The scientific community turned toward solid-state electronic materials as alternative solutions to overcome the challenges of vacuum tube-based devices that were widely used as amplifiers and rectifiers prior to 1940s. The invention of semiconductor devices in the late 1950s led to the development of solid-state devices which are smaller, efficient, reliable, and cost-effective than the vacuum tube-based devices (Figure 1.4). Furthermore, the solid-state devices were associated with low heat dissipation and fast response. The solid-state devices work on the principle of electronic conduction in the material in contrary to the vacuum tube devices which operate through thermionic

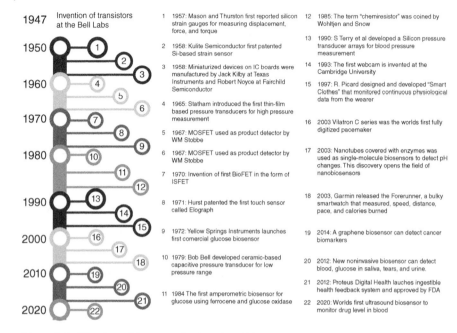

1947 Invention of transistors at the Bell Labs

1950
1
2
3

1960
4
5
6

1970
7
8
9

1980
10
11
12

1990
13
14
15

2000
16
17

2010
18
19
20
21

2020
22

1 1957: Mason and Thurston first reported silicon strain gauges for measuring displacement, force, and torque

2 1958: Kulite Semiconductor first patented Si-based strain sensor

3 1958: Miniaturized devices on IC boards were manufactured by Jack Kilby at Texas Instruments and Robert Noyce at Fairchild Semiconductor

4 1965: Statham introduced the first thin-film based pressure transducers for high pressure measurement

5 1967: MOSFET used as product detector by WM Stobbe

6 1967: MOSFET used as product detector by WM Stobbe

7 1970: Invention of first BioFET in the form of ISFET

8 1971: Hurst patented the first touch sensor called Elograph

9 1972: Yellow Springs Instruments launches first comercial glucose biosensor

10 1979: Bob Bell developed ceramic-based capacitive pressure transducer for low pressure range

11 1984 The first amperometric biosensor for glucose using ferrocene and glucose oxidase

12 1985: The term "chemiresistor" was coined by Wohltjen and Snow

13 1990: S Terry et al developed a Silicon pressure transducer arrays for blood pressure measurement

14 1993: The first webcam is invented at the Cambridge University

15 1997: R. Picard designed and developed "Smart Clothes" that monitored continuous physiological data from the wearer

16 2003 Vilatron C series was the worlds first fully digitized pacemaker

17 2003: Nanotubes covered with enzymes was used as single-molecule biosensors to detect pH changes. This discovery opens the field of nanobiosensors

18 2003, Garmin released the Forerunner, a bulky smartwatch that measured, speed, distance, pace, and calories burned

19 2014: A graphene biosensor can detect cancer biomarkers

20 2012: New noninvasive biosensor can detect blood, glucose in saliva, tears, and urine.

21 2012: Proteus Digital Health lauches ingestible health feedback system and approved by FDA

22 2020: Worlds first ultrasound biosensor to monitor drug level in blood

Figure 1.4 Timeline of key contributions in solid-state sensors.

emission. With the advent of microfabrication techniques of devices, the demand for solid-state (semiconductor) devices increased manifold due to their batch fabricability which cuts down processing cost. Thus, the rise of semiconducting materials and solid-state devices provided a new dimension to the field of electronics and associated devices. The invention of transistors by American Physicists John Bardeen, Walter Brattain, and William Shockley in 1947 at the Bell Labs virtually redefined the modern-day electronics [35, 36]. This transistor (Figure 1.5a) was based on point contact configuration and paved the way for cheaper radios, televisions, calculators, computers, and other devices. Bardeen, Walter, and Shockley (Figure 1.5b) were awarded with the Nobel Prize in Physics in the year 1956 for their study on semiconductors and the discovery of the transistor effect, and later in the year 2009 this feat was acknowledged on the list of IEEE milestones in electronics. Shockley continued his studies on semiconductor and along with Gordon Teal and Morgan Sparks successfully demonstrated the working of N-P-N bipolar transistor in the year 1950 (Figure 1.5c). The striking advantage of transistors lies in the potential use as amplifiers and switches. Due to the small size and energy efficiency of the transistors, it was widely used in the design of complex electronic circuits with high switching speeds and energy efficiency.

Figure 1.5 The foundation and pillars of modern-day solid-state electronics. (a) The first transistor of the world, born in Bell Labs in 1947, invented by (b) the American Physicists John Bardeen (L), Walter Brattain, and William Shockley (R). (c) Sparks N-P-N bipolar transistor of 1950. (d) Jack Kilby with his invented first pocket calculator. (e) Robert Noyce with his invented monolithic IC-based motherboard. (f) The first monolithic (2D) integrated circuit developed at Fairchild Semiconductor in 1959 by R. Noyce. Kilby and Noyce are recognized as the inventors of the microchip. Kilby received the Nobel Prize in year 2000 for his contribution to integrated circuit. *Sources:* (b) AT&T/Wikipedia Commons/public domain. (d) The Library of Congress. (e) Intel Free Press/Wikipedia Commons/CC BY-SA 2.0. (f) Fairchild/Courtesy of the Computer History Museum.

The need for the miniaturization of electrical components arose when they were required to be assembled on a single chip. Miniaturization further enhances the energy efficiency of the circuit and reduces the time lag in response. Miniaturized devices on IC boards were manufactured by Jack Kilby at Texas Instruments, and Robert Noyce at Fairchild Semiconductor in 1958 and 1959, respectively. Researchers were successful in fabricating not only transistors but also resistors, capacitors, and diodes on a single monolithic layout to form an electronic chip. Kilby used germanium and Noyce used silicon for the semiconductor material. In 1959, both parties applied for patents. Jack Kilby (Figure 1.5d) and Texas Instruments received US patent #3,138,743 for miniaturized electronic circuits [37]. Robert Noyce (Figure 1.5e) and the Fairchild Semiconductor Corporation received US patent #2,981,877 for a silicon-based integrated circuit [38]. In 1961, the first commercially available integrated circuits (Figure 1.5f) came from the Fairchild Semiconductor Corporation. All computers then started to be made using chips instead of the individual transistors and their accompanying parts. Jack Kilby holds patents on over 60 inventions and is also well-known as the inventor of the portable calculator (1967) [39, 40] (Figure 1.5d).

With the advent of miniaturization and advanced microfabrication technique, researchers designed new devices aimed at different applications. As evident from published literature, the researchers have extensively worked on different devices and sensors and classified them in accordance with the transduction principle used in the device. Although the terms "sensor" and "transducer" are often used as synonyms, the American National Standards Institute (ANSI) standard in 1975 preferred the latter over the former [41]. However, the scientific community has adopted the term "sensor" to define *a device which provides a usable output in response to a specific measurand* and thus is commonly used in scientific articles. The output of a sensor may be any form of energy. Many early sensors converted (by transduction) a physical measurand to mechanical energy; for example, pneumatic energy was used for fluid controls and mechanical energy for kinematic control. However, the introduction of solid-state electronics created new avenues for the development of sensor systems aided by computer-based controls, archiving/recording, and visual display. The modern-day sensor system is often associated with algorithms, AI, and other techniques which require electrical interfacing with microchips and computer-aided controls, thereby broadening the definition of a sensor to include the systems interface and signal conditioning features that form an integral part of the sensing system. With progress in optical computing and information processing, a new class of sensors, which involve the transduction of energy into an optical form, is likely. Thus, the definition of a "sensor" is likely to continue evolving with time.

1.2.3 Emergence of New Technologies

The evolution of solid-state electronics and its contribution in the field of solid-state sensors is largely attributed to the emergence of new technologies which facilitated the transformation of different aspects of solid-state sensors since its origin. Emergence of new technologies and their advancements over the years (Figure 1.6) have led to technological convergence of their key features toward the evolution and transformation of solid-state electronics that have forced scientists to setup new benchmarks in defining solid-state sensors. This technological convergence has shifted the paradigm of research in solid-state sensors from semiconductor electronics to a more open-ended approach. In this section, we will discuss about the key technologies that helped transform solid-state research. The key new technologies with their evolutionary timeline are represented in Figure 1.6.

(a)

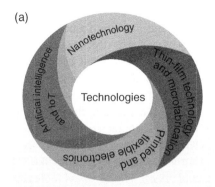

(b)

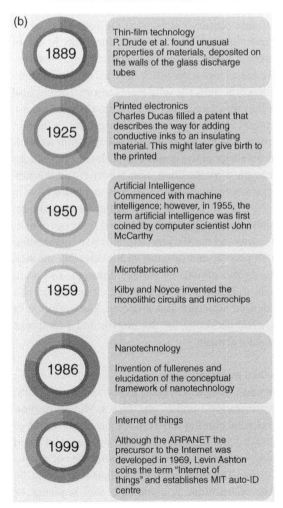

1889 — Thin-film technology
P. Drude et al. found unusual properties of materials, deposited on the walls of the glass discharge tubes

1925 — Printed electronics
Charles Ducas filled a patent that describes the way for adding conductive inks to an insulating material. This might later give birth to the printed

1950 — Artificial Intelligence
Commenced with machine intelligence; however, in 1955, the term artificial intelligence was first coined by computer scientist John McCarthy

1959 — Microfabrication
Kilby and Noyce invented the monolithic circuits and microchips

1986 — Nanotechnology
Invention of fullerenes and elucidation of the conceptual framework of nanotechnology

1999 — Internet of things
Although the ARPANET the precursor to the Internet was developed in 1969, Levin Ashton coins the term "Internet of things" and establishes MIT auto-ID centre

Figure 1.6 (a) Emerging technologies and their (b) timeline of events, which lead to the evolution of solid-state sensors.

1.2.3.1 Thin-Film Technology

The thin-film technology evolved at the end of the nineteenth century when Drude et al. [42] in 1889 found unusual properties of materials deposited on the walls of the glass discharge tubes, which were different from their bulk counterpart. Over the years, thin films of different materials deposited by wide range of techniques including physical (1940s) and chemical (1970s), vapor deposition, atomic layer deposition, molecular beam epitaxy (1980s), spin coating, dip coating, sol gel, and Langmuir Blodgett method have been well characterized and understood by electron diffraction method (1927), electron microscopy (1940), surface analytical methods (1960s), atomic resolution surface imaging techniques like STM and AFM (1980), and ultrahigh-resolution TEM (1990). The advent of more advanced thin-film deposition and characterization techniques such as UHV TEM, fast STM, and synchrotron in early twenty-first century has boosted the solid-state sensor industry due to the thickness controllability, thus providing tunability in sensor performance. Thin films have taken a dominant role in revolutionizing the development of new active and passive sensor elements which have led to a drastic metamorphosis of the electronic devices especially the solid-state sensors. The physical and chemical properties of the thin film largely depend on the material, microscopic structure, and the parameters (e.g. temperature and pressure) utilized to generate the desired microstructure. The microstructure in the thin-film technology relates to the phase state, morphology of the grains and surfaces, orientations of the crystals planes, texture, chemical composition, homogeneity, and the substrate and thin-film interface. The key feature of the thin-film technology lies in the evolution of self-organized structure by atom by atom adding process at temperatures far from dynamic equilibrium which allows controlled synthesis of metastable phases and artificial structures and multilayers [43]. This self-organization in thin film is a thermal activated process and initiates with nucleation through migration of atoms, followed by crystal growth by surface diffusion, and grain growth and restructuring by bulk diffusion [44]. However, contaminations can produce adverse effect on the desired structure and property. The evolution of the thin-film technology is truly the backbone for the growth and widespread acceptance of modern-day solid-state sensors.

1.2.3.2 Advancements in Micro- and Nanofabrication

The inventions of the transistors and integrated circuit in the 1940s and 1950s, respectively, gave rise to a miniaturization trend in electronics. Miniaturization through microfabrication and micromachining has revolutionized the world of solid-state sensors. These microfabrication processes have evolved and been modified into advanced technologies that are pushing the boundaries of resolution, feature sizes, and aspect ratios. The term microfabrication was coined by the semiconductor industry. Microfabrication is a multiple step sequence of photolithographic and chemical processing which are used to make microscopic devices and electronic circuits on silicon, polymer, and other biomaterials substrates (rigid or flexible).

With the improvement in microfabrication techniques, smaller and more complex integrated circuits were built, which facilitated the dense packing of electronic components on a given area of the microchip. Due to the extremely small dimension of the structures certain high-tech tools must be used when performing microfabrication work. The major concepts and principles of microfabrication are photolithography, doping, deposition, etching, bonding, and polishing. A typical microdevice is fabricated by a sequence of microfabrication steps including micromachining, deposition, followed by patterning the film with various microfeatures using photolithography and subsequently etching parts of the undesired area. These processes are detailed in Chapter 2. Industrial microfabrication process is a cumbersome sequence of events where a typical memory chip requires approximately 30 lithography, 10 oxidation, 20 etching, and 10 doping steps to fabricate. Microfabrication is used in the development of integrated circuits on a chip, most of the miniaturized sensors including the semiconductors, microfluidic devices, solid-state sensors, MEMS, and BioMEMS.

Thin film plays a key role in microfabrication where microchips are typically created using multiple thin films. For electronic devices, the patterned films contain conductive metals that allow for the flow of electricity, while for optical devices, thin films are in the form of reflective or transparent films to improve visibility and clarity. The thin film may also be chemical and biological materials in the form of active sensing or encapsulation material often found in medical devices and gas sensors. Dielectric thin films of polymers are used in capacitive sensors and gate dielectric as the insulation material.

The process of microfabrication is not only limited to the deposition patterning and etching of polymers and semiconducting materials, but also offers a way to produce homogeneous monodisperse particles that are not only spherical, but having controlled or asymmetrical shapes and architectures with a specific size to be fabricated, which is not possible with other methods [45, 46]. Using microfabrication techniques such as particle replication in nonwetting templates [47], microcontact hot printing [48], step and flash imprint lithography [49], and hydrogel template [50] achieve such feat. Microfabrication techniques are used to generate patterns of cells on surfaces. This cellular patterning is a necessary component for cell-based biosensors, cell culture analogs, tissue engineering, and fundamental studies of cell biology [51]. Moreover, alternative techniques, such as microcontact printing [52, 53], microfluidic patterning using microchannels [54, 55], and laminar flow patterning [56], have been developed for use in biological applications. Microfabrication is even used in advanced manufacturing processes for engineered neural prosthetics where high-resolution neuroelectrodes are fabricated with the aim of reducing the size of the electronic component. This is achieved by 2D printing of neural arrays or fabricating topographical and biochemical features on the surface of engineered neuroelectrode [57].

1.2.3.3 Emergence of Nanotechnology

The emergence and subsequent evolution of nanomaterials have revolutionized research in the field of materials science. The convergence of nanomaterial and several microfabrication techniques have led to diverse sensors with different applications [58–63]. The reduction of particle size and tuning the particle morphology of nanomaterials from micro to nano size has led to unique properties with versatile applications. The reason for the nanomaterials to exhibit unique/ enhanced properties is due to the large surface-to-volume ratio and quantum confinement effect. The word nanoscience refers to the study, manipulation, and engineering of matter, particles, and structures on the nanometer scale (one millionth of a millimeter). Due to the quantum mechanical effects, the properties of materials in nanoscale differ from the properties of the same material in bulk form. Material properties, such as electrical, optical, thermal, and mechanical properties, are governed by the morphology and particle size of the nanomaterial, where properties of materials larger than 100 nm tend to show bulk properties. Nanotechnology is the application of nanoscience leading to the use of new nanomaterials in sensors and devices. Due to tunable properties of nanomaterials, nanotechnology has the capability and promise to provide us with custom-made materials and products with new enhanced properties, new nanoelectronics components, new types of "smart" medicines and sensors, and even interfaces between electronics and biological systems.

Although the field of nanoscience and nanotechnology is relatively new and scientific developments in these fields bloomed after 1990, the key concepts of these branches of science were practiced over a longer period of time where the earliest evidence dates back to 600 BCE. Pottery shards excavated from Keeladi in India's southern state Tamil Nadu and reported in the year 2020 showed unique black coating on the inner surface of the pottery, which was investigated by the Indian researchers at the Vellore Institute of Technology, India, found to be carbon nanotubes [64] (Figure 1.7a). The use of metal nanoparticles in the fourth century Roman era was evidenced in the Lycurgus Cup which is kept in the British Museum. The glass contains gold–silver nanoparticles which are distributed in such a way that it appears green in daylight, but changes to red, when illuminated from the inside. The Damascus steel sword from the Mesopotamian civilization made between 300 and 1700 CE contained nanoscale wires and tube-like structures which exceptionally enhanced the sharp cutting edge of the sword. However, the re-emergence of nanotechnology in the 1980s was attributed jointly to the invention of advanced experimental tools such as scanning tunneling microscope,

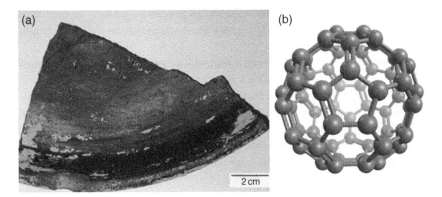

Figure 1.7 (a) Artifacts of pottery shards dated to 600 BCE unearthed from an archaeological site in Keeladi, Tamil Nadu showed evidences of carbon nanotubes. (b) Molecular structure of Fullerene, an allotrope of carbon discovered in 1986. *Sources:* (a) Springer Nature Limited. (b) The Trustees of the British Museum.

invention of fullerenes (Figure 1.7b), and elucidation of the conceptual framework of nanotechnology in 1986 [65]. The growing awareness on nanotechnology led to the discovery of carbon nanotubes in 1991 [66], which triggered multiple avenues of research and led to understanding of peculiar material properties in nanoscale, opening up applications in diverse fields.

Systematic research developments on the experimental as well as the conceptual aspects of nanoscience and nanotechnology resulted in vivid understanding of the quantum size effects and its role on material properties. The findings opened up new and exciting possibilities to tailor the chemical and physical properties of a material through exploiting and varying their nanoscopic properties. Unusual properties of nanomaterials led to the evolution of new devices which are aimed at smart sensor application. Controlled particle size and consistent yield of nanostructured material facilitated the widening of the realm of nanotechnology and encouraged researchers to integrate nanomaterials into smart devices and produce advanced and cost-effective sensor systems.

1.2.3.4 Printed Electronics on Flexible Substrates
Research on printed and flexible electronics is constantly evolving with extensive scope for innovation and has led to new avenues for the development of new age sensors. Printed electronics is an all-encompassing term for the printing method used to create electronic devices by printing on a variety of substrates. Printed circuitry is widely used in organic or plastic electronics, where carbon-based and carbon–metal composite inks are extensively utilized. The advent of flexible keyboards, antennas, and electronic skin patches has boosted the demand for

printed electronics. Printed electronics is one of the fastest growing technologies today and the importance is already realized in several industries including health care, aerospace, and media.

The evolution of printed electronics initiated in 1903 when famous German Inventor Albert Hanson filed a British patent for a device described as a flat, foil conductor on an insulating board having multiple layers. He called the conductor as "Printed Wires," and showed that we can punch a hole into the two layers and had perpendicular wires to establish electrical connectivity. In 1925, Charles Ducas filed a patent that describes the way for adding conductive inks to an insulating material (Figure 1.8a). This later gave birth to the printed wiring board

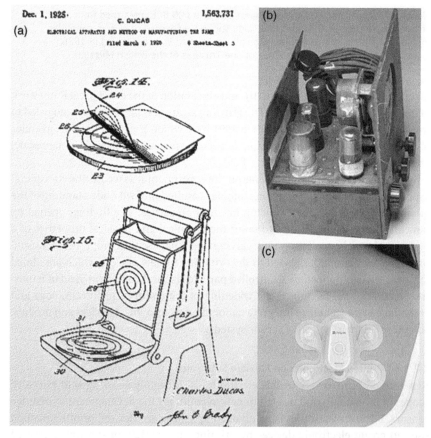

Figure 1.8 (a) Charles Ducas patent drawing. (b) A radio made by Paul Eisler that uses first printed circuit board (PCB). (c) The Life Signals Biosensor Patch 1AX enable remote wireless mass population monitoring for COVID-19. *Sources:* (b) By Georgi Dalakov - http://history-computer.com/ModernComputer/Basis/printed_board.html, CC BY-SA 3.0, https://commons.wikimedia.org/w/index.php?curid=19448601. (c) Courtesy of how2electronics.com.

(PWB). This was used as a flat heating coil and showed the way for printed circuitry. The period from 1930 to 1945 is marked as the most significant time in the history of printed circuit board (PCB). The PCB was used in proximity fuse invented by the British for high-velocity artillery shells, which needed to fireside precisely over massive distances in either sky or land. In 1936, Paul Eisler, an Austrian living within the United Kingdom, filed a patent for copper foil circuit on a nonconductive base of a glass terming it as printed circuit [67–69]. Eisler later took this concept one step further for large-scale production of radios (Figure 1.8b) by the United States during World War II in 1943, which might pave the way for future military applications. Printed circuit technology was released for commercial use in the United States in 1948 (Printed Circuits Handbook, 1995). Widespread production of printed electronics for household use began in the 1960s when the PCB became the foundation for all consumer electronics. Over a half-century since its inception, printed electronics have evolved from PCBs to today's use of solution-based conductive inks on flexible substrates for the development of RFID, photovoltaic, and electroluminescent technologies. At present, printed electronics have revolutionized the solar cell industry and wearable sensors market. In 2011, researchers from MIT created a flexible solar cell by inkjet printing on normal paper [70]. In 2018, researchers at Rice University have developed organic solar cells which can be painted or printed onto surfaces [71]. In 2020, researchers have successfully placed wearable circuits directly onto the surface of human skin to monitor health indicators, such as temperature, blood oxygen, heart rate, and blood pressure. This printed wearable electronics for health monitoring combines soft on-body sensors with flexible PCBs (FPCBs) for signal readout and wireless transmission to healthcare workers [72]. In 2020, following COVID-19 pandemic, some companies have developed electronic skin patches for monitoring a patient's temperature remotely, the benefit being that the patient does not need to come into contact with healthcare professionals to take their temperature which may expose them to the virus (Figure 1.8c). Again conductive ink has been used in sensor test systems as part of a system to detect infections, including COVID-19 [73, 74]. Furthermore, Cu nanoparticles due to their antiviral benefits were 3D printed onto surfaces such as door handles to combat spread of infection [75]. Printed force sensors have been applied into mats to alert shoppers where they are standing too close to each other.

Owing to the advantages of printed electronics, the sensors and devices have become light weight, flexible, and cost-effective, making them more appealing to a wide cross-section of industries. Printed circuits have the potential to reduce costs and technical constraints typically associated with mass producing electronics. In addition, printed electronics have paved the way for flexible sensors and allied wearable devices which have opened up new set of applications. Printed electronics holds promise to change the photovoltaic industries due to less expensive polymer electronics.

The benefits of printed electronics are listed as:

1) Use of custom-made organic printable inks
2) Use of various different substrates including bendable and biodegradable
3) Cost-effective and fast fabrication
4) Attractive and flexible form factor
5) Ease of production
6) Ease of integration

Although printed electronics have been around since 1960s, a major burst in total revenue was seen in the hybrid age after year 2000. According to IDTechEx publication, the total printed electronic revenue in 2011 was reported to be at $12.385 (billion) and expected to reach $330 (billion) in 2027. This huge increase in revenue is attributed to the incorporation of printed electronic into cell phones. Despite the fast growing market of printed and flexible electronics, researchers are still working on the development of best performance conductive inks for on-fabric use and its use on different substrates. Printed electronics continue to face challenges in batch fabrication as the processes are so new that trial and error and testing are critical to success.

1.2.3.5 Smart Devices with Artificial Intelligence

Smart sensors are expected to perform intelligently like humans and possess multifunctional and flexible sensing systems involving vision, handwriting recognition, speech recognition, human–robots interactions, gaming control, touch, smell, gesture control, and taste. These sensor systems can be made to operate more effectively and make smart decisions using AI. AI is a way of providing human-like intelligence and decision-making ability to machines, sensor systems, computer-controlled robots, and software with an aim to create expert systems and to implement human intelligence in machines. AI emerged in the mid-1950s, and evolved as a powerful tool for the development of smart sensor systems for automatically solving problems that would normally require human intelligence. The AI tools which are frequently used in sensor system are knowledge-based systems, fuzzy logic, automatic knowledge acquisition, neural networks, genetic algorithms, case-based reasoning, and ambient intelligence. These tools have minimal computation complexity and can be implemented with small sensor systems, single sensors, or system arrays with low-capability microcontrollers. Effective use of AI tools will contribute to the development of more competitive sensor systems and applications. Other technological developments in AI that will impact sensor systems include data mining techniques, multiagent systems, and distributed self-organizing systems. Ambient sensing involves integrating many microelectronic processors and sensors into everyday objects to make them "smart." This method is called sensor fusion. AI is useful when smart sensors are expected to perform advanced and complex task other than just reading raw data from the sensors.

Using AI, sensor fusion can be done easier and more accurately than with classical algorithms. Neural networks can cope with unknown situations much better. A combination of sensor systems and AI algorithms can empower intelligent robots with advanced capabilities in many areas such as environmental monitoring, gas leakage detection, food and beverage production and storage, and especially disease diagnosis through detection of different types and concentrations of target gases with the advantages of portability, low-power consumption, and ease of operation. Sensors equipped with AI in wearable and flexible smart electronics will meet the rising demand and growth in smart watch and sport bracelets in the market. Such devices as a part of IoT or as a part of our daily life can greatly enhance life quality of people and promote performance and interactivity of infrastructure in modern life. Contributions from various disciplines of research will help in the growth of smart AI-aided sensor systems.

1.2.3.6 IoT-Enabled Sensors

The internet of things, or IoT, is a system of interrelated devices, mechanical and digital machines, sensors or people (Figure 1.9a) that are provided with unique identifiers (UID) and transfers data over a network in the absence of any human-to-human or human-to-system interaction. The word "thing" in IoT refers to devices and sensors like wearable medical devices, an antenna connected to implantable device, built in sensors in tire of automobiles, RFID tags, or any device that has been assigned an IP protocol address and is able to transfer data over a network. IoT-enabled sensors share the sensor data they collect by connecting to an IoT gateway or other devices where data are processed in the cloud or analyzed locally. Such devices communicate with other connected devices in the network and also can act on the information received from another. The IoT enabled device/sensor at the receiving end collects the raw data and performs preprocessing in order to present relevant results. In advanced sensor systems, the raw data received from IoT-enabled device needs to be processed using AI for smart and intelligent decision making which can act as an alert for the users. In near future, the processing of the data from multiple sensors that are remotely connected to each other is believed to take place in cloud where the role of IoT-enabled devices will be effective. This will allow more complex sensor solutions through sensor data fusion for more advanced systems. The schematic working of IoT devices is shown in Figure 1.9b. The IoT-enabled devices together with AI can revolutionize the field of sensor systems, remote diagnosis, alerts for thefts, climate variations, natural disasters, medical emergency, and fire and thereby combat with appropriate measures without human interferences, thus making life more dynamic and smoother. The IoT helps people live and work smarter, as well as gain complete control over their lives. In addition to offering smart devices to automate homes, IoT is essential to business. IoT provides businesses with a real-time look

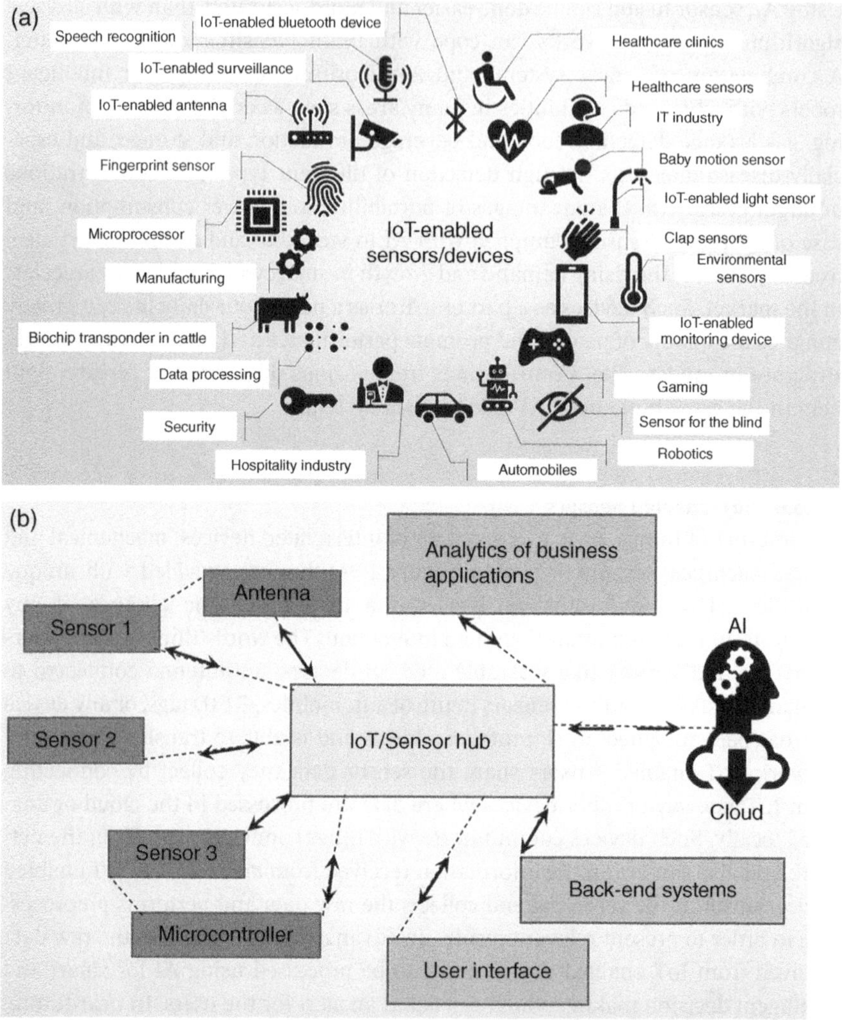

(a)

Speech recognition	IoT-enabled bluetooth device
IoT-enabled surveillance	
IoT-enabled antenna	Healthcare clinics
Fingerprint sensor	Healthcare sensors
	IT industry
Microprocessor	Baby motion sensor
Manufacturing	IoT-enabled light sensor
Biochip transponder in cattle	Clap sensors
Data processing	Environmental sensors
Security	IoT-enabled monitoring device
Hospitality industry	Gaming
	Sensor for the blind
	Robotics
	Automobiles

IoT-enabled sensors/devices

(b)

Analytics of business applications

Antenna

Sensor 1

AI

Sensor 2

IoT/Sensor hub

Cloud

Sensor 3

Back-end systems

Microcontroller

User interface

Figure 1.9 (a) Different IoT-enabled devices and systems. (b) Schematic representation of working of AI-supported IoT-enabled smart sensor system.

into how their systems really work, delivering insights into everything from the performance of machines to supply chain and logistics operations.

1.2.4 Paradigm Shift in Solid-State Sensor Research

With the advent of new technologies discussed in detail in Section 1.2.3, a distinct shift in paradigm is noticed in solid-state sensor research especially after the year 2000. The solid-state sensor research in the present day focuses on the development of devices which mostly uses nanomaterials, biomolecules, and organic polymers, are mechanically bendable, fabricated using solution process routes and thin-film technology, and cost-effective, makes them commercially attractive and integrable with microchips for the realization of smart sensing system with AI and IoT support. Some of the research products emerging out from the new technologies are listed below but are not limited to the following.

1.2.4.1 Organic Devices

Organic electronics is a newly emerging branch of modern electronics which deal with carbon-based materials often in the form of polymers and small molecules and molecules of living beings. Organic materials are known to be excellent insulators and used in wide range of applications until Heeger, MacDiarmid, and Shirakawa in the 1970s discovered that polymer polyacetylene when doped iodine can be made conductive with electrical conductivity enhanced many folds. Semiconducting behavior of organic polymer materials was first demonstrated by Tsumura et al. [76] in 1986 when they used polythiopene as semiconducting material in the channel of the first solid-state field effect transistor. The discovery of electrically conductive and semiconducting polymers drastically changed our perspective on polymer materials and formed the foundation of organic electronics. Over the years, the organic electronics have found their way from research laboratories to industrial applications and is due to replace Si electronics in commercial market in near future. The organic electronic market includes applications in the fields of light-emitting diodes (LEDs) [77], FET [78], and solar cells [79] and also open new avenues for other technologies.

Organic semiconductors based on polymers or small molecules consist of π-conjugated bonds with delocalized electrons which affect the electrical properties of the material [80, 81]. The molecules are held together by π–π noncovalent stacking interactions, which are weak Van der Waals, and dipole–dipole forces, which despite being feeble in nature, are sufficiently strong when large amount of interactions are involved [82]. The delocalized electrons in the π orbitals of the molecules constitute the π way through which the transport of electrons takes place under the influence of potential bias [83]. The amount of π-conjugated bonds and hence the number of delocalized electrons can be regulated by modifying the molecular structure using chemical methods. Thus, material property of organic conjugated polymers can be modified by substituting aliphatic side chains to aromatic part which may lead to a formation of distinct superstructure driven by the local phase separation between the flexible aliphatic part and rigid aromatic fraction or changes the molecular packing of the conjugated polymer. This affects the availability of delocalized electrons in the π way and thus on the transport of charge carriers from one molecule to another. An increased π-stacking distance reduces the hopping rate of charge carriers.

The principle advantage of using the conjugated polymers lies in the solution-based processability to fabricate devices, thereby allowing large area low-cost fabrication [84, 85]. Moreover, organic polymer materials are mechanically flexible and thus suitable for application in bendable devices [86]. The mass fabrication of the devices occurs by high-speed and inexpensive methods at low temperatures including printable circuits using inkjet and screen-printing techniques. Due to the low cost processing, the electronic elements can be used in the realization of radiofrequency identification (RFID) tags and sensors in which FETs play a key role [87]. Due to the low-cost fabrication process with conjugated polymers and their mechanical bendability [88] and strength, the organic materials [4, 89] and biopolymers [1, 90] finds their way into medical and healthcare applications.

1.2.4.2 Wearable Devices

The prehistory of wearable technology starts with the watch, which was worn by people to tell time. In 1500, the German inventor Peter Henlein created small watches that were worn as necklaces. A century later, men began to carry their watches in their pockets as the waistcoat became a fashionable item, which led to the creation of pocket watches. Wristwatches were also created in the late 1600s, but were worn mostly by women as bracelets. Today at the advent of modern-day technology supported by advanced microfabrication tools and techniques, wearable devices have encroached into our daily lives in the form of smart watches, medical devices, safety alarms, and fitness equipment. Driven by the increasing scientific and technological interest and its huge commercial promise, wearable sensors will continue to evolve over the next few decades in providing a deeper understanding of the subjects' environment [91]. As the name suggests, the wearable sensors are packaged into wearable objects which can be directly worn on the body to help monitor health and/or provide clinically relevant data for care. Due to the burst in technological advancements (Figure 1.10), the recent focus has shifted to wearable sensing platforms, using stretchable and flexible electronics. However, these sensors are fabricated on bendable substrates which make microfabrication challenging, but opens up new avenues in research and applications.

Nowadays, various techniques such as transfer printing, screen printing, photolithography, microchannel molding, filling, and lamination are utilized in the realization of flexible sensors (Sections 3.4 and 3.7). Here, deformable or stretchable conducting electrodes are patterned onto a bendable/stretchable substrate [92]. However, these methods have some limitations such as high cost, multistep fabrication protocols, poor durability, and challenges in prototyping and

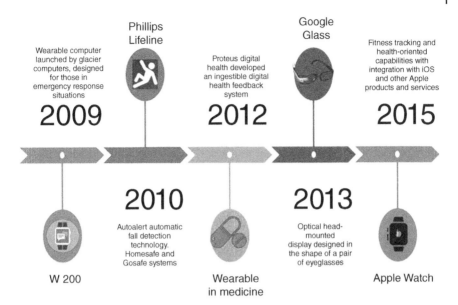

Figure 1.10 Timeline of wearable devices during the last decade.

scalability. The real challenge in wearable devices mainly lies in the acceptability of these technologies by the users/patients. For instance, monitoring of elderly people is one application that has been tentatively addressed since the beginning of the era of wearable monitoring, with the main motivation being the decrease in the costs of assistance for the chronic diseases typically affecting this category of patients. On the other hand, the limited familiarity of these patients with technology immediately highlighted the main limit of these systems. To turn this threat/challenge into an opportunity, several actions are needed.

1.2.4.3 Implantable Sensors

Implantable solid-state sensors play an important role especially in medical science not only to provide real-time treatment facilities to patients, but also in monitoring vital physiological statistics in humans. The ease of commercial manufacture, storage, and sensor performance in terms of accuracy and repeatability along with operations that are not dependent on user handling make the solid-state implantable solid-state sensors ideal for medical applications. Much before the burst of applications of implantable sensors in the medical industry in 1960s, they were developed in mid-1940s for space applications to monitor and broadcast real-time biological parameters of dogs and chimps [93]. With the advent of new materials and fabrication processes, the implantable sensors show promises for wider scope of application but are presently in their early stages of medical trials. In solid-state implantable sensors, the sensor element is invasive and communicates sensor data to an external system, worn either on the body or located remotely, for processing and display. In most implantable sensors, the active sensor element generates chemical signals or undergoes biological changes, which produces signals that have to be transformed into electrical signals for electronic detection. Even if piezoelectric sensors are used as detection systems, the electrical signal generated from the piezoelectric crystals must be modulated and amplified to a signal that can be exported from the body. Smart implantable sensors can collect data and store the information in buffers until the unit is interrogated by an external device.

Implantation of biomaterials and devices into soft tissues often causes foreign body response (FBR), which may interfere with the satisfactory working of the implanted device and/or lead to failure [94, 95]. The FBR involves overlapping acute and persistent inflammatory phases associated with collagenous coverage and proteins adsorption, which have no therapeutic remedies. The use of appropriate biomaterials for encapsulated packaging of implantable devices is an established way to protect devices from FBR. Such encapsulating materials include variety of synthetic polymers [96], such as PU, poly (2-methoxyethyl acrylate), and PVA [97], but found to have limited durability. Recently, polyester fabric coated with a biodegradable PGA sheet was developed and implanted in a rat model [98]. As the PGA degraded, cells deposited their extracellular matrix, much of which is collagen, on the polyester fabric. In terms of design, the implantable sensors must be capable of unobstructive and uninterferred data acquisition from the patient, be lightweight and small in size, its battery should last many years as it would be impractical to replace it frequently, not cause any kind of discomfort or damage to human tissues through overheating [99], and the sensor element need to be reliable and sensitive [100]. Present advancements in implantable sensors can be seen in sensor miniaturization and in the design of low-power circuitry [99].

Despite the recent developments in fabrication testing evaluation and statistical analysis of sensor performance, the acceptability and adoption of implantable sensors on the commercial scale are far from realization. This is because any failure of the devices could potentially lead to patient death or develop other forms of disease. Thus, to reform and revolutionize the field of implantable sensors, dedicated effort in engineering and technology coupled with public awareness of such devices need to be undertaken. Once the technology shows promises in comfort and effectiveness in real-time tracking of physiological parameters, there will be more patients willing to consent to monitoring, only to eliminate other painful and expensive methods of treatments [101].

1.3 Outline

There has been a rapid growth in the field of solid-state sensors in the last decade, mostly attributed to the fast advancements in monolithic fabrication processes, computer-controlled remote automation, and signal processing tools. The field of solid-state sensor technology has expanded its realm so much that it finds wide applications in industrial processes, healthcare systems, and household use. Its cost-effectiveness and portability has made it affordable, convenient, and user-friendly to the wide section of the global population and thus forms the highest trading market of this century. With the rise in integrated and automated remote sensing technology, sensor terminology, such as "sensor," "sensor element," and "sensor system," has been introduced and explained to facilitate coherent and consistent analysis of sensor technologies. The consistency and cost control in sensor system depend on advancements in fabrication procedures and tools. In this chapter, we discuss various fabrication procedures and tools including micromachining and lithography which lead to the advancements in the development of microsensor systems. Since modern solid-state sensors encompass much more than a transduction material, there are many opportunities for introducing novel materials in sensor systems which are discussed in subsequent chapters.

List of Abbreviations

AFM	Atomic Force Microscopy
AI	Artificial Intelligence
CAGR	Compound Annual Growth Rate
COVID-19	Coronavirus Disease 2019
ECG	Electrocardiogram
FPCB	Flexible Printed Circuit Boards
MEMS	Micro-electro-mechanical Systems
PCB	Printed Circuit Board
PWB	Printed Wiring Board
RFID	Radiofrequency Identification
SOC	System on Chip
STM	Scanning Tunneling Microscope
TEM	Transmission Electron Microscopy

References

1 Kafi, M.A., Paul, A., Vilouras, A., and Dahiya, R. (2020). Mesoporous chitosan based conformable and resorbable biostrip for dopamine detection. *Biosensors and Bioelectronics* 147: 111781.

2 Kim, J.Y., Park, J., Lee, M.J. et al. (2020). Diffusion-dependent graphite electrode for all-solid-state batteries with extremely high energy density. *ACS Energy Letters* 5 (9): 2995–3004.

3 Jeong, Y., Park, J., Lee, J. et al. (2020). Ultrathin, biocompatible, and flexible pressure sensor with a wide pressure range and its biomedical application. *ACS Sensors* 5 (2): 481–489.

4 Hosseini, E.S., Manjakkal, L., Shakthivel, D., and Dahiya, R. (2020). Glycine-chitosan-based flexible biodegradable piezoelectric pressure sensor. *ACS Applied Materials & Interfaces* 12 (8): 9008–9016.

5 Yogeswaran, N., Navaraj, W.T., Gupta, S. et al. (2018). Piezoelectric graphene field effect transistor pressure sensors for tactile sensing. *Applied Physics Letters* 113 (1): 014102.

6 Yan, C., Deng, W., Jin, L. et al. (2018). Epidermis-inspired ultrathin 3D cellular sensor array for self-powered biomedical monitoring. *ACS Applied Materials & Interfaces* 10 (48): 41070–41075.

7 Taube Navaraj, W., García Núñez, C., Shakthivel, D. et al. (2017). Nanowire FET based neural element for robotic tactile sensing skin. *Frontiers in Neuroscience* 11 (501): 40.

8 Lopez-Marzo, A.M. and Merkoci, A. (2016). Paper-based sensors and assays: a success of the engineering design and the convergence of knowledge areas. *Lab on a Chip* 16 (17): 3150–3176.

9 Nag, A., Mukhopadhyay, S.C., and Kosel, J. (2017). Wearable flexible sensors: a review. *IEEE Sensors Journal* 17 (13): 3949–3960.

10 Xu, Y., Hu, X., Kundu, S. et al. (2019). Silicon-based sensors for biomedical applications: a review. *Sensors* 19: 2908.

11 Luan, E., Shoman, H., Ratner, D.M. et al. (2018). Silicon photonic biosensors using label-free detection. *Sensors (Basel)*. 18 (10): 3519.

12 PRNewswire (2020). *Global Automotive Sensor Market by Type (Temperature Sensors, Pressure Sensors, Motion Sensors, Speed Sensors, and Gas Sensors) and Application (Powertrain, Chassis, Body Electronics, Safety & Security, and Telematics): Global Opportunity Analysis and Industry Forecast, 2020–2027.* Portland: Allied Market Research.

13 McIntosh, J. (2008). *The Ancient Indus Valley: New Perspectives.* ABC-CLIO.

14 Lewis, M.J.T. (2001). *Surveying Instruments of Greece and Rome.* Cambridge University Press.

15 Northrup, C.C. (ed.) (2014). *Encyclopedia of World Trade: From Ancient Times to the Present*, vol. 4. Armonk, NY: Sharpe Reference.

16 Sleeswyk, A.W. and Sivin, N. (1983). Dragons and toads. The Chinese seismoscope OF A.D. 132. *Chinese Science* 6: 1–19.

17 H. Selin, "*Encyclopaedia of the History of Science, Technology, and Medicine in Non-western Cultures* Ed. 2," vol. 1, Berlin; New York Springer, 2008, pp. 197.

18 Lewis, O. (2017). Studies in ancient medicine, fragments and interpretation. In: *Praxagoras of Cos on Arteries, Pulse and Pneuma*, vol. 48, 350. Brill.

19 Hajar, R. (2018). The pulse from ancient to modern medicine: part 3. *Heart Views* 19: 117–120.

20 Lawrence, C. (1978). Physiological apparatus in the Wellcome Museum. 1. The Marey sphygmograph. *Medical History* 22 (2): 196–200.

21 Fung, Y.C. and Yih, C.-S. (1971). Biomechanics: a survey of the blood flow problem. In: *Advances in Applied Mechanics*, vol. 11, 65–130. Elsevier.

22 Mulcahy, R. (1997). *Medical Technology: Inventing the Instruments, Innovators Series*, 144. Oliver Press.

23 Reichert, P. (1978). A history of the development of cardiology as a medical specialty. *Clinical Cardiology* 1: 5–15.

24 Barold, S. (2003). Willem Einthoven and the birth of clinical electrocardiography a hundred years ago. *Cardiac Electrophysiology Review* 7 (1): 99–104.

25 Klooster, J.W. (2009). *Icons of Invention: The Makers of the Modern World from Gutenberg to Gates*, vol. 1. Greenwood Press: ABC-CLIO.

26 History.com Editors (2010). *First radio transmission sent across the Atlantic Ocean.* https://www.history.com/this-day-in-history/marconi-sends-first-atlantic-wireless-transmission (accessed 2 August 2023).

27 Fitzgerald, M. (2005). *Runner's World The Cutting-Edge Runner: How to Use the Latest Science and Technology to Run Longer, Stronger, and Faster*, 63. Rodale.

28 Tolmie, N. and Burich, M. (2005). *Textronics(TM) Introduces Heart Sensing Sports Bra; NuMetrex(TM) Sports Bra Monitors Heart Rate with Innovative Built-in Heart Monitor.* Wilmington: BusinessWire.

29 Islam, M.M., Rahaman, A., and Islam, M.R. (2020). Development of smart healthcare monitoring system in IoT environment. *SN Computer Science* 1 (3): 185.

30 Emerson, D.T. (1997). The work of Jagadis Chandra Bose: 100 years of millimeter-wave research. *IEEE Transactions on Microwave Theory and Techniques* 45 (12): 2267–2273.

31 Braun, A., Braun, E., and MacDonald, S. (1982). *Revolution in Miniature: The History and Impact of Semiconductor Electronics*. London: Cambridge University Press.

32 Graff, K.F. (1982). *A History of Ultrasonics, in Physical Acoustics*, vol. XV. New York: Academic Press.

33 Matté, J.A. (1996). *Forensic Psychophysiology Using the Polygraph: Scientific Truth Verification, Lie Detection*, vol. 1. Williamsville, New York: J.A.M. Publications.

34 Fleming, J.A. (1905). Instruments for converting alternating current into continuous current. *United States Patent Office*. United States Patent 803,684, U. F. Government, USA, 7 November 1905.

35 Bardeen, J. and Brattain, W. (1948). The transistor, a semi-conductor triode. *Physical Review B* 74: 230–231.

36 Bardeen, J. and Brattain, W. (1950). Three-electrode circuit element utilizing semiconductor materials. U. S. Patent 2,524,035.

37 Kilby, J. (1964). Miniaturized electronic circuits. US Patents, U. S. P. Office, Texas Instruments.

38 Noyce, R.N. (1959). Semiconductor device-and-lead structure. US2981877A, U. P. Office, Fairchild Semiconductor Corp.

39 Bellis, M. (2020). *The History of the Integrated Circuit (Microchip)*. ThoughtCo. thoughtco.com/history-of-integrated-circuit-aka-microchip-1992006.

40 Kilby, J., Merryman, J., and Van, T.J. (1972). Miniature electronic calculator. Google Patents, US3819921A, U. S. P. Office, Texas Instruments Inc.

41 Instrument Society of America (1975). *Electrical Transducer Nomenclature and Terminology. ANSI Standard MC6.1*. Research Triangle Park, North Carolina: Instrument Society of America.

42 Drude, P. (1889). Ueber Oberflächenschichten. *Annalen der Physik* 272 (2): 532–560.

43 Grzegorz, G., Ivan, P., Greene, J.E., and Lars, H. (2019). Paradigm shift in thin-film growth by magnetron sputtering: From gas-ion to metal-ion irradiation of the growing film. *Journal of Vacuum Science & Technology A* 37 (6): 060801.

44 Burnside, S.D., Shklover, V., Barbe, C. et al. (1998). Self-organization of TiO2 nanoparticles in thin films. *Chemistry of Materials* 10 (9): 2419–2425.

45 Luo, H., Cardinal, C.M., Scriven, L.E., and Francis, L.F. (2008). Ceramic Nanoparticle/Monodisperse Latex Coatings. *Langmuir* 24 (10): 5552–5561.

46 Nagao, D., Goto, K., Ishii, H., and Konno, M. (2011). Preparation of asymmetrically nanoparticle-supported, monodisperse composite dumbbells by protruding a smooth polymer bulge from rugged spheres. *Langmuir* 27 (21): 13302–13307.

47 Xu, J., Wong, D.H.C., Byrne, J.D. et al. (2013). Future of the particle replication in nonwetting templates (PRINT) technology. *Angewandte Chemie International Edition* 52 (26): 6580–6589.

48 Koide, Y., Such, M.W., Basu, R. et al. (2003). Hot microcontact printing for patterning ITO surfaces. methodology, morphology, microstructure, and OLED charge injection barrier imaging. *Langmuir* 19 (1): 86–93.

49 Resnick, D.J., Sreenivasan, S.V., and Willson, C.G. (2005). Step & flash imprint lithography. *Materials Today* 8 (2): 34–42.

50 Acharya, G., Shin, C.S., McDermott, M. et al. (2010). The hydrogel template method for fabrication of homogeneous nano/microparticles. *Journal of Controlled Release* 141 (3): 314–319.

51 Hou, S., Yang, K., Qin, M. et al. (2008). Patterning of cells on functionalized poly(dimethylsiloxane) surface prepared by hydrophobin and collagen modification. *Biosensors and Bioelectronics* 24 (4): 912–916.

52 Fernandes, T.G., Diogo, M.M., Cabral, J.M.S. et al. (2013). Microscale technologies for stem cell culture. In: *Stem Cell Bioprocessing*, 143–175. Woodhead Publishing.

53 Otsuka, H., Ohshima, H., and Makino, K. (2014). Chapter 11 – Micropatterning of cell aggregate in three dimension for in vivo mimicking cell culture. In: *Colloid and Interface Science in Pharmaceutical Research and Development* (ed. H. Ohshima and K. Makino), 223–241. Amsterdam: Elsevier.

54 Tan, W. and Desai, T.A. (2003). Microfluidic patterning of cells in extracellular matrix biopolymers: effects of channel size, cell type, and matrix composition on pattern integrity. *Tissue Engineering* 9 (2): 255–267.

55 Suri, S., Singh, A., Nguyen, A.H. et al. (2013). Microfluidic-based patterning of embryonic stem cells for in vitro development studies. *Lab on a Chip* 13 (23): 4617–4624.

56 Berthier, E., Warrick, J., Casavant, B., and Beebe, D.J. (2011). Pipette-friendly laminar flow patterning for cell-based assays. *Lab on a Chip* 11 (12): 2060–2065.

57 Kelly, A., Ballerini, L., Lowery, M. et al. (2017). Engineering the neural interface. In: *Comprehensive Biomaterials II* (ed. P. Ducheyne), 642–660. Oxford: Elsevier.

58 Kong, J., Franklin, N.R., Zhou, C. et al. (2000). Nanotube molecular wires as chemical sensors. *Science* 287 (5453): 622–625.

59 Harris, K.D., Huizinga, A., and Brett, M.J. (2002). High-speed porous thin film humidity sensors. *Electrochemical and Solid-State Letters* 5 (11): H27–H29.

60 Besteman, K., Lee, J.-O., Wiertz, F.G.M. et al. (2003). Enzyme-coated carbon nanotubes as single-molecule biosensors. *Nano Letters* 3 (6): 727–730.

61 Ito, Y. and Fukusaki, E. (2004). DNA as a 'nanomaterial'. *Journal of Molecular Catalysis B: Enzymatic* 28 (4–6): 155–166.

62 Penza, M., Antolini, F., and Antisari, M.V. (2004). Carbon nanotubes as SAW chemical sensors materials. *Sensors and Actuators B: Chemical* 100 (1–2): 47–59.

63 Mahar, B., Laslau, C., Yip, R., and Sun, Y. (2007). Development of carbon nanotube-based sensors – a review. *IEEE Sensors Journal* 7 (2): 266–284.

64 Kokarneswaran, M., Selvaraj, P., Ashokan, T. et al. Discovery of carbon nanotubes in sixth century BC potteries from Keeladi, India. *Scientific Reports* 10: 19786.

65 Newton, M.D. and Stanton, R.E. (1986). Stability of buckminsterfullerene and related carbon clusters. *Journal of the American Chemical Society* 108 (9): 2469–2470.

66 Iijima, S. (1991). Helical micro-tubules of graphitic carbon. *Nature* 345: 56–58.

67 Tracy, M. (2019). The legendary life of Paul Eisler, the father of pcb circuit boards. In: *Oneseine Enterprise Building A*. Shajing, Baoan, Shenzhen, China: Shixiaganglian Industrial Park.

68 Paul, E. (1944). Manufacture of electric circuit components. US Patent Office, in *Google Patents*, US2441960A. USA: Eisler Paul.

69 Paul, E. (1949). Electric capacitor and method of making it. US Patent Office, *Google Patents*, US2607825A. USA: Eisler Paul.

70 Kraemer, D., Poudel, B., Feng, H.-P. et al. (2011). High-performance flat-panel solar thermoelectric generators with high thermal concentration. *Nature Materials* 10 (7): 532–538.

71 Mok, J.W., Hu, Z., Sun, C. et al. (2018). Network-stabilized bulk heterojunction organic photovoltaics. *Chemistry of Materials* 30 (22): 8314–8321.

72 Zhang, L., Ji, H., Huang, H. et al. (2020). Wearable circuits sintered at room temperature directly on the skin surface for health monitoring. *ACS Applied Materials & Interfaces* 12 (40): 45504–45515.

73 Laghrib, F., Saqrane, S., El Bouabi, Y. et al. (2021). Current progress on COVID-19 related to biosensing technologies: new opportunity for detection and monitoring of viruses. *Microchemical Journal* 160: 105606.

74 Singh, V.K., Chandna, H., Kumar, A. et al. (2020). IoT-Q-band: a low cost internet of things based wearable band to detect and track absconding COVID-19 quarantine subjects. *Transactions on Internet of Things* 6 (21): 4.

75 Xue, X., Ball, J.K., Alexander, C., and Alexander, M.R. (2020). All surfaces are not equal in contact transmission of SARS-CoV-2. *Matter* 3 (5): 1433–1441.

76 Tsumura, A., Koezuka, H., and Ando, T. (1986). Macromolecular electronic device: field-effect transistor with a polythiophene thin film. *Applied Physics Letters* 49 (18): 1210–1212.

77 Tang, C.W. and VanSlyke, S.A. (1987). Organic electroluminescent diodes. *Applied Physics Letters* 51 (12): 913–915.

78 Tsumura, A., Fuchigami, H., and Koezuka, H. (1991). Field-effect transistor with a conducting polymer film. *Synthetic Metals* 41 (3): 1181–1184.

79 Dyer-Smith, C., Nelson, J., Li, Y., and Kalogirou, S.A. (2018). Chapter I-5-B – organic solar cells. In: *McEvoy's Handbook of Photovoltaics*, 3e, 567–597. Academic Press.

80 Jager, E.W.H. and Inganas, O. (2016). Electrochemomechanical devices from conjugated polymers. In: *Reference Module in Materials Science and Materials Engineering* (ed. S. Hashmi). Elsevier.

81 Peng, H., Sun, X., Weng, W. et al. (2017). 2 – Synthesis and design of conjugated polymers for organic electronics. In: *Polymer Materials for Energy and Electronic Applications*, 9–61. Academic Press.

82 Wang, Y.J. and Yu, G. (2019). Conjugated polymers: from synthesis, transport properties, to device applications. *Journal of Polymer Science Part B: Polymer Physics* 57 (23): 1557–1558.

83 Paul, A. and Bhattacharya, B. (2010). DNA functionalized carbon nanotubes for nonbiological applications. *Materials and Manufacturing Processes* 25 (9): 891–908.

84 MacFarlane, L.R., Shaikh, H., Garcia-Hernandez, J.D. et al. (2020). Functional nanoparticles through TT-conjugated polymer self-assembly. *Nature Reviews Materials* 6: 7–26.

85 Zeglio, E., Rutz, A.L., Winkler, T.E. et al. (2019). Conjugated polymers for assessing and controlling biological functions. *Advanced Materials* 31 (22): 1806712.

86 Che, B., Zhou, D., Li, H. et al. (2019). A highly bendable transparent electrode for organic electrochromic devices. *Organic Electronics* 66: 86–93.

87 Pankow, R.M. and Thompson, B.C. (2020). The development of conjugated polymers as the cornerstone of organic electronics. *Polymer* 207: 122874.

88 Kafi, M., Paul, A., Vilouras, A., and Dahiya, R. (2018). Chitosan-Graphene Oxide based ultra-thin conformable sensing patch for cell-health monitoring. *Presented at IEEE Sensors* (28–31 October 2018). New Delhi: IEEE.

89 Lima, C.G.A., de Oliveira, R.S., Figueiró, S.D. et al. (2006). DC conductivity and dielectric permittivity of collagen–chitosan films. *Materials Chemistry and Physics* 99 (22): 284–288.

90 Vilouras, A., Paul, A., Kafi, M.A., and Dahiya, R. (2018). Graphene oxide-chitosan based ultra-flexible electrochemical sensor for detection of serotonin. *Presented at 2018 IEEE Sensors* (28–31 October 2018).

91 An, B.W., Gwak, E.-J., Kim, K. et al. (2016). Stretchable, transparent electrodes as wearable heaters using nanotrough networks of metallic glasses with superior mechanical properties and thermal stability. *Nano Letters* 16 (1): 471–478.

92 Hu, W., Niu, X., Li, L. et al. (2012). Intrinsically stretchable transparent electrodes based on silver-nanowire-crosslinked-polyacrylate composites. *Nanotechnology* 23 (34): 344002.

93 Martin, S., Duncan, E., Inmann, A., and Hodgins, D. (2013). 8 – Sterilisation considerations for implantable sensor systems. In: *Implantable Sensor Systems for Medical Applications*, 253–278. Woodhead Publishing.

94 Malik, A.F., Hoque, R., Ouyang, X. et al. (2011). Inflammasome components Asc and caspase-1 mediate biomaterial-induced inflammation and foreign body response. *Proceedings of the National Academy of Sciences* 108 (50): 20095.

95 Goswami, R., Arya, R.K., Biswas, D. et al. (2019). Transient receptor potential vanilloid 4 Is required for foreign body response and giant cell formation. *The American Journal of Pathology* 189 (8): 1505–1512.

96 Li, R., Wang, L., Kong, D., and Yin, L. (2018). Recent progress on biodegradable materials and transient electronics. *Bioactive Materials* 3 (3): 322–333.

97 Mulinti, P., Brooks, J.E., Lervick, B. et al. (2018). 10 – Strategies to improve the hemocompatibility of biodegradable biomaterials. In: *Hemocompatibility of Biomaterials for Clinical Applications*, 253–278. Woodhead Publishing.

98 Yekrang, J., Semnani, D., Zadbagher, A. et al. (2017). A novel biodegradable micro-nano tubular knitted structure of PGA braided yarns and PCL nanofibres applicable as esophagus prosthesis. *Indian Journal of Fibre & Textile Research* 42: 264–270.

99 Islam, S.U., Ahmed, G., Shahid, M. et al. (2017). Chapter 2 – Implanted wireless body area networks: energy management, specific absorption rate and safety aspects. In: *Ambient Assisted Living and Enhanced Living Environments* (ed. C. Dobre, C. Mavromoustakis, N. Garcia, et al.), 17–36. Butterworth-Heinemann.

100 Alam, M.M. and Ben Hamida, E. (2014). Surveying wearable human assistive technology for life and safety critical applications: standards, challenges and opportunities. *Sensors (Basel, Switzerland)* 14 (5): 9153–9209.

101 Eltorai, A.E.M., Fox, H., McGurrin, E., and Guang, S. (2016). Microchips in medicine: current and future applications. *BioMed Research International* 2016: 1743472–1743472.

2

Classification and Terminology

The classification of solid-state sensors provides a framework to understand their operation, application domains, and performance characteristics. By considering the physical phenomenon sensed, the sensing mechanism employed, and the materials used, engineers and researchers can select the most suitable sensor for their specific needs. This knowledge facilitates the development of innovative sensor technologies and their integration into a wide range of devices and systems, enhancing our ability to sense and interact with the world around us. The classification of solid-state sensors is crucial for understanding their diverse applications and functionalities. These sensors can be categorized based on various criteria, including the type of physical phenomenon they detect, their sensing mechanism, and the materials used in their construction. Each classification offers insights into the sensor's operation, performance characteristics, and suitability for specific applications. Again, understanding the terminology associated with solid-state sensors is crucial for effective communication and comprehension. Understanding these fundamental terms associated with solid-state sensors will aid in effective communication, troubleshooting, and selecting the most suitable sensor for a given application.

2.1   Sensor Components

The definition of a sensor does not precisely define what physical elements constitute the sensor [1]. Thus, a knowledge of sensor terminology will facilitate us in understanding the different functionaries of a working device. The definitions adopted in this book are as follows.

Sensor element: A component of the sensor that detects or measures a physical property and records, indicates, or responds to it. It forms a fundamental

Solid-State Sensors, First Edition. Ambarish Paul, Mitradip Bhattacharjee, and Ravinder Dahiya.
© 2024 The Institute of Electrical and Electronics Engineers, Inc.
Published 2024 by John Wiley & Sons, Inc.

transduction mechanism or material constituting as sensor in which one form of energy converts into another. Some sensors may incorporate more than one sensor element (e.g. a compound sensor).

Sensor: A sensor element including its physical packaging and external connections (e.g. electrical or optical).

Solid-state sensor: An integrated electronic device with no movable part, generating sensory response by the virtue of various transducing actions occurring within the active sensory matrix of the device is termed as the solid-state sensor.

Sensor system: An integrated sensor and its assorted signal-processing hardware (analog or digital) with the processing either in or on the same package or discrete from the sensor itself.

Sensor systems offer cheap and effective remedy for controlling various automatized industrial processes and provide continuous monitoring of clinical parameters in healthcare applications. For any given application, the strength of a physical quantity should be measured. A solid-state sensor is designed in such a way that the measurand, the physical property to be sensed, exploits a physical phenomenon within the sensor element. This physical phenomenon leads to an electrical response that can be detected using a sensor and magnified with integrated electronics in the form of sensor system. However, the design of such an integrated sensor system must be implemented using optimized dimensions and under technical restrictions.

2.2 Classification of Solid-State Sensors

Solid-state sensors may be classified in terms of the input and output energy to and from the sensor, respectively, taking into consideration the transduction principle and disregarding the internal and compound transduction effects. Table 2.1 lists the six energy forms or signal domains generally encountered with examples of typical properties that are measured using those energy forms. Thus, the world of solid-state sensors was divided into self-generating and modulation type transduction. Self-generating solid-state sensor operates without an auxillary energy source, whereas that which works on the modulation principle requires an auxillary energy source. The self-generating and the modulating solid-state sensors are depicted schematically in Figure 2.1a, b, respectively. An example of a self-generating sensor is a piezoelectric pressure sensor. In this case, the mechanical energy form (strain or pressure) creates electrical signal (an electrical charge) as a result of the fundamental material behavior of the sensor element. An example of a modulating sensor is a fiber optic magnetic field sensor in which a magnetostrictive jacket is used to convert a magnetic field into an induced strain in the optical fiber. The resulting change in the gauge length of the fiber is measured using

Table 2.1 List of various measurands for different forms of energy.

Energy forms	Example measurands
Mechanical	Length, area, volume, all time derivatives such as linear/angular velocity, linear/angular acceleration, mass flow, force, torque, pressure, acoustic wavelength, and acoustic intensity
Thermal	Temperature, specific heat, entropy, heat flow, state of matter
Electrical	Voltage, current, charge, resistance, inductance, capacitance, dielectric constant, polarization, electric field, frequency, dipole moment
Magnetic	Field intensity, flux density, magnetic moment, permeability
Optical	Intensity, phase, wavelength, polarization, reflectance, transmittance, refractive index
Chemical	Composition, concentration, reaction rate, pH, oxidation/reduction potential

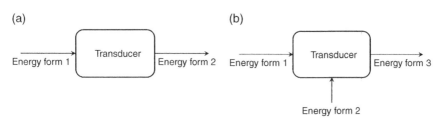

Figure 2.1 Block diagram of (a) self-generating and (b) modulating sensors showing the transformation of energy.

interferometry. Schematic representations of a piezoelectric pressure sensor and a fiber optic magnetic field sensor are depicted in Figure 2.2a, b, respectively.

The field of sensor technology is diverse and constantly evolving, with new sensor types and applications emerging regularly. Solid-state sensors can be classified based on various criteria, including their applications, operating principles, and sensing mechanisms. Based on operating principle, solid-state sensors can be classified as:

1) *Resistive sensors*: These sensors measure changes in electrical resistance, such as strain gauges and thermistors.
2) *Capacitive sensors*: These sensors measure changes in capacitance, such as touch sensors and humidity sensors.
3) *Piezoelectric sensors*: These sensors generate an electrical signal in response to mechanical pressure or vibration, such as pressure sensors and accelerometers.
4) *Chemical sensors*: These sensors detect and measure specific chemical species or changes in chemical properties, such as gas sensors, pH sensors, and biosensors.

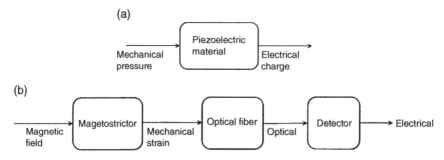

Figure 2.2 Schematic representations of (a) piezoelectric pressure sensor and (b) fiber optic magnetic field sensor.

5) *Biological sensors*: These sensors are designed to interact with biological substances or processes, such as biosensors, DNA sensors, and biomedical sensors.
6) *Optical sensors*: These sensors use light or photons to detect changes, such as photodiodes, phototransistors, and optical encoders.
7) *Magnetic sensors*: These sensors measure changes in the magnetic field, such as magnetic field sensors and current sensors.

On the other hand, the solid-state sensors can also be classified in terms of potential applications as:

1) *Environmental sensors*: These sensors are used to monitor and measure environmental parameters like temperature, humidity, air quality, and radiation levels.
2) *Automotive sensors*: These sensors are used in vehicles for various applications, including engine management, airbag deployment, tire pressure monitoring, and proximity detection.
3) *Medical sensors*: These sensors are used in medical devices and healthcare applications, such as patient monitoring, diagnostic tools, and drug delivery systems.
4) *Industrial sensors*: These sensors are employed in industrial processes and machinery for monitoring parameters like temperature, pressure, flow, and position.
5) *Consumer electronics sensors*: These sensors are integrated into various consumer devices like smartphones, wearables, gaming consoles, and home appliances for functions such as touch sensing, motion detection, and environmental monitoring.

The field of sensor technology is vast and continually evolving, leading to the development of new types of sensors and innovative applications across various industries.

The solid-state sensors can also be conveniently classified by their input energy form or signal domain of interest. However, this form of classification does not emphasize the underlying mechanisms and would be more understandable when the sensors are classified in accordance with the science-based taxonomy. Classification based on the input energy may infuse confusion among readers when multiple response interactions and transduction mechanism takes place. In order to have a scientific understanding on the sensors, we have followed the transduction principle-based classification and divided the regime of solid-state sensors into seven categories, viz. (i) piezoelectric, (ii) piezoresistive, (iii) capacitive, (iv) chemical, (v) biomedical, (vi) optical, and (vii) magnetic as depicted in Figure 2.3. Alternative sensor taxonomy includes sensor material, cost application,

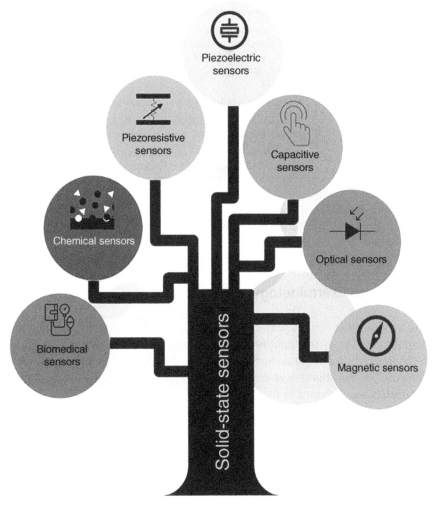

Figure 2.3 Schematic representation portraying the main classes of solid-state sensors.

Table 2.2 Various sensor applications for different classes of solid-state sensors.

S. no	Classification of solid-state sensors	Typical sensor applications
1	Physical sensor	Pressure, temperature, strain, piezoresistive, sensors for blood pressure monitoring
2	Chemical sensor	Gas, clinical analysis, environment and process control, CHEM-FET and ISFET, integrated biosensors for online monitoring through wearable sweat monitoring
3	Piezoelectric	Vibration and impact sensing, acoustic and ultrasonic transducers, force and load measurement, energy harvesting, piezoelectric generators, ultrasound imaging, medical diagnostics
4	Capacitive	Touch screens and human–machine interfaces, proximity and object detection, position and displacement measurement, level sensing and liquid-level detection, humidity and moisture sensing, pressure and force sensing, level sensing and object detection
3	Optical sensor	Photodiodes, phototransistors, color sensor, CCDs
4	Acoustic	electrocardiography, ultrasonic sensors, environmental sensors, surface acoustic wave microsensors, corrosion sensors
5	Magnetic sensor	Hall sensors, magnetotransistors, thermogalvanometric sensors, CMOS magnetic field detectors

and output signal-based classification, which are also dealt with in respective chapters. Table 2.2 lists the different types of solid-state sensors and their possible applications.

2.3 Sensor Terminology

The knowledge of sensors characteristics is imperative in correctly mapping the output signal versus the input of the measurand. The sensor characteristics are determined by various parameters that describe the operation of the sensor under different external stimuli and environmental conditions [2]. Some of the sensor parameters are explained as follows.

2.3.1 Accuracy

Accuracy of a sensing system represents the correctness of its output in comparison to the actual value of a measurand. To assess the accuracy, either the system is benchmarked against a standard measurand or the output is compared with a measurement system with superior accuracy. Accuracy can be closely defined for a given set of operating conditions, but it is vitally important to take into account all contributing factors if the best results are required. It should be appreciated that there are two main aspects to these potential sources of error. First, there is the inherent performance of the transducer, and second, the quality of the means of measurement of that performance, i.e. the calibration equipment.

2.3.2 Precision

Precision represents capacity of a sensing system to give the same reading when repetitively measuring the same measurand under the same conditions. The precision is a statistical parameter and can be assessed by the standard deviation (or variance) of a set of readings of the system for similar inputs. The difference between accuracy and precision is depicted in Figure 2.4.

2.3.3 Calibration Curve

A sensing system has to be calibrated against a known measurand to assure that the sensing results in correct outputs. The relationship between the measured variable (x) and the signal variable generated by the system (y) is called a calibration curve. When the relation between the measured variable and the signal variable is linear the calibration curve is represented as $y = mx + c$.

2.3.4 Sensitivity

It is the ratio of the incremental change in the sensor's output (Δy) to the incremental change of the measurand in input (Δx). The sensitivity can be graphically determined from the slope of the calibration curve $y = f(x)$, and for a linear curve,

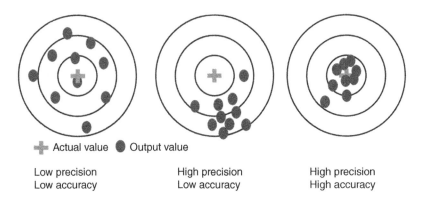

Figure 2.4 Dart illustration representing the concept of precision and accuracy.

the sensitivity is simply the slope m ($=\Delta y/\Delta x$) of the straight line. For a nonlinear calibration curve, the sensitivity of the sensor system depends on the operation range in which the measurand is detected. For example, with increasing measurand value, a calibration curve that initially follows a linear curve but saturates at a certain value of the measurand, the sensitivity is constant in the linear regime but gradually decreases in the vicinity of the saturation regime of the sensor. At saturation, the sensor no longer responds to any changes.

2.3.5 Threshold/Minimum Detectable Limit

In a sensing system, minimum detectable signal (MDS) is the minimum signal increment that can be observed, when all interfering factors are taken into account. When the increment is assessed from zero, the value is generally referred to as threshold or detection limit. If the interferences are large relative to the input, it will be difficult to extract a clear signal and a small MDS cannot be obtained. The MDS can be calculated by solving for x after setting $y = 0$.

2.3.6 Null Offset

The null offset is the electrical output present when the sensor system is not exposed to any measurand.

2.3.7 Dynamic Range

All sensing systems are designed to perform over a specified range. Signals outside of this range may be unintelligible, cause unacceptably large inaccuracies, and may even result in irreversible damage to the sensor. The maximum and minimum values of the measurand that can be measured with a sensing system are called the dynamic range. This range results in a meaningful and accurate output for the sensing system.

2.3.8 Nonlinearity

The closeness of the calibration curve to a specified straight line shows the linearity of a sensor. Linearity error is the deviation of the sensor output curve from a specified straight line over a desired output range. This linearity error is also defined as nonlinearity. There are two common ways of specifying the linearity error, viz. (i) best-fit straight-line method and the (ii) end-point linearity method. In the former method, the error is specified as the maximum deviation $\pm x\%$ of the output range from the straight line. The end-point linearity is determined by drawing a straight line perpendicular to the line joining the end data points on the output curve. The maximum value of this perpendicular line obtainable in the output range gives the linearity error.

2.3.9 Hysteresis

This represents the difference in output from a transducer when any particular value of the measurand is approached from the low and the high sides. Hysteresis may cause false and inaccurate readings. In general, hysteresis occurs when a sensing technique relies on the stressing of a particular material such as elastomers. The hysteresis caused by loosening of mechanical joints is called the backslash. These errors are fixed physical characteristics of a particular device and are generally independent of temperature.

2.3.10 Selectivity

Selectivity is the sensing system's ability to measure a target measurand in the presence of other interferences.

2.3.11 Repeatability

It is the ability of a transducer to consistently reproduce the same output signal for repeated application of the same value of the measurand. Repeatability is closely related to precision. Both long-term and short-term repeatability estimates can be important for a sensing system. The repeatability error is calculated using the relation, $\delta_{rep} = SD/\Omega$, where SD is denoted as the standard deviation of the output values over successive measurements and Ω is the mean of the dataset over which the SD was calculated.

2.3.12 Reproducibility

It is the sensing system's ability to produce the same responses after measurement conditions have been altered.

2.3.13 Resolution

Resolution is the lowest change of the measurand that can produce a detectable increment in the output signal. Resolution is strongly limited by any noise in the signal.

2.3.14 Stability

Stability is a sensing system's ability to produce the same output value when measuring the same measurand over a period of time. The stability is classified into mechanical stability and electrical stability. Since there are no movable parts in solid-state sensors, the term stability mostly refer to electrical stability. However, for bendable sensors the stability of output signal against physical bending should also be taken into consideration. The stability error is calculated by the relation, $\delta_{stab} = SD_{stab}/\Omega$, where SD is denoted as the standard deviation of the output values over repetitive measurements and Ω is the mean of the dataset over which the SD was calculated.

2.3.15 Noise

The unwanted fluctuations in the output signal of the sensing system, when the measurand is not changing, are referred to as noise. The standard deviation value of the noise strength is an important factor in measurements. The mean value of the signal divided by this value gives a good benchmark, as how readily the information can be extracted. As a result, signal-to-noise ratio (S/N) is a commonly used figure in sensing applications where S is the mean value of signal and N is the SD of noise.

2.3.16 Response and Recovery Time

Response time t_{res} is defined as the time required to reach 90% of the final and stable value when the sensing system is exposed to a measurand. Conversely, the recovery time t_{rec} is defined as the time required for the final output signal (obtained during exposure) to fall to 10% of its value after the measurand is completely withdrawn. The response and recovery times are depicted in Figure 2.5.

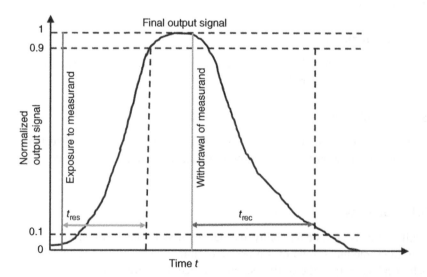

Figure 2.5 Graphical representation of the response and recovery times of the sensor obtained from the sensor output signal versus time plot.

2.3.17 Drift

Drift is the undesirable change in system output over a period of time when the measurand and the sensor position are actually constant. It is considered a systematic error, which can be attributed to interfering parameters such as mechanical instability and temperature instability, contamination, mechanical and electrical aging, environmental changes, and the sensor's materials degradation. The drift in sensor output is determined with respect to a sensor's baseline. The baseline refers to the value of output signal obtained when the sensor system is not exposed to any measurand.

2.4 Conclusion

In conclusion, solid-state sensors play a crucial role in various industries and applications by enabling the measurement and detection of physical, chemical, and biological parameters. They provide valuable data for monitoring, control, and decision-making processes. Sensors can be classified in terms of their operating principles, sensing mechanisms, and applications. The classification of sensors helps in understanding their underlying principles and choosing the most suitable sensor for a specific application. It also aids in the development of new sensor technologies and applications. As technology advances, new types of sensors and sensing techniques continue to emerge, expanding the capabilities and possibilities of sensing in various fields. The future holds the promise of more advanced, miniaturized, and integrated sensors that will further revolutionize industries and improve the quality of life.

List of Abbreviations

CCD	Charge-coupled Device
CHEM-FET	Chemical Field-effect Transistors
CMOS	Complementary Metal Oxide Semiconductor
ISFET	Ion-selective Field-effect Transistors
MDS	Minimum Detectable Signal
SD	Standard Deviation

References

1 McGrath, M.J. and Scanaill, C.N. (2013). *Sensing and sensor fundamentals.* In: *Sensor Technologies: Healthcare, Wellness, and Environmental Applications*, 15–50. Berkeley, CA: Apress.

2 Nyce, D.S. (2016). *Glossary of sensor terminology.* In: *Position Sensors* (ed. D.S. Nyce), 355–361. Hoboken, New Jersey: John Wiley & Sons, Inc.

3

Fabrication Technologies

3.1 Introduction

The solid-state sensor regime varies from simple two-electrode structure bridged by active sensor material, or stacked layers of different sensor materials sandwiched between electrodes, to complex array architecture where an $(n \times n)$ matrix of similar simple sensor elements are formed to generate dedicated response to change the localized environment. The individual sensors in the array are integrated on a common chip and controlled by suitable and separate integrated circuits (ICs). The size of the devices also varies from less than one micron to several millimeters. Smart sensors involve an integration of sensing device consisting of microstructure transducers and supporting microelectronics to control the total functioning of the device. In the last century, solid-state sensors of different types were realized covering all realms of piezoelectric, piezoresistive, chemical, biological, optical, and magnetic-based technology. These sensors in their miniaturized scale outperform the macrocounterpart in terms of sensitivity, power consumption, and user-friendliness. Successful realization of miniaturized sensors is attributed to their (i) advanced fabrication process comprising different forms of lithography, state-of-the-art etching techniques, and simple solution process routes and (ii) batch processing which provided commercial boost. The combination of low production cost and better efficiency has provided numerous commercial market opportunities.

Miniaturization open up new avenues in interdisciplinary research associated with physics, materials science, chemistry, engineering, and equipment design. The miniaturized solid-state sensors can be realized by a wide cross section of advanced microfabrication techniques and tools involving microlithography, deposition, doping, etching, polishing, and printing. To fabricate a microsensor, different processes are undertaken one after another, and many times repeatedly. The processes include film deposition, patterning the film with the desired

Solid-State Sensors, First Edition. Ambarish Paul, Mitradip Bhattacharjee, and Ravinder Dahiya.
© 2024 The Institute of Electrical and Electronics Engineers, Inc.
Published 2024 by John Wiley & Sons, Inc.

microfeatures, and etching portions of the film. The complexity of microfabrication processes can be described by performing number of times patterning step. Microfabrication resembles multiple photolithographic exposures where many patterns are aligned to each other to create the final structure.

Microfabricated devices may be fabricated on rigid substrates like Si or on flexible substrates like polyimide (PI), polyvinyl alcohol (PVA), chitosan, and polymethyl methacrylate (PMMA). Flexible sensors are not generally fabricated on freestanding bendable substrates, but are usually formed over or in a thicker support called the carrier substrate. The IC chips for electronic applications are fabricated on semiconducting substrates such as silicon wafers; optical devices or flat panel displays are fabricated on transparent substrates such as pyrex, glass, or quartz; and wearables and bendable devices are fabricated on flexible substrates deposited on carrier wafers. The substrate enables easy handling of the microdevice through many fabrication steps. Often many individual devices are made together on one substrate and then isolated into separated devices toward the end of fabrication.

The principal fabrication techniques include (i) deposition, (ii) patterning, (iii) etching, and (iv) doping, which will be discussed in this chapter. These fabrication techniques and their different types are listed in Figure 3.1.

3.2 Deposition

Microfabricated devices are typically constructed using single deposited layer or stacking of multiple thin films. The purpose of these thin films depends on the type of device and its functionalities. Electronic devices may have thin films which are conductors (metals), insulators (dielectrics), or semiconductors. Optical devices may have films which are reflective, transparent, light guiding, or scattering, while that of bendable devices use elastomers or/and flexible polymers. On the other hand, wearable devices and implantable devices utilize biocompatible materials as substrates which may be deposited by spin coater and spray coating. Films used as substrates provide mechanical stability in wearables and bendable devices and as active sensor elements in chemical, optical, and biosensors.

Many different kinds of bulk materials and thin films are used in the fabrication of Si-based micromachined devices. The bulk materials are predominantly semiconducting. The most important semiconductors in the micromachining fabrication technique are silicon and gallium arsenide. There are four important thin-film materials (or class of materials) in micromachining technique: (i) thermal silicon oxide, (ii) dielectric layers, (iii) polycrystalline silicon (poly-Si), and (iv) metal films (predominantly aluminum).

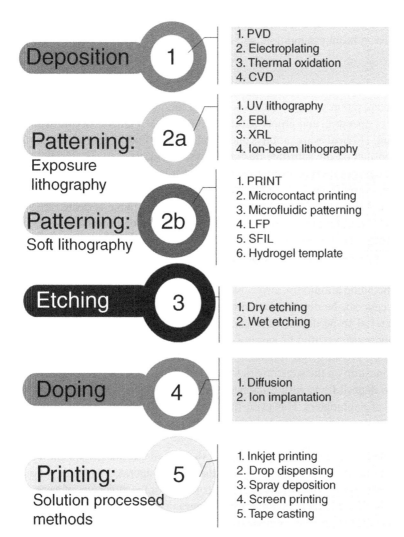

Figure 3.1 Different types of fabrication techniques and their types.

3.2.1 Physical Vapor Deposition

Physical vapor deposition (PVD) is a technique used to deposit thin films of materials onto a substrate through a physical process, typically under vacuum conditions. PVD involves the evaporation or sputtering of material from a solid source and the subsequent condensation of the vapor onto a substrate to form a thin film. There are several methods of PVD, including, thermal evaporation, sputtering, electron-beam evaporation, and laser ablation.

3.2.1.1 Thermal Evaporation

Thermal evaporation is one of the most popular among the deposition methods. Simplicity of operation and proper speed are the notable strengths of this deposition method. Thermal evaporation is one of the PVD methods during which a thin film is deposited on the substrate during a physical process.

In the process of thermal evaporation, the target material is located inside an evaporation source (boat, coil, and basket) which is heated by the passage of electric current. The target material inside the evaporation source is heated to the evaporation point. Because heat generation is due to electrical resistance of the evaporation source, this method is also called resistive evaporation. After evaporation, the molecules of the target material move to the substrate and form a thin film on the surface of the substrate. Many materials can be deposited using this method, including aluminum, silver, nickel, chromium, and magnesium.

The process can be described according to the following sequence of steps: (i) the material to be deposited is converted into a vapor by physical means (high-temperature vacuum or gaseous plasma), (ii) the vapor is transported to a region of low pressure from its source to the substrate, and (iii) the vapor undergoes condensation on the substrate to form a thin film. Typically, PVD processes are used to deposit films with thicknesses in the range of a few nanometers to thousands of nanometers. However, they can also be used to form multilayer coatings, graded composition deposits, very thick deposits, and freestanding structures.

One of the important features of vacuum evaporation systems is their ultimate pressure. In the process of thermal evaporation, if the evaporated molecules of the target material in their path to the substrate collide with the molecules of gases in the chamber, they will undergo undesired changes in their path and this will negatively affect the quality of the sample coverage. To prevent this, the thermal evaporative deposition process must be performed in a high vacuum environment. At a pressure of 10–5 Torr, the mean free path length of the molecules is 1 m. This means that the average distance traveled by the molecules in the vacuum chamber, before colliding with another molecule, is 1 m. As a result, it can be said that the molecules of the vaporized material can reach the substrate in a straight line without colliding with another molecules.

On the other hand, the presence of background gases in the deposition chamber may contaminate the deposited thin film. For example, the presence of a small amount of oxygen or moisture during thermal evaporation in the process of organic light-emitting diode (OLED) and organic photovoltaic devices fabrication makes the active functional parts of these devices inefficient. By reducing the pressure of the deposition chamber to 10–6 Torr, the purity of the deposited thin film and the performance of the devices will increase significantly.

Some of the advantages of thermal evaporation deposition include:

1) *High-purity films*: Thermal evaporation allows for the deposition of high-purity films, as the process can be carried out under high-vacuum conditions, which minimizes the introduction of impurities.
2) *Good adhesion*: Films deposited using thermal evaporation exhibit good adhesion to the substrate due to the high-energy particles that are generated during the process.
3) *Versatility*: Thermal evaporation can be used to deposit a wide range of materials, including metals, semiconductors, and organic compounds.
4) *Cost-effectiveness*: The equipment required for thermal evaporation is relatively simple and inexpensive, compared to other deposition techniques.
5) *Control over film thickness*: The thickness of the deposited film can be controlled precisely by adjusting the deposition time and rate.
6) *Uniformity*: Thermal evaporation can produce highly uniform films over large areas, making it suitable for industrial-scale production.

However, the thermal evaporation technique also suffers from limitations:

1) *Limited deposition rate*: The deposition rate for thermal evaporation is typically lower than other deposition techniques, such as sputtering or chemical vapor deposition (CVD).
2) *Limited control over film composition*: Thermal evaporation generally results in a stoichiometric ratio of the deposited material, which may not be suitable for certain applications that require precise control over the composition of the film.
3) *Limited control over film morphology*: Thermal evaporation can result in films with poor surface morphology, such as roughness, pinholes, or cracks, which can affect the film's performance.
4) *Susceptibility to substrate contamination*: Although thermal evaporation can be carried out under high-vacuum conditions, there is still a risk of substrate contamination from residual gases or impurities in the evaporation source material.
5) *Limited material choices*: While thermal evaporation can be used to deposit a wide range of materials, it may not be suitable for depositing some materials, such as insulators, that are not compatible with the process conditions.
6) *Limited film thickness*: The thickness of the deposited film is limited by the amount of material that can be evaporated from the source, which can be a drawback for certain applications that require thicker films.

Thermal evaporation deposition has variety of uses, including the following.

- Creating metal bonding layers in devices such as OLEDs, solar cells, and thin-film transistors.
- Create a thin layer of indium on the back of the ceramic sputtering targets to connect the metal backing plate.
- Deposition of metal thin film (often aluminum) on polymers for use in food packaging and heat and sound insulation.

Metalized polymer films are polymer films that are coated with a thin layer of metal (usually aluminum) using thermal evaporation deposition method. These films have a shiny, metallic appearance like aluminum foil, but are lighter and cheaper than aluminum foil. These films are mostly used for decorating purpose and packaging food products.

Polypropylene and polyethylene terephthalate (PET) polymers are commonly used for metallization. Metalized PET films are used in NASA spacesuit to reflect heat to keep astronauts warm. PET in firefighter's uniforms is also used to reflect the heat radiated from the side of the fire. Other applications of these metalized films, which are deposited by thermal evaporation coating, include emergency aluminum blankets, which are used to conserve the shock patients body heat. Metalized PET films are also used in the construction of antistatic electricity and antiheat and soundproof enclosures in aircraft.

3.2.1.2 Sputter Deposition

The word "sputtering" originates from the Latin "Sputare," which means "to spit out noisily." Sputtering crudely resembles a spray fountain made up of tiny moist particle flux. Peter J. Clarke presented the first sputtering device in 1970 when he used electron and ion collisions to deposit an atom scale coating on a target surface. Half a century later, thin film applications proved to be an inseparable part of research and development and indispensable part of various industries. Sputter deposition is a versatile technique that can be used to deposit a wide range of materials including metals, alloys, ceramics, and semiconductors onto different kinds of rigid and flexible substrates like silicon, polymers, and naturally obtained biopolymers. Sputter deposition is a PVD technique used to deposit thin films of various materials onto substrates. This deposition technique uses plasma to bombard high-energy ions on the source to eject atoms of the source material that travel to the substrate to form a thin film of source material by bonding at the atomic level. Thus, source material on bombardment by high-energy ions changes the state of matter from solid to vapor, traverses in vacuum to reach the substrate where the state returns back to solid in the form of thin film. The schematic representation of the sputtering deposition process is illustrated in Figure 3.2a. It is commonly used in the semiconductor industry to deposit thin films for

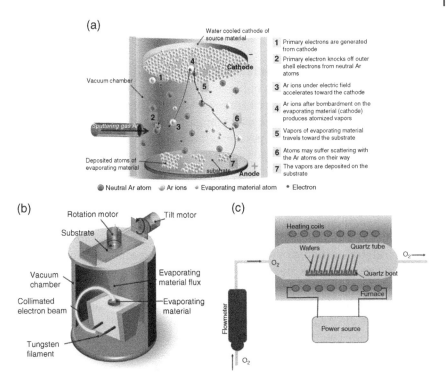

Figure 3.2 Schematic representation of the (a) sputtering deposition process and (b) EBPVD technique, and (c) basic furnace arrangement for the thermal oxidation of silicon wafers.

microelectronics and photovoltaic applications. Sputtering is also used in the manufacturing of magnetic storage media, optical coatings, and decorative films.

The process of sputter deposition initiates when the substrate that needs to be coated with the desired material is placed in a vacuum chamber ($\sim 8 \times 10^{-2}$ to 2×10^{-2} mbar) filled with inert gas atom, commonly argon. The substrate is biased negatively relative to the source material with which the substrate needs to be coated. Thus, the substrate serves as the cathode, while the source is used as the anode in the sputtering process. The inert gas filled inside the sputtering chamber is called the sputtering gas. For efficient momentum transfer, the atomic weight of the sputtering gas should be close to the atomic weight of the source material. Although argon is often used as the sputtering gas during gold deposition, neon is preferred for sputtering light elements, while for heavy elements gases such as krypton or xenon are used. Reactive gases can also be used to sputter compounds which can be formed on the surface of source material in-flight or on the substrate depending on the process parameters. The availability of many parameters that control sputter deposition not only makes it a complex process, but also allows experts a large degree of control over the growth and microstructure of the film.

When the source and the substrate are appropriately biased as cathode and anode, respectively, the electrons of the source material tear away from the surface, gains kinetic energy due to the high potential bias (~100 V–3 KV) and collides with the outer shell electrons of the sputtering gas atoms and knocks them off from the outer shell. Once the neutral gas atoms lose electrons from their outer shell, they turn into positively charged ions. In a nut shell, the primary electrons from the source material on collision with the sputtering gas atoms produces secondary electrons which knocks off outer shell electrons from other sputtering gas atoms thereby creating an avalanche of electrons which gives rise to copious supply of gas ions. These ionized gas ions form the gaseous plasma and appear as glow discharge in the sputtering chamber. The positively charged gas ions travel toward the source material (cathode) with high kinetic energy (and momentum depending on the mass of the sputtering gas atom) when the atomized source material get "sputtered" or detached from the parent source. The ionized gas ion collects an electron after bombardment on the source material and transform into neutral gas atoms before getting ionized again. Sputtered atoms ejected from the target source have a wide energy distribution, typically up to tens of eV (100,000 K). The sputtered atoms are ionized and can ballistically fly from the source target through the vacuum chamber and impact energetically on the substrates producing deposition of atomized source material on the substrate.

The advantages of sputter deposition are as follows:

1) *Precise control over film properties*: Sputter deposition allows for precise control over film thickness, composition, and microstructure. The process parameters such as target material, gas pressure, target–substrate distance, and applied voltage can be adjusted to achieve specific film properties such as optical or electrical conductivity.
2) *High uniformity*: Sputter deposition results in a highly uniform film deposition, which is important for ensuring consistent film properties across large areas or complex shapes.
3) *Versatility*: Sputter deposition is a versatile technique that can be used to deposit a wide range of materials, including metals, alloys, ceramics, and semiconductors.
4) *High deposition rates*: Sputtering offers high deposition rates, which makes it a faster process than other PVD techniques.
5) *Low substrate temperatures*: Sputtering can be carried out at low substrate temperatures, which is important for depositing films onto temperature-sensitive substrates.

6) *Compatibility with various substrate materials*: Sputtering can be carried out on a variety of substrate materials, including glass, metals, and semiconductors.
7) *Minimal waste*: Sputtering is a relatively clean process that generates minimal waste, making it an environmentally friendly technique.
8) *Good adhesion*: Sputtered films typically exhibit good adhesion to substrates, making them suitable for various applications, including microelectronics, photovoltaics, and magnetic storage media.

However, sputtering does have some limitations as listed below:

1) *Limited deposition rate for thick films*: Sputter deposition is not suitable for depositing thick films, as the target material can become depleted over time, resulting in a decrease in deposition rate.
2) *Limited scalability*: Sputter deposition may not be suitable for large-scale production of thin films, as the deposition rate can be slow for large areas.
3) *Expensive equipment*: Sputter deposition requires specialized equipment, such as a vacuum chamber and power supply, which can be expensive to purchase and maintain.
4) *High-purity target materials*: Sputter deposition requires high-purity target materials to ensure film quality and consistency, which can be expensive.
5) *Limited film uniformity in complex shapes*: Sputter deposition may result in non-uniform film deposition in complex shapes, such as those with high aspect ratios or sharp corners.
6) *Limited substrate temperatures*: While sputter deposition can be carried out at low substrate temperatures, high-temperature sputtering may be necessary for certain materials, which can limit the choice of substrates.
7) *Sputtered particles can damage sensitive substrates*: The high-energy sputtered particles can damage sensitive substrates or devices, which may require additional processing steps to protect them.

Sputter deposition is a highly effective technique for depositing thin films with precise control over film properties. Some of the variants of sputter deposition include ion-beam deposition, reactive sputtering, and ion-assisted deposition. Their versatility and efficiency make it an important tool in various industries, from microelectronics to renewable energy.

3.2.1.3 Electron-Beam PVD

Electron-beam PVD (EBPVD) is a form of PVD technique of thin-film deposition where an electron beam is used to vaporize a material for deposition onto a substrate. In this process, a high-energy electron beam is bombarded onto a target material, causing it to vaporize and form a gaseous plasma which is then directed toward a substrate, where it condenses to form a thin film. The schematic representation of the process is illustrated in Figure 3.2b. EBPVD is expensive to implement in part because of the low efficiency with which materials are evaporated and the low deposition rate. EBPVD can be used to deposit a variety of materials on substrates. EBPVD is a powerful technique for depositing high-quality thin films with excellent control over film composition and uniformity, making it useful in a wide range of applications, including electronics, optics, and aerospace.

In an EBPVD system, the deposition chamber needs to be evacuated to a pressure of $\sim 7.5 \times 10^{-5}$ Torr (10^{-2} Pa) to allow passage of electrons generated from the electron gun either by thermionic emission, field electron emission, or the anodic arc method to reach the evaporation material. However, advanced EBPVD system, which has magnetron sputtering support, can operate at vacuum levels as low as 5.0×10^{-3} Torr. The low pressure in the deposition chamber is maintained to increase the mean free path of the electrons and reduce electron collision in the chamber and allow passage of electrons from the electron gun to the evaporation material without losing significant amount of energy during flight. These high-energy swarm of electrons are directed toward the evaporating (source) material using an array of strong electromagnets and are bombarded onto the evaporating material ingot or rod. Upon striking the evaporation material, the electrons lose their kinetic energy very rapidly. This kinetic energy of the electrons is predominantly converted into thermal energy through electron–matter interactions. Although some of the incident electron energy is lost through the creation of X-rays and secondary electron emission, most of the electrons kinetic energy is converted to thermal energy when the accelerating voltage is held at 20–25 kV and the beam current is a few amperes. The thermal energy that is produced heats up the evaporation material causing them to evaporate into the gaseous phase. These atoms then precipitate on the substrate as thin film where the atoms form closed packed structures using metallic bond where the outermost electron shell of each of the metal atoms overlaps with a large number of neighboring atoms.

A clear advantage of this process is it permits direct transfer of energy to source during heating and very efficient in depositing pure evaporated material to substrate. Also, deposition rate in this process can be as low as 1 nm/min to as high as few μm/min. The material utilization efficiency is high relative to other methods and the process offers structural and morphological control of films. To add to these, the benefits of the EBPVD are manifold as described below:

1) *High-quality films*: EBPVD produces films that are dense, uniform, and of high quality. This is because the process allows for precise control over the deposition rate, which results in films that are free from defects and have excellent adhesion to the substrate.

2) *Wide range of materials*: EBPVD can be used to deposit a wide range of materials, including metals, ceramics, and semiconductors. This makes it a versatile process that can be used in many different applications.
3) *High-purity films*: EBPVD produces films that are of high purity and stress free, making it an ideal process for applications that require high purity and uniformity, such as in the semiconductor industry.
4) *Low contamination*: EBPVD is a vacuum-based process, which means that there is little or no contamination of the deposited films. This is because there is no exposure to the atmosphere during the deposition process.
5) *High flexibility*: EBPVD is a highly flexible process that can be used to deposit films onto a variety of substrates including bendable substrates, architectured surfaces, and complex shapes.

Despite several disadvantages, the EBPVD suffers from various limitations such as:

1) EBPVD is a line-of-sight deposition process when performed at a low enough pressure (roughly <10−4 Torr). Thus, metal deposition on fabrication templates of complex 3D geometries may not yield desirable results, even with the translational and rotational motion of the substrate. The inner surfaces of microfabricated structures may not be uniformly coated.
2) The filament degradation in the electron gun results in a nonuniform evaporation rate and hence nonuniform thickness of the thin film on substrates.

EBPVD is a powerful technology for the fabrication of sensors due to its ability to deposit high-quality thin films with precise control over film thickness, uniformity, and composition, making it an ideal process for creating complex and highly sensitive sensors for a wide range of applications. EBPVD can be used to deposit thin films of metal oxides, such as zinc oxide or titanium dioxide, which are sensitive to various gases. These films can be patterned using photolithography to create a gas-sensitive layer, which can be used to detect and measure gas concentrations. Again, this technique can be used to deposit thin films of various metals, such as gold or platinum, which can be used as electrodes for biosensors. These films can be patterned using photolithography to create an array of electrodes that can detect and measure biological molecules such as proteins, DNA, or cells. Moreover, EBPVD can be used to deposit thin films of various metals or alloys, such as platinum or nickel–chromium, which exhibit changes in resistance with changes in temperature. These films can be patterned using photolithography to create a temperature-sensitive layer, which can be used to measure temperature changes. Furthermore, EBPVD can be used to deposit thin films of piezoelectric materials, such as zinc oxide or aluminum nitride, which exhibit changes in electrical properties with changes in strain. These films can be patterned using photolithography to create a strain-sensitive layer, which can be used to measure changes in strain.

3.2.1.4 Laser Ablation

Laser ablation is a technique used to remove material from a solid surface by irradiating it with a laser. This process can also be used to deposit thin films of materials onto a substrate through the vaporization and subsequent condensation of the material. In laser ablation, a high-powered laser is directed onto the surface of a target material, causing it to vaporize and form a plume of atoms or molecules. This plume then condenses onto a nearby substrate, forming a thin film. The process can be carried out under vacuum or in a controlled gas environment.

Laser ablation has several advantages, including:

1) *High precision*: Laser ablation allows for precise control over the removal or deposition of material, allowing for the creation of intricate patterns and structures.
2) *High-quality films*: Films deposited using laser ablation tend to have high purity, low defect density, and good adhesion to the substrate.
3) *Versatility*: Laser ablation can be used to deposit a wide range of materials, including metals, semiconductors, and ceramics.
4) *Low-heat input*: Laser ablation typically involves low-heat input, which reduces the risk of damage to the substrate and the deposited film.
5) *Speed*: Laser ablation can be a fast process, allowing for high throughput and efficient production of thin films.

Some of the disadvantages of laser ablation include:

1) *Limited thickness*: The thickness of the deposited film is limited by the amount of material that can be vaporized from the target, which can be a drawback for certain applications that require thicker films.
2) *Limited scalability*: Laser ablation may not be suitable for large-scale production due to the cost and complexity of the equipment.
3) *Limited material choices*: While laser ablation can be used to deposit a wide range of materials, it may not be suitable for depositing some materials, such as insulators, that are not compatible with the process conditions.

Laser ablation offers a precise and flexible approach to sensor fabrication, allowing for the creation of complex patterns and structures with high reproducibility and accuracy. Laser ablation can be used to pattern electrodes on a substrate, such as a silicon wafer or a flexible polymer. The electrodes can be used to detect changes in conductivity, capacitance, or impedance, which can be indicative of various chemical or physical parameters. It can also be used to create microfluidic channels on a substrate, which can be used to transport fluids containing analytes to the sensing area. The channels can be designed to have specific dimensions and shapes, which can influence the flow rate and mixing properties. Moreover, laser ablation can be used to modify the surface of a substrate, creating a rough or porous texture that can increase the surface area and enhance the binding of molecules. This can be useful for sensing applications that rely on the interaction between the sensor surface and the analyte.

3.2.2 Electroplating

Electroplating also called the electrochemical process was invented in 1805 by chemist and inventor Luigi Valentino Brugnatelli. This process is used to deposit a layer of metal on another metallic object using an electrolytic cell. This is an electrochemical deposition process where the metallic coat on a solid metallic substrate is achieved through the reduction of cations of the former metal by means of a direct electric current. The basic electroplating process requires two electrodes – a cathode and an anode, an electrolyte or solution, and a direct current electrical source. The metallic object desired to be coated serves as the cathode, while the anode is usually a metallic stock block or inert conductive material with which the target object is to be electroplated. The electrolyte is a solution of a salt of the metal to be electroplated on the object. An external power supply is connected between the metal object and the cathode to drive an electric current through the solution. The versatility, efficiency, and cost-effectiveness of the electroplating process make it a popular choice for a variety of applications in industries such as electronics, automotive, and jewelry. It is also used to deposit copper and other conductors in forming printed circuit boards, and copper interconnects in ICs.

Let us consider a metallic object Me which needs to be electroplated by metallic copper Cu. To perform Cu coating on object Me, the object is immersed in an electrolytic chamber filled with a solution containing dissolved salt of Cu, such as aqueous solution of $CuSO_4$, which dissociates into Cu^{2+} cations and SO_2^{-4} anions in the solution. The negative and the positive terminals of the power supply are connected to the object Me (cathode) and the metallic Cu electrode (anode), respectively. Under potential bias, the dissociated Cu^{2+} cations in the electrolyte migrate toward the cathode and subsequently reduced to metallic Cu at the cathode by gaining two electrons. The metallic Cu atoms are deposited onto the metal object, forming a thin layer of metal coating on it. On the other hand, at the metallic Cu anode, the Cu atoms are oxidized by losing two electrons to turn into dissolved Cu^{2+} ions in the solution. Here the rate at which the anode is dissolved is equal to the rate at which the cathode is plated and thus the ions in the electrolyte bath are continuously replenished by the anode. The net result is the effective transfer of metal from the anode to the cathode. If this electrochemical process is attempted with lead or carbon anode which resists electrochemical oxidation, oxygen, hydrogen peroxide, or some other byproducts may be produced at the anode instead. In such a case, fresh $CuSO_4$ solution must be periodically replenished in the bath as Cu^{2+} ions are exhausted in the electrolyte solution.

Electroplating has several advantages, including:

1) *Versatility*: Electroplating can be used to coat a wide range of metal objects with various metals such as copper, nickel, gold, and silver.
2) *Enhanced properties*: Electroplating can improve the mechanical, electrical, and corrosion resistance properties of the coated metal object.
3) *High efficiency*: Electroplating can be a highly efficient process, with a high deposition rate and low material waste.
4) *Cost-effective*: Electroplating can be a cost-effective process, especially for high-volume production.
5) *Precise control*: Electroplating allows for precise control over the thickness and composition of the plated layer.

Some of the disadvantages of electroplating include:

1) *Limited to conductive surfaces*: Electroplating can only be used to coat metal objects with conductive surfaces, and nonmetallic objects require pretreatment to create a conductive surface.
2) *Limited to simple shapes*: Electroplating is not suitable for complex-shaped objects or objects with intricate features.
3) *Environmental concerns*: Electroplating can involve the use of toxic chemicals, and disposal of the waste generated during the process can be environmentally challenging.
4) *Limited to small objects*: Electroplating is typically used for small objects, and it may not be suitable for coating larger objects.

In sensor fabrication, electroplating is typically used to deposit a thin layer of metal onto a substrate or template. The substrate or template can be made from a variety of materials, including silicon, glass, ceramics, or polymers, depending on the application requirements. The electroplated metal layer can be used as the sensing element or as a conductive layer for attaching other components such as electrodes or wires. The metal layer can also be patterned or etched to create specific shapes or features that are important for the sensing function. Electroplating can also be used to create multilayered structures for more complex sensors. For example, a multilayered sensor may consist of a substrate, an intermediate layer, and a sensing layer. Each layer can be electroplated separately, allowing for precise control over the composition, thickness, and properties of each layer. One of the advantages of electroplating in sensor fabrication is the ability to create uniform and repeatable structures. The process can also be scaled up for high-volume production, making it suitable for mass production of sensors. However, electroplating for sensor fabrication also has some limitations. The process is typically limited to planar surfaces, making it difficult to fabricate sensors with complex 3D structures. Electroplating also requires precise control over the electroplating conditions, such as the current density, plating solution, and temperature, to ensure the quality and reproducibility of the plated layer.

3.2.3 Thermal Oxidation

Thermal oxidation is the method by which a thin film of SiO_2 is grown on top of a silicon wafer. The basic thermal oxidation apparatus is shown in Figure 3.2c. The apparatus comprises resistance-heated furnace, a cylindrical fused quartz tube that contains the silicon wafers held vertically in slotted quartz boat, and a source of either pure dry oxygen or pure water vapor. The loading end of the furnace tube protrudes into a vertical flow hood, wherein a filtered flow of air is maintained. The hood reduces dust in the air that surrounds the wafers and minimizes contamination during wafer loading.

Thermal oxidation of silicon in oxygen or water vapor can be described by the following two chemical reactions:

$$Si\ (solid) + O_2\ (gas) \xrightarrow{900 - 1200°C} SiO_2(solid) \tag{3.1}$$

and

$$Si\ (solid) + O_2\ (gas) \xrightarrow{900 - 1200°C} SiO_2(solid) + 2H_2\ (gas) \tag{3.2}$$

The silicon–silicon dioxide interface transverses the silicon during the oxidation process. Using the densities and molecular weights of silicon and SiO_2, it can be shown that growing an oxide of thickness x consumes a layer of silicon that is $0.44x$ thick. The basic structural unit of thermal SiO_2 is a silicon atom surrounded tetrahedrally by four oxygen atoms, as shown in Figure 3.3a. The silicon–oxygen and

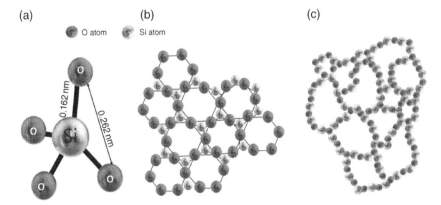

(a) (b) (c)

● O atom   ◉ Si atom

0.162 nm
0.262 nm

Figure 3.3 Schematic representation of (a) the molecular structure of SiO_2 unit cell, (b) crystalline SiO_2 in quartz, and (c) disordered array of amorphous SiO_2 in glass.

oxygen–oxygen interatomic distances are 1.6 and 2.27 Å, respectively. This SiO_4 unit cell forms the basic building block of crystalline quartz (Figure 3.3b). Thermally grown oxides are usually amorphous in nature (Figure 3.3c). One advantage of high-pressure oxide growth is that oxides can be grown at significantly lower temperatures and at acceptable growth rates. Thermally grown oxide layer has excellent electrical and mechanical properties and is used for insulation between conducting layers.

3.2.4 Chemical Vapor Deposition

Chemical gases or vapors react on the surface of solid, produce solid byproduct on the surface in the form of thin film. Other byproducts are volatile and leave the surface. There are three deposition methods that are commonly used to form a thin film on a substrate. These methods are all based on CVD and are as follows: (i) atmospheric pressure CVD (APCVD), (ii) Low-pressure CVD (LPCVD), and (iii) Plasma-enhanced CVD (PECVD) or plasma deposition.

3.2.4.1 Atmospheric Pressure Chemical Vapor Deposition
An APCVD reactor is shown in diagram Figure 3.4a. This reactor is used to deposit silicon dioxide. The samples reach the reactor through a conveyor belt. Due to the

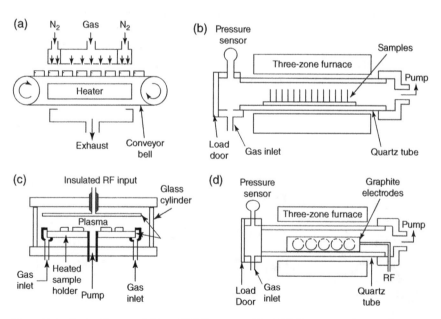

Figure 3.4 Typical layout of (a) an APCVD reactor, (b) a LPCVD reactor, (c) a parallel-plate PECVD reactor, and (d) hot-wall PECVD reactor.

high-speed flow of nitrogen, gas curtains are formed which cover the reactant gases flowing through the center of the reactor. The samples are heated by convection. The advantages of APCVD reactors are – (i) high throughput and (ii) Ability to handle large diameter wafers. However, this process suffers from the disadvantages of contamination of film and requirement of fast gas flow.

3.2.4.2 Low-Pressure Chemical Vapor Deposition

Figure 3.4b shows an LPCVD reactor that is used to deposit polysilicon, silicon dioxide, and silicon nitride. The reactor consists of a quartz tube heated by a three-zone furnace. The gas is introduced through one end of the furnace and pumped out at the opposite end. The pressure inside the reaction chambers varies from 30 to 250 Pa, with a temperature range between 300 and 900°C. Wafers are kept in a quartz holder and are kept to stand in the vertical position, perpendicular to the gas flow. Such a reactor can easily hold 150 mm diameter wafers. Each run processes 50–200 wafers, with thickness uniformities of the deposited films within $\pm5\%$. The type of LPCVD reactor shown in Figure 3.4b is a hot-wall LPCVD reactor, in which the quartz tube wall is hot because it is adjacent to the furnace; this is in contrast to a cold-wall LPCVD reactor, such as the horizontal epitaxial reactor that uses radiofrequency (RF) heating. The advantages of LPCVD process are – (i) Achievement of pure and highly uniform film and (ii) Ability to accommodate large diameter wafers. This process suffers from the disadvantages of low deposition rate of film and frequent use of toxic corrosive or flammable gases.

3.2.4.3 Plasma-Enhanced Chemical Vapor Deposition

Figure 3.4c shows a plasma-enhanced CVD (PECVD) or plasma deposition reactor which is a radial-flow, parallel-plate type. The reaction chamber is a cylinder, with aluminum plates on the top and bottom. Samples lie on the grounded bottom electrode. An RF voltage is supplied to the top electrode so as to create a glow discharge between the two plates. This causes the gases to flow radially through the discharge. These gases begin at the outer edge and take the direction toward the center.

However, if needed, the pattern of the flow can also be reversed. Resistance heaters or high-intensity lamps heat the bottom, grounded electrode to a temperature between 100 and 400°C. Due to its low temperature deposition, this reactor finds its application in the plasma-enhanced deposition of silicon dioxide and silicon nitride. However, the disadvantages of this reactor are – (i) Its limited capacity is limited, especially for large diameter wafers, (ii) Requirement of individual loading and unloading of wafer, and (iii) Contamination of wafer.

Figure 3.4d shows a PECVD or plasma deposition reactor of hot-wall type. This reactor will help in solving most of the problems that occurred in radial-flow reactor. The reaction takes place in a quartz tube heated by a furnace. The samples should be held vertically, and that too parallel to the gas flow. The samples should be supported with the help of materials like long graphite or aluminum slabs, kept in the electrode assembly. A discharge is produced in the space between the electrodes. This discharge is produced by the alternating slabs that are connected to the power supply. The advantages of hot-wall PECVD reactor are – (i) Its high holding capacity and (ii) Low deposition temperature. However, this process still suffers from problem of individual handling of wafer during its loading and unloading.

3.3 Exposure-Based Lithography Techniques

The word lithography derived from the word "Lithos," which in Greek means stone, was invented by Alois Senefelder in the Kingdom of Bavaria in 1796. In earlier days, lithography worked on limestone using mutual repulsion of oil and water to create images. However, in modern times, this technology in its microscale was used to fabricate miniaturized devices with ICs and compact circuitry.

Microlithography is the process of imprinting a geometric micropattern from a mask onto a thin layer of material called a resist, which is a radiation sensitive material. The pattern-transfer process is accomplished by using a lithographic exposure tool that emits radiation. The performance of the tool is determined by three properties: (i) resolution, (ii) Registration, and (iii) Throughput. They are described as:

1) *Resolution*: Resolution is defined in terms of the minimum feature size that can be repeatedly exposed and developed in the photoresist layer.
2) *Registration*: Registration is a measure of how accurately patterns on successive masks can be aligned with respect to previously defined patterns on wafer.
3) *Throughput*: Throughput is defined as number of wafers exposed per unit time.

Depending on the resolution, several types of radiation, including electromagnetic (e.g. ultraviolet [UV] and X-rays) and particulates (e.g. electrons and ions), may be employed in lithography. Based on the types of radiation used in lithography technique, it can be classified as (i) UV lithography, (ii) electron-beam lithography (EBL), (iii) X-ray lithography, and (iv) ion-beam lithography. Optical lithography uses UV radiation ($\lambda = {\sim}0.2$–$0.4\,\mu m$). Optical exposure tools are capable of approximately $1\,\mu m$ resolution, $0.5\,\mu m$ registration, and a throughput of 50–100 wafers/h. Because of backscattering, EBL is limited to a 0.5-μm resolution with 0.2-μm registration. Similarly, X-ray lithography typically has 0.5-μm resolution with 0.2-μm registration. However, the electron-beam, X-ray, and ion-beam lithographies require more complicated masks that of the photolithography. Thus, the optical lithography used for both bulk and surface micromachining techniques are described in this chapter in detail.

3.3.1 UV Lithography

Photo lithography requires (i) exposure tool, (ii) Masks containing design information, and (iii) Photosensitive resist (positive and negative photoresists).

3.3.1.1 Exposure Tool

Optical lithography uses two methods for imprinting the desired pattern on the photoresist. These two methods are shadow printing and projection printing. In shadow printing, the mask and wafer are in direct contact during the optical exposure (contact printing is shown in Figure 3.5a) or are separated by a very small gap g that is on the order of 10–50 μm (proximity printing is shown in Figure 3.5b).

The minimum line width (L_{min}) that can be achieved by using shadow printing is given by,

$$L_{min} = \sqrt{\lambda g} \tag{3.3}$$

The intimate contact between the wafer and mask in contact printing offers the possibility of very high resolution, usually better than 1 μm. However, contact printing often results in mask damage caused by particles from the wafer surface that gets attached to the mask. These particles may end up as opaque spots in regions of the mask that are supposed to be transparent.

Projection printing is an alternative exposure method in which the mask damage problem associated with shadow printing is minimized. Projection printing exposure tools are used to project images of the mask patterns onto a resist-coated wafer several centimeters away from the mask as shown in Figure 3.6. To increase resolution in projection printing, only a small portion of the mask is exposed at a time. A narrow arc-shaped image field, about 1 mm in width, serially transfers the

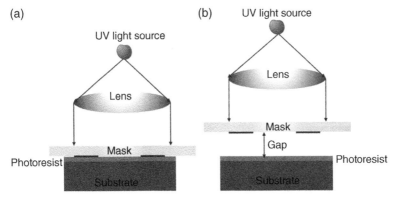

Figure 3.5 Basic lithographic mask arrangements: (a) shadow printing and (b) proximity printing (not to scale as chrome layer on glass mask is exaggerated).

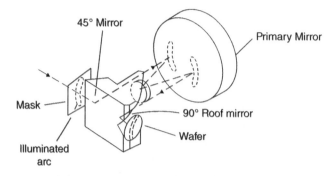

Figure 3.6 Basic lithographic arrangement for mask projection.

slit image of the mask onto the wafer. Typical resolutions achieved with projection printing are on the order of 1 μm.

3.3.1.2 Mask

Generate 2D physical layout of device/circuit – consists of different layers. Layout data are given as input to the mask making system (shown in Figure 3.7). Mask is an optically flat glass/quartz substrate consists of transparent and opaque regions. Based on the material used to fabricate the opaque region of the mask, it can be classified as – (i) photo emulsion-coated mask plate and (ii) Chrome mask plate

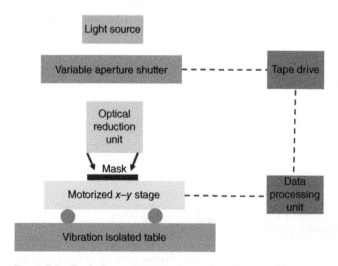

Figure 3.7 Typical arrangement of a mask making machine.

(higher resolution). Again, the mask is of two types depending on photoresist used in the lithography process – (i) bright-field mask and (ii) Dark field mask.

3.3.1.3 Photoresist

Photoresist is an organic chemical consisting of resin, photoactive compound, and solvent (to maintain proper viscosity). The characteristics of a good photoresist are – good adhesion and resistance to dry and wet etch process. The photoresist is of two types – (i) negative and (ii) positive photoresists.

A) Negative Photoresist

On exposure of negative photoresist to light, photoactive agent reacts with rubber to form cross-links between rubber molecules, making the rubber less soluble in an organic developer. This type of resist is composed of (i) resin: polyisoprene rubber, (ii) Photoactive component: *bis*-aryl-diazides, and (iii) Solvent: aliphatic and aryl hydrocarbons. For negative photoresist, the pattern is formed by masking away the unpolymerized resist with solvent (hydrocarbon mixture).

B) Positive Photoresist

Positive photoresist does not rely on polymerization. The photoactive component is dissolution inhibitor. When destroyed by exposure to light, the resin becomes more soluble in aqueous developer solution. This type of resist is composed of (i) Resin: phenol formaldehyde novolac, (ii) Photoactive component: diazo-oxides, and (iii) Solvent: ethylene glycol monoethyl ether. For positive photoresist, the pattern is formed by washing away the exposed resist in aqueous developer solution. A comparative study on the characteristics of negative and positive photoresists is given in Table 3.1.

Figure 3.8 shows schematically the lithographic process that is used to fabricate a microdevice. First, the oxide film deposited on the surface of the silicon wafer as shown Figure 3.8a. In the following step, a positive photoresist is spin coated or sprayed onto the oxide layer (Figure 3.8b). After that, the resist is exposed to

Table 3.1 A comparative study on the characteristics of negative and positive photoresists.

	Negative photoresist	Positive photoresist
1.	Limited ability to resolve fine line because of (i) size of molecules and (ii) chain reaction	Higher resolution
2.	Swelling of polymerized resist	Does not swell
3.	Aspect ratio is low	Aspect ratio is high
4.	Large throughput	Lower throughput

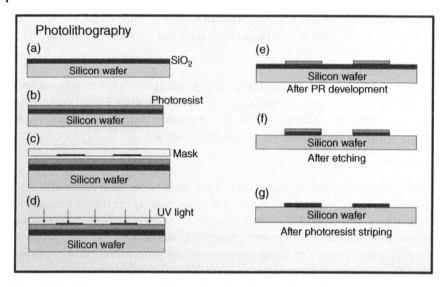

Figure 3.8 Fabrication process flow during photolithography. (a) Deposition of SiO_2 on Si by thermal oxidation; (b) spin coating of photoresists on the oxide layer and dried; (c) the photoresists are covered by pattern carrying mask and (d) exposed to ultraviolet (UV) light; (e) exposed parts of the photoresists are soluble in the developer creating a pattern of photoresist on SiO_2 layer; (f) etching of the SiO_2 layer; (g) the unexposed parts of the photoresists are removed.

UV light (using exposure tool) through a pattern carrying mask (Figure 3.8c, d). The positive resist is then rendered soluble in a developer when it is exposed to radiation. Thus, to develop the positive photoresist, its exposed areas are washed away by using a developer and pattern is formed in the resist as shown in Figure 3.8e. After that, the etching of oxide layer (except the under-lining layer of patterned photoresist) is performed (Figure 3.8f). Finally, the dry plasma strip of O_2 is used to remove the remaining photoresist and the pattern of the mask is transferred into the oxide layer (Figure 3.8g). This oxide layer serves as hard mask for the silicon wafer.

3.3.2 Electron-Beam Lithography

EBL refers to a direct writing maskless lithographic process that uses a focused beam of electrons to form nanoscale patterns by material modification, material deposition (additive), or material removal (subtractive). The schematic diagram of EBL is shown in Figure 3.9.

Similar to photolithography, substrates for EBL are coated with Na resist that either cross-link when struck by electrons, rendering it less soluble in developer

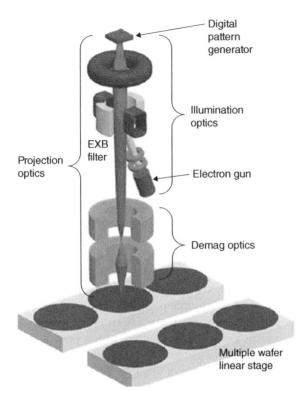

Digital
pattern
generator

Illumination
optics

EXB
filter

Projection
optics

Electron gun

Demag optics

Multiple wafer
linear stage

Figure 3.9   Schematic representation of EBL set up.

solution (negative EBL), or alters the resist to become more soluble (positive EBL). Again, the EBL utilizes an electron beam instead of a visible or UV light beam in the lithography process. Just as UV, energetic electrons can change the chemical characteristics or the solubility of the resist, enabling selective removal of either the exposed or nonexposed regions of the resist. As opposed to conventional UV photolithography, the resolution of EBL can reach precision levels down to 1 nm. Since the electron beam radiation is of a much shorter wavelength than that used in direct-write photolithography, the spatial resolution of EBL is greater than direct-write photolithography. For this reason, EBL is a very powerful technique for creating patterns at the nanoscale, with feature sizes of 10–100 nm easily fabricated by EBL.

The advantages of EBL are listed as:

- *Mode of fabrication*: maskless direct-write format.
- *Focusing*: much more tightly than light, which makes much finer patterns with sub-10 nm resolution.

- *Resolution*: Higher resolution than light, limited by lateral scattering of (secondary) electrons from the beam-defined location.
- *Proximity effect*: Spilling over exposure of an adjacent region into the exposure of the currently written feature, effectively enlarging image, and reducing its contrast.
- *Electron energy*: Fast (higher energy) electrons provide higher resolution. Slow (low energy) electrons especially in the range <0–5 eV compromise resolution as they have significant range (as far as 100 nm), suffering very little loss on colliding resist molecules. Low-energy electrons are weakly interacting because their kinetic energy falls below the ionization potential.

EBL, however, has several limitations: first, although existing SEM systems can be easily configured for EBL, they are expensive, so EBL is not a routine technique that can be set up in a laboratory, and hence requires access to a dedicated instrument that is typically located in a shared instrumentation facility. Second, EBL requires high vacuum, which limits its utility for patterning biological molecules. Third, EBL is slow and can only pattern relatively small areas, though larger areas can be "stitched together" by moving the substrate using a motorized stage. The high resolution, low throughput, and high cost make EBL useful largely as a method for high-resolution fabrication of nanoscale patterns over a small area.

EBL scans a focused beam of electrons in a patterned fashion across a surface in order to create very small structures, such as ICs or other nanostructures. There are two basic ways to scan an electron beam – raster scanning and vector scanning. In raster scanning, the image is partitioned in pixels that are printed in a left-to-right/top-to-bottom sequence. The beam sweeps the substrate horizontally left to right at a steady rate, turning on when surface exposure is needed and turning off when exposure is not required. After one line is finished, the beam blanks and then moves back to the left where it starts sweeping the next line. In raster scanning, the beam scans the whole surface even in those areas where no features are present. In vector scanning, the image is partitioned in features (vectors). The electron beam is directed only to the specific positions where features are present, and hops from feature to feature. In vector scanning, time is saved because the beam is able to move in any direction and does not scan the whole surface. In EBL, the minimum obtainable feature dimension is not limited by diffraction because the quantum mechanical wavelength of high-energy electrons is exceedingly small. The resolution of EBL is related to the spot size of the focused beam, but it is also affected by forward scattering of the electron beam inside the resist and by backscattering from the substrate and the resist itself. A major commercial application of EBL systems is photomask production. These masks are made on a quartz substrate, which provides ideal transmission characteristics in the ultraviolet region, a wavelength range normally used in photolithography. Masks are made by depositing a chrome layer on the quartz glass plate. An electron beam-sensitive resist layer coated on the top will subsequently be exposed and developed to generate the required pattern.

3.3.3 X-Ray Lithography

X-ray lithography (XRL) is a process used in electronic industry to selectively remove parts of a thin film. It uses X-rays to transfer a geometric pattern from a mask to a light-sensitive material, chemical photoresists, or simply "resist," on the substrate. A series of chemical treatments then engraves the produced pattern into the material underneath the photoresist. XRL uses X-rays, characterized by a wavelength of 0.4–4 nm, usually generated by a synchrotron source, to expose a resist-coated sample through a proximity shadow mask. The advantage of using X-rays in this regime of wavelength is to achieve enhanced resolution and a large depth of focus. However, these advantage features of X-rays in XRL come with a cost largely based on its absorption coefficient. In general, the absorption coefficient is proportional to $\rho Z^4 \lambda^3$ over a wide range of wavelengths, where ρ is the density, Z is the atomic number, and λ is the wavelength. Abrupt decrease in the absorption coefficient is attributed to the energy required to ionize inner shell electrons of a given material. The difference in transparencies of materials is not very distinct in the regime over which X-ray lithography is practiced. Thus, it is not possible to generate an X-ray mask structure analogous to the familiar optical mask structure: a thin absorbing layer deposited on a thick, rigid, glass plate that is very transmissive. Hence, the membrane material must be relatively thin, with the absorber thickness selected to provide an acceptable mask contrast. Mask contrast can be defined as the ratio of the X-ray intensity through the membrane relative to that through the membrane and absorber. The most challenging point is mask fabrication due to their unit magnification and the necessity of constructing the mask from materials that are adequately X-ray absorbing. In extreme ultraviolet (EUV) light, sources with $\lambda = 13.5$ nm are used. The main problem of this technique is the throughput. The throughput of EUV can be improved by increasing the light source power and the resist sensitivity. Despite the high-resolution capabilities, this technique failed to provide an economically attractive lithographic process owing to the high tool costs.

3.3.4 Ion-Beam Lithography

Ion-beam lithography (IBL) or focused IBL (FIBL) refers to a direct writing process that uses a narrow scanning ion-beam source (e.g. 20 nm in diameter) typically of gallium ions. IBL is employed for several nanofabrication processes involving milling, etching, ion implantation, and resist exposure. Since ions have much heavier mass than electrons by 3–5 orders of magnitude, they have much less back scattering, which results in less proximity effects. Second, because of the heavier particles and more momentum, they suffer less forward scattering; hence, it affords higher resolution patterning than UV, X-ray, or EBL. The large mass and momentum give the ion beam a smaller wavelength than even an electron beam and therefore almost no diffraction. As compared with EBL, IBL requires only about 1–10% of the particle dose to expose a resist. The resist can be an ordinary PMMA resist that absorbs most of the ions during exposure. Therefore, radiation damage to sensitive underlying structures is minimized and small as compared with that of EBL.

FIBL can be used in direct writing format, or it can be used with a mask. In masked IBL, the beam passes through ion-transparent membrane, patterned with absorber material, positioned close to the resist-coated substrate. The most common resists used in this application is PMMA. The advantages and disadvantages of IBL processing can be summarized as follows:

- High exposure sensitivity.
- Two or more orders of magnitude higher resolution than that of EBL.
- Negligible ion scattering in the resist and low backscattering from the substrate.
- Penetration of the particle beam is small, compared to electron beam, reducing blurring resulting from beam scattering (reduce proximity effect, increase localization, precision).
- Can be used as physical sputtering etch, milling, and chemical-assisted etch (see Chapter 4 for use in preparation of plasmonic nano tweezer tips).
- Can also be used as direct deposition or chemical-assisted deposition, or doping.

However, FIBL suffers from lower throughput and extensive substrate damage.

Ion-beam techniques such as ion projection can be used to structure materials on micrometer and nanometer scales. It can complement current optical or EBL techniques for micro- or nanodevice fabrication. Besides IBL, also direct ion-beam structuring methods that do not rely on resist materials for the pattern transfer have been investigated. Local sputtering, magnetic nanopatterning, and inducement of selective electroplating are examples for such resistless ion-beam patterning techniques. A new type of instrument based on a parallel multibeam concept will allow combining high resolution with high throughput.

3.4 Soft Lithography Techniques

Soft lithography technique is a family of methods used for fabricating and patterning microstructures in submicron scale by replicating the same microstructural features using an elastomeric stamp, molds, microfluidic channels, and conformable photomask. The family of techniques are termed "soft" as they are performed using soft elastomeric materials like the polydimethylsiloxane (PDMS).

The advantages of soft lithographic techniques are as follows:

1) Cost-effective alternative to photolithography which used costly exposure tools and instruments.
2) The use of patterned elastomer as mask, stamp, or mold enhances large area batch fabrication and cuts down processing time.
3) Basis in self-assembly tends to minimize certain types of defects.
4) Many soft lithographic processes are additive and minimize waste of materials.
5) Isotropic mechanical deformation of PDMS mold or stamp provides routes to complex patterns.
6) Optical transparency of the mask allows through-mask registration and processing.
7) Embrace chemical concepts of self-assembly, templating, and crystal engineering, with soft lithographic techniques of microcontact printing and micromolding.
8) Shape materials over different length scales from 1 to 500 mm.
9) Pattern 2D and 3D structures on planar and curved surfaces.
10) Different material can be processed simultaneously and provides more pattern-transferring methods, thus offering versatility in device fabrication process.
11) Does not need a photoreactive surface to create a nanostructure.
12) Higher resolution than conventional photolithography, no diffraction limit, and features as small as 30 nm have been fabricated.
13) Good control over surface chemistry, very useful for interfacial engineering.
14) Best suited for biological patterning, bendable electronics, and microfluidics.

However, soft lithographic techniques suffer from a few disadvantages such as:

1) Patterns in the stamp or mold may distort due to the deformation (pairing, sagging, swelling, and shrinking) of the elastomer used.
2) Difficulty in achieving accurate registration with elastomers.
3) Compatibility with current integrated circuit processes and materials must be demonstrated.
4) Defect levels higher than photolithography.

The different techniques included in the family of soft lithography include (i) particle replication in nonwetting templates (PRINT), (ii) microcontact printing, (iii) microfluidic patterning, (iv) laminar flow patterning (LFP), (v) step and flash lithography, and (vi) hydrogel template patterning, and are discussed as follows.

3.4.1 Particle Replication in Nonwetting Templates

PRINT is a continuous, roll-to-roll, high-resolution molding technology which allows the design and synthesis of precisely defined micro- and nanoparticles [1]. This technology blends the lithographic-based fabrication technique with the roll-to-roll processes to obtain control over particle size, shape, chemical composition, and surface properties. The PRINT process initiates with the fabrication of Si master template used for the preparation of the elastomeric molds. For this, the Si wafers are patterned with desired features using standard photolithographic techniques and forms the master templates in the PRINT process. The liquid precursor of the elastomer when poured over the Si master template completely wets the Si wafer with micro- and nanosized patterns because of its positive spreading coefficient on almost all surfaces and can be cured to generate an elastomeric PRINT mold with complementary micro/nanofeatures derived from the patterns on the Si master template. A liquid-based patternable material is filled into the cavities (complementary micro/nanofeatures) without wetting the noncavity area using a roll-to-roll process. The roll-to-roll process uses the film-split technique against a high-surface-energy PET counter sheet which passed through a roller

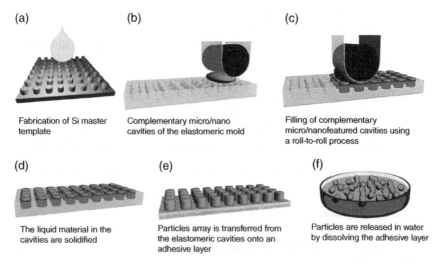

(a) Fabrication of Si master template

(b) Complementary micro/nano cavities of the elastomeric mold

(c) Filling of complementary micro/nanofeatured cavities using a roll-to-roll process

(d) The liquid material in the cavities are solidified

(e) Particles array is transferred from the elastomeric cavities onto an adhesive layer

(f) Particles are released in water by dissolving the adhesive layer

Figure 3.10 Schematic representation of PRINT technique. (a) Fabrication of Si master template. (b) Preparation of complementary micro/nano cavities of elastomeric mold using the Si master template. (c) Filling of complementary micro/nano cavities with liquid material using a roll-to-roll process. (d) Solidification of liquid material in the micro/nano cavities. (e) Solidified array of micro/nano particles is transferred from the elastomeric cavities onto an adhesive layer. (f) Particles are dissolved in water by dissolving the adhesive layer. *Source:* Adapted from Xu et al. [1] and Jeong et al. [2].

and applies pressure to the mold, thereby spreading the liquid over the microcavities. The liquid in the mold cavities is solidified using one of the following ways: (i) photocuring, (ii) vitrification by filling at an elevated temperature and cooling down, and (iii) solvent evaporation. When the liquid in the mold cavities is solidified, the array of particles can be transferred from the elastomeric cavities by bringing the mold in contact with an adhesive layer (yellow) which can collect the particles from the low-surface-energy mold. The transferred array of particles can be released in water by dissolving the adhesive layer. The schematic fabrication process flow is shown in Figure 3.10a–f.

The stochiometric composition and size of particles fabricated using PRINT technology can be modified under desired demand. The particles prepared using PRINT includes: (i) hydrogels such as cross-linked poly(ethylene glycol)s (PEG) and poly(silyl ether)s [3, 4]; (ii) thermoplastic polymers such as poly(lactic acid) (PLLA) and poly(lactic-co-glycolic acid) (PLGA) [3, 5]; (iii) biologics such as insulin and albumin; and (iv) pure small-molecule compounds including sugars and small molecular drugs [6]. Multiphasic and region specifically functionalized particles can also be fabricated using PRINT. Zhang et al. [7] developed a strategy to prepare end-labeled particles, biphasic Janus particles, and multiphasic shape-specific particles by integrating two compositionally different chemistries into a single particle. The surface properties of the PRINT particles can be easily modified, and the porosity, texture, and modulus of the particles can be altered through careful design of matrix formulation.

3.4.2 Microcontact Printing

Microcontact printing is a process of transferring materials from a prepatterned elastomeric stamp or mold with micro/nanofeatures onto a receptor substrate by contacting the elastomeric stamp with the substrate as shown in Figure 3.11. Microcontact printing is a particular type of soft lithography and replica molding procedure which consist of two parts: stamp inking and printing. The patternable ink material ink solution is typically composed of proteins, protein mixtures, or small molecules. The stamp is a complementary elastomeric mold and produced using soft lithography approaches by pouring and subsequently curing liquid elastomeric resin on Si master template. Generally, the PDMS stamp is replica molded by casting a solution of PDMS prepolymer onto a photolithographically fabricated patterned Si master template. This stamp is immersed into a solution of the ink material by which the ink adheres to the embossed regions of the stamp. Alternatively, the ink can also be collected using the stamp from a donor substrate on which the ink is predeposited. The inked pattern on the stamp is transferred on the receptor substrate at the interface of stamp and the donor substrate through

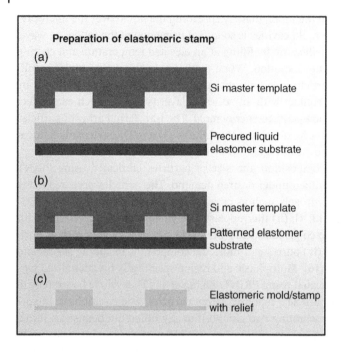

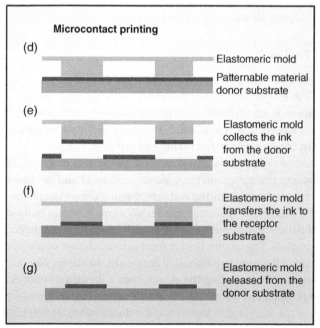

Figure 3.11 Fabrication process flow for the preparation of (a) Si master template and (b, c) elastomeric stamp followed by microcontact printing process (d–g).

conformal contact between the two. This microcontact printing process yields a pattern of inked material on the surface of the receptor substrate, allowing the pattern and deposition of geometrically defined and spatially addressable materials onto flat surfaces. Microcontact printing has been used to create microscale patterns of adhesive biomolecules, like collagen and fibronectin, in cell-repellent substrates [8]. This results in cellular arrays in which cell growth is restricted to the islands of adhesive materials [9]). Thus, *in vitro* culture of geometrically organized cell communities is possible using such patterning techniques. Alkanethiols are often microprinted onto a gold surface to prepare cell micropatterns [10]. Here the microprinted substrates are backfilled with PEG-thiol followed by immersion in a fibronectin solution. This process yields selective adsorption of fibronectin onto alkanethiol-stamped regions, which allows the selective attachment of cells onto the stamped regions.

3.4.3 Microfluidic Patterning

The microfluidic patterning is a process closely related to microcontact printing of particles, organic molecules, and biological cells [11] on a target substrate. However, the striking difference in microcontact printing and microfluidic printing is that in the former, the embossed regions elastomeric stamp forms the relief of the Si master template, while the same regions are attached to the target substrate in the latter and the gap forms the microchannels. When the materials solution/biological cells are injected through the microchannels the target regions on the substrate are exposed to the flow, resulting in patterning of the material. Although this method is less frequently used than microcontact printing, it is more useful to micropattern multiple components on a surface because single microcontact printing divides the surface into only two regions unless a multilevel stamp is used. The schematic representation of the microfluidic patterning process is shown in Figure 3.12. Microfluidic patterning assists a directed delivery of cells onto a surface, in addition to selective deposition of materials [12].

The major advantages of the microfluidic patterning are minimal wastage of materials since the material is directly fed into the microchannels for patterning, multiple materials can be patterned on the target substrate using a single elastomeric stamp, reusability of the elastomeric stamp, and low cost and flexibility of the stamp to serve as a versatile, cost-effective, high-throughput patterning technique. However, the potential drawback of this scheme lies in its mechanical instabilities imposed by the soft stamp – such as lateral collapse or pairing of narrowly spaced features, or sagging of the region between two features widely apart. This restricts the pattern geometries that can be produced; blurring of the pattern caused by diffusion of the solution of molecules stamped; the risk of contamination of the surface to be patterned by transference of molecules or fragments from the

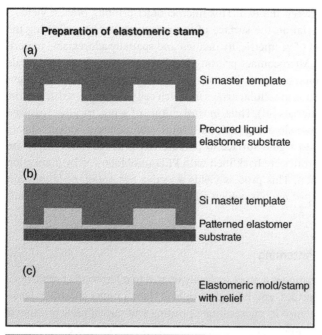

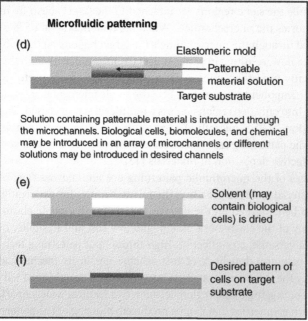

Figure 3.12 Fabrication process flow for the preparation of prerequisites – (a) Si master template and (b, c) elastomeric stamp followed by microfluidic process (d–f).

stamp material with consequences for the wetting; and adhesion properties of the patterned substrate, multiple printing, and difficulty of achieving precise alignment of the stamp with previous printing steps and the resulting pattern inhomogeneity along large surfaces. Microfluidic patterning was successfully used in the patterning of cells [13], proteins [14], DNA arrays [15], organoids [16], or gold nanoparticles [17] on substrates such as glass slides, polymers [15], hydrogels [18], silicon nitride [19], and SiO_2.

3.4.4 Laminar Flow Patterning

LFP is a unique case of microfluidic patterning first developed by Takayama et al. [20], where the flow configuration must be highly controlled. LFP technique is used to pattern surfaces with particles and biomaterials including cells and also position them so as to develop a gradient along its path. The LFP utilizes laminar flow principle in liquid which is achieved when it flows through microfluidic channels where the ratio of inertial to viscous forces of the flow configuration is low, thereby limiting the effect of turbulence. Such flow configuration through capillaries is termed as the low Reynold number (Re) flow [21, 22] and is the primary prerequisite of achieving laminar flow-based patterning. Re is a dimensionless parameter relating the ratio of inertial to viscous forces in a specific fluid flow configuration, and is a measure of the tendency of the liquid to develop turbulence [23]. Laminar flow allows two or more streams of fluid to flow side by side in a channel without convective mixing. Diffusion of the constituent molecules of fluids present in respective channels cannot be ruled away across the boundaries when the fluid flows side by side in the main channel [24]. However, in such cases, the diffusion of particulate matter, for example cells across the boundaries, is very slow and can be easily addressed during fabrication [25]. This ability to generate and sustain laminar streams of different solutions in capillaries provides a unique opportunity to pattern cells not only by positioning them along the stream, but also chemically modifying the surface on which the cells are attached, allowing other particulates to flow alongside in its vicinity and the composition of the fluid medium surrounding the patterning cells [20]. LFP is blessed with the unique ability that allows patterning of the culture medium itself in a highly controlled fashion [26].

The steps for LFP initiate with the fabrication of elastomeric template for realization of network of microfluidic capillaries with multiple inlets that converge into a single main channel. This elastomeric template prepared by casting precured liquid elastomer (PDMS) on suitable complementary Si master mold and subsequently processed for polymerization. The as-prepared elastomeric template with the network of microcapillary channels is attached to the flat surface of the substrate with the relief side in contact with the substrate. The network of relief

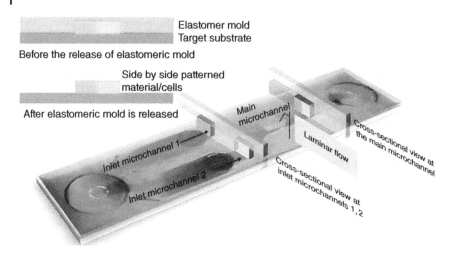

Figure 3.13 Schematic diagram representing the laminar flow patterning.

patterns forms continuous microfluidic channel. By allowing different patterning fluid components to flow from different inlets of microfluidic channels, patterns of parallel stripes of desired materials are created alongside each other in the main channel where the different fluids under laminar flow converges. The process is schematically represented in Figure 3.13. LFP has some features that make it complementary to other patterning techniques used for biological applications. It takes advantage of the easily generated, multiphase laminar flows to pattern fluids and to deliver components for patterning. This mild delivery method allows the use of cells themselves as the patterning component. LFP is experimentally simple. Multiple-component patterns can be made in a short sequence of steps, without the need for multiple stages of pattern transfer with registration required by other methods. The method is applicable to the patterning of metals, organic polymers, inorganic crystals, and ceramics on the inner walls of preformed capillaries, using both additive and subtractive processes.

3.4.5 Step and Flash Imprint Lithography

Step and flash imprint lithography (SFIL) was developed at The University of Texas at Austin in the late 1990s and since then it has grown over the years [27]. SFIL is a nanomolding process where the pattern created on the target substrate is defined by the topography of the template mold with resolution in the sub-50 nm regime [28–30]. Here a transparent material such as quartz which allows the passage of UV rays during exposure is used as a mold for the process.

The advantage of this process over conventional photolithography technique lies in its simplicity and cost-effectiveness as it does not require expensive projection optics or advanced illumination sources and the quartz template can be used multiple times. SFIL is most suitable in the fabrication processes which require several lithography steps and suffers from registration issues. These problems are minimized in SFIL through the use of transparent fused quartz/silica template, facilitating the viewing of alignment marks on the template and wafer simultaneously. Moreover, the imprint process is performed at low pressure and room temperature, minimizing magnification and distortion errors. SFIL process is found to be robust with an inherent self-cleaning mechanism for removing particle contamination which is in contrast with other contact printing methods where fabrication related defects are quite common [31]. In addition, surface treatment during template fabrication has improved the lifetime of the template and helps to reduce the process-generated defects during fabrication. SFIL shows promise as a low-cost manufacturing tool for a wide variety of semiconductor, microelectromechanical, optoelectronic, microfluidic, and patterning hard disk substrates [32].

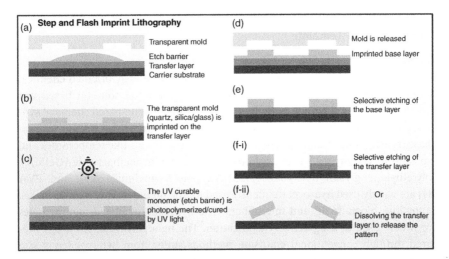

Figure 3.14 Fabrication process flow of the step and flash imprint lithography (SFIL). (a) Transparent quartz mold/template is brought in contact with the low viscous monomer to be patterned; (b) monomer liquid spread across the surface and fill the relief structures of the template when the transparent template conforms over sacrificial layer; (c) monomer liquid is photopolymerized/cured by UV light; (d) transparent template is separated from the substrate leaving a complementary solid replica of the template on the substrate surface; (e) etching of the residual layer of the monomer; (f-i) transfer layer is etched away leaving behind the target features over a layer of sacrificial layer; (f-ii) dissolving the transfer layer thereby releasing the separated array of relief patterns.

In the SFIL process, a low-viscosity UV-curable monomer (known as the etch barrier) is dispensed onto a removable sacrificial substrate deposited over carrier surface [33]. The transparent quartz mold/template is brought in contact with the low viscous monomer to be patterned. The low viscosity of the monomer liquid allows it to spread across the surface and fill the relief structures of the template when the later conforms over sacrificial layer. The monomer liquid trapped inside the transparent template is photopolymerized/cured by UV light before the transparent template is separated from the substrate leaving a complementary solid replica of the template on the substrate surface. Postprocessing consists of a breakthrough etch of the residual layer of the monomer where the target features created using through the relief are separated on the substrate but not yet released. The desired pattern is obtained either through (i) transfer etch where the sacrificial layer that is exposed after breakthrough etching is selectively etched away leaving behind the target features over a layer of sacrificial layer or (ii) dissolving the sacrificial layer thereby releasing the separated array of relief patterns. The SFIL process is schematically depicted in Figure 3.14. This process has the potential to become a high-throughput means of producing high aspect ratio and high-resolution patterns without projection optics.

3.4.6 Hydrogel Template

The hydrogel template approach for the preparation of nanoparticles, nanowires, microcapsules, and fibers was developed to make morphology-oriented chemical synthesis simpler, cost-effective, and faster than the other solution based wet chemical and laser ablation methods. This hydrogel template-based chemical synthesis technique utilizes unique properties of physical gels which undergo sol-gel phase transition upon changes in environmental conditions like temperature and exposure to UV rays. Alginate, gelatin, agarose, gelatin methacrylate (GelMA), poly(ethylene glycol) diacrylate (PEGDA), poly(N-vinylimidazole), and PVA [34] are widely used hydrogel frequently used for the preparation of nanoparticles, nanowires, bioscaffolds, and microfibers. The hydrogel-templated nanoparticles are widely used in drug delivery application. The nanoparticle-based drugs prepared through conventional emulsion methods suffer from heterogeneous size distribution with suboptimal drug loading and release properties. The use of hydrogel-templated nanoparticles ensures the preparation of monodispersed uniformly sized nanoparticles, where solid template-based methods seldom work. This hydrogel template method is capable of preparing nanoparticles with resolution of 50–200 nm and offer flexibility in controlling the size in drug delivery formulations [35]. In a nut shell, the hydrogel template approach provides a new strategy of preparing nano/microstructures of predefined size and shape with homogeneous size distribution most suitable for drug delivery applications.

In this process, the nanoparticle precursor is loaded in the prepolymer solution to form the hydrogel. The precursor can be reduced by UV rays or using suitable reducing agents mixed in the prepolymer. The hydrogel traps the nanoparticles produced by UV reduction or by chemical reduction to form nucleus of the seed nanoparticles. The prepolymer of the hydrogel is subsequently dissolved in water to release the nanoparticles. Wunnemann et al. [36] used microstructured hydrogel to develop a new template-guided method to obtain conductive Au-nanowire arrays on a large scale. To generate the hydrogel template, wrinkled PDMS was used as a master mold to form complementary microstructured template of hydrogel by imprint lithography method. Here polymer-based hydrogel was prepared by PDMS stamp for the fabrication of parallel Au-nanowire array. The polymer hydrogel was preloaded with reducing agent *N*-vinylimidazole which reduces the Au precursor tetrachloroaurate(III) in the grooves of the hydrogel. The aqueous solution of the Au precursor when coordinated in the groove of the hydrogel and reduced yields parallel array of Au nanowires. It was found that the lateral wire to wire distance was shortened as compared to that of the PDMS wrinkles. This method can be regarded as a facile nonlithographic top-down approach from micrometer-sized structures to nanometer-sized features. Tamayol et al. [37] used the alginate hydrogel template to develop polymeric network of fibers and 3D structures. This was achieved by syringe-based wet spinning a mixture of Na-alginate and polymeric solution in a $CaCl_2$ coagulation bath where rapid and reversible ionic gelation of sacrificial Na alginate template takes place when it subsequently cross-links with $CaCl_2$. Here the Na-alginate was cross-linked rapidly by exchange of Na^+ with Ca^{2+}, which resulted in the formation of hydrogel networks. The polymer fibers trapped in the hydrogel was released by dissolving the alginate template polymer fibers in ethylenediaminetetraacetic acid (EDTA) solution. However, mechanical strength of the fibers formed through hydrogel template method contributes toward the success of the method. Chen et al. [38] reported a simple way to fabricate custom-shaped microcapsules using hydrogel templates. Nanoparticle containing microcapsules were prepared by coating hydrogel particles with single layer of poly-L-lysine followed by one-step core degradation and capsule cross-linking procedure.

3.5 Etching

Etching is used extensively in material processing for delineating patterns, removing surface damage and contamination, and fabricating 3D structures. Etching is a chemical process wherein material is removed by a chemical reaction between the etchants and the material to be etched.

Etching processes are characterized by three parameters: (i) etch rate, (ii) etch selectivity, and (iii) etch uniformity.

I) *Etch rate*: The etch rate is defined as the material thickness etched per unit time.

II) *Etch selectivity*: Etch selectivity is a measure of how effective the etch process is in removing the material to be etched without affecting other materials or films present in the wafer. Quantitatively, etch selectivity can be expressed as the ratio between the etch rate of the material to be etched and etch-mask materials on the wafer.

III) *Etch directionality*: Based on the etching directionality, the etching process can be classified in two categories – (i) isotropic and (ii) anisotropic etching process. Process flow of isotropic and anisotropic etching is shown in Figure 3.15.

1) *Isotropic etching*: Isotropic etching proceeds in all directions at the same rate. Two-dimensional view of a structure consisting of semiconductor substrate and insulating layer with a photoresist mask is shown Figure 3.15a. Figure 3.15b, exhibits 2D view of the side wall profile of insulating film obtained by its isotropic etching.

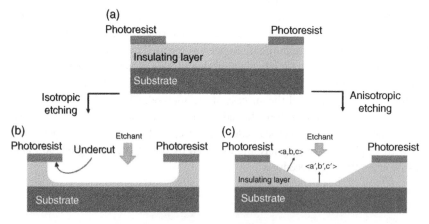

Figure 3.15 Process flow of isotropic and anisotropic etching. (a) Sample with protective mask on insulating layer before etching. (b) Isotropic etching process where the etchant dissolves the exposed region of the insulation (etching) layer equally in all directions creating an undercut. (c) Anisotropic etching process where the etchant dissolves the insulating layer at different rates in different crystallographic directions. The etching rate is higher along <a′,b′,c′> than along <a,b,c>.

2) *Anisotropic etching*: Anisotropic etching is preferential in one direction. Figure 3.15c, exhibits 2D view of the side wall profile of insulating film obtained by its anisotropic etching.

The etching process is of two types – (i) wet and (ii) dry etching.

3.5.1 Wet Etching

Wet etching is a method of fabricating devices or structures by selectively removing materials using a liquid chemical solution. It is commonly used in the semiconductor industry and other fields of microfabrication to create patterns, remove unwanted layers, or shape the surface of a material. It is important to note that wet etching is a subtractive process, where the material is removed, rather than adding or depositing it. The selectivity and control of the etching process are crucial to achieving the desired patterns or structures with high precision. However, wet etching has certain limitations, including limited resolution and isotropic etch profiles (where the etch rate is the same in all directions). However, it remains a valuable technique for many applications, especially in cases where high throughput, simplicity, or cost-effectiveness is desired.

The wet etching process typically involves the following steps:

1) *Substrate preparation*: The material to be etched, usually a silicon wafer or a thin film on a substrate, is thoroughly cleaned to remove any contaminants or particles that could interfere with the etching process.
2) *Masking*: A masking layer is deposited or patterned on the surface of the material to protect certain areas from the etchant. The mask is typically made of a resistant material such as photoresist or metal.
3) *Etchant selection*: An appropriate etchant solution is chosen based on the material being etched and the desired selectivity. The etchant selectively attacks the material to be removed while minimally affecting the masking material or other surrounding materials.
4) *Etching*: The substrate is immersed in the etchant solution, and the etching process is allowed to proceed for a specific duration. During this time, the etchant chemically reacts with the exposed regions of the material, dissolving or breaking down the material, and removing it from the surface.
5) *Rinse and clean*: After the etching process is complete, the substrate is rinsed thoroughly with deionized water or a rinsing solution to remove any residual etchant and reaction byproducts.
6) *Mask removal*: If a temporary masking layer (e.g. photoresist) was used, it is typically removed using an appropriate solvent or stripping process. This step exposes the underlying material and completes the wet etching process.

Different etchants and process parameters are employed for the removal of various material. For example, hydrofluoric acid (HF) is commonly used for etching silicon dioxide (SiO_2), while a mixture of sulfuric acid (H_2SO_4) and hydrogen peroxide (H_2O_2) may be used for etching metals like aluminum (Al). The etching process of Si, SiO_2, and Al are briefly described next:

1) Semiconductor (Silicon [Si])

The most widely used isotropic etchant for silicon is a mixture of nitric acid (HNO_3) and hydrofluoric acid (HF), commonly referred to as the "piranha solution." The etching rate and selectivity of silicon can be influenced by various factors, including the etchant composition, temperature, concentration, and etching time. The process parameters need to be carefully controlled to achieve the desired etch depth and pattern resolution. Wet etching of silicon can be used to create various structures, such as trenches, wells, or other features on the silicon wafer.

The substrate is suitably cleaned to remove the contaminants and then masked with photoresists on the Si wafer at appropriate sites where etching is not desired. The etching process initiates with the preparation of the piranha solution by mixing HNO_3 and HF in the appropriate ratio. The Si wafer is immersed in the piranha solution, and the etching process is allowed to proceed for a specific duration. The etchant selectively reacts with the exposed silicon surface, dissolving the silicon and removing it. Wet chemical etching usually proceeds by oxidation. Initially, silicon is oxidized in the presence of holes as follows:

$$Si + 2H^+ \rightarrow Si^{2+} + H_2. \tag{3.4}$$

Water dissociates according to the reaction:

$$H_2O \rightarrow (OH)^- + H^+. \tag{3.5}$$

The hydroxyl ions $(OH)^-$ recombine with positively charged silicon ions to form SiO_2 in two steps:

$$Si^{2+} + 2(OH)^- \rightarrow Si(OH)_2 \tag{3.6}$$

and

$$Si(OH)_2 \rightarrow SiO_2 + H_2. \tag{3.7}$$

SiO_2 dissolves in HF according to the reaction:

$$SiO_2 + 6HF \rightarrow H_2SiF_6 + 2H_2O, \tag{3.8}$$

where H_2SiFe is soluble in water. The reactions of (3.4)–(3.8) may be represented with HNO_3 by the following overall reaction:

$$Si + HNO_3 + 6HF \rightarrow H_2SiF_6 + HNO_2 + H_2O + H_2 \tag{3.9}$$

However, this process yields isotropic etch profiles. For anisotropic etching of Si, other etchants such as KOH solution, TMAH (tetramethylammonium hydroxide), and ethylenediamine-based etchants are used. Etching Si using a potassium hydroxide (KOH) solution is a widely used method for shaping and etching Si wafers. KOH is an alkaline etchant that selectively reacts with Si, allowing for controlled etching of the material. This process is often referred to as "anisotropic etching" as the etch rates are significantly different in different crystallographic directions of Si. The etching rate and selectivity of Si in KOH solution are influenced by various factors such as (i) temperature, (ii) KOH concentration, and (iii) doping level of the Si. The orientation of the Si crystal also plays a significant role, as the etch rate varies depending on the crystallographic planes of the Si. The (110) plane has the fastest etching primary surface as it has more corrugated atomic structures than the (100) and (111) primary surfaces. The (111) plane is an extremely slow etching plane that is tightly packed, has a single dangling bond per atom, and is overall atomically flat. KOH etching of silicon is employed for creating features with specific crystallographic orientations, such as V-grooves, pyramids, or other structures. The anisotropic nature of the etch enables precise control over the shape and dimensions of the etched features. The KOH etching is primarily suited for bulk etching of silicon and may not be suitable for etching through thin films or high-aspect ratio structures. In such cases, other etching methods like plasma etching or deep reactive ion etching (DRIE) may be more appropriate.

2) Insulator (Silicon Dioxide [SiO_2])

Etching SiO_2 using HF is a common method for selectively removing or patterning SiO_2 layers. HF is highly reactive toward SiO_2 and can be used to etch SiO_2 while leaving other materials, such as silicon, unaffected. The etch rate and selectivity of SiO_2 in HF solution can be influenced by various factors, including the concentration of HF, temperature, and exposure time. Higher concentrations of HF or longer etching times generally result in faster etch rates. The etching process can be controlled to achieve specific patterns or features in the SiO_2 layer. For example, a photolithographic technique can be used to define a pattern on the masking layer, and the SiO_2 can be selectively etched away using HF to transfer the pattern onto the SiO_2 surface.

The Si substrate with SiO_2 layer is thoroughly cleaned to remove any contaminants or particles that may interfere with the etching process. A photoresist (materials resistant to HF) is applied or patterned on the SiO_2 surface to protect certain areas from the HF etchant. When the substrate is immersed in the HF solution and allowed to proceed for a specific duration, the HF selectively reacts with the SiO_2, dissolving and removing it from the surface according to reaction.

$$SiO_2 + 6HF \rightarrow H_2 + SiF_6 + 2H_2O. \tag{3.10}$$

The reaction produces silicon fluoride gas (SiF_6) as a byproduct. The etch rate and selectivity of SiO_2 in HF solution can be influenced by various factors, including the concentration of HF, temperature, and exposure time. The etching process can be controlled to achieve specific patterns or features in the SiO_2 layer. For example, a photolithographic technique can be used to define a pattern on the masking layer, and the SiO_2 can be selectively etched away using HF to transfer the pattern onto the SiO_2 surface.

3) Metal (Aluminum [Al])

Etching aluminum (Al) using phosphoric acid (H_3PO_4) is a common method for selectively removing or patterning aluminum layers. Phosphoric acid is a relatively mild etchant for aluminum and offers good control over the etching process. The reaction between H_3PO_4 and Al forms byproducts $AlPO_4$ which dissolves in water and H_2 escapes. Etch rate of Al is of 2000 Å/min at 25 °C. The etch rate and selectivity of aluminum in phosphoric acid solution can be influenced by the concentration of phosphoric acid, temperature, and exposure time. Higher concentrations of phosphoric acid or longer etching times generally result in faster etch rates. The etching process can be controlled to achieve specific patterns or features in the aluminum layer. However, H_3PO_4 is a relatively slow etchant for aluminum. For faster and more selective etching of aluminum, other etchants such as hydrochloric acid (HCl) or mixtures like $HCl–HNO_3$ (aqua regia) are commonly used.

In Table 3.2, we have listed some materials and their etchants with desired composition and etch rate.

Table 3.2 Wet etchants used in etching and some selected electronic materials.

	Material	Etchant composition	Etch rate (Å/min)
1.	Si	3 ml HF + 5 ml HNO_3	3.5×10^5
2.	GaAs	8 ml H_2SO_4 + 1 ml H_2O_2 + 1 ml H_2O	0.8×10^5
3.	SiO_2	28 ml HF + 170 ml H_2O + 113 g NH_4F or 15 ml HF + 10 ml HNO_3 + 300 ml H_2O	1000 or 120
4.	Si_3N_4	Buffered HF or H_3PO_4	5 or 100
5.	Al	1 ml HNO_3 + 4 ml CH_3COOH + 4 ml H_3PO_4 + 1 ml H_2O	350
6.	Au	4 g KI + 1 g I_2 + 40 ml H_2O	1×10^5

3.5.2 Dry Etching

Plasma-assisted etching is generally referred to as dry etching, and the term dry etching is now used to denote several etching techniques that use plasma in the form of low-pressure discharges. A glow discharge is used to generate chemically reactive species (atoms, radicals, and ions) from a relatively inert molecular gas. The etching gas is chosen so as to produce species that react chemically with the material to be etched to form a reaction product that is volatile. The etch product then desorbs from the etched material into the gas phase and is removed by the vacuum pumping system. The most common example of the application of plasma etching is in the etching of carbonaceous materials, for example, resist polymers, in oxygen plasma – a process referred to as plasma ashing or plasma stripping. In this case, the etch species are oxygen atoms and the volatile etch products are CO, CO_2, and H_2O gases.

In etching silicon and silicon compounds, glow discharges of fluorine-containing gases, such as CF_4, are used. In this case, the volatile etch product is SiF_4 and the etching species are mainly fluorine atoms. In principle, any material that reacts with fluorine atoms to form a volatile product can be etched in this way (e.g. W, Ta, C, Ge, Ti, Mo, B, U). Chlorine-containing gases have also been used to etch some of the same materials, but the most important uses of chlorine-based gases have been in the etching of aluminum and poly-Si. Both aluminum and silicon form volatile chlorides. Aluminum is not etched in fluorine-containing plasmas because its fluoride is nonvolatile.

Plasma etching is predominantly an isotropic process. However, anisotropy in dry etching can be achieved by means of the chemical reaction preferentially enhanced in a given direction to the surface of the wafer by some mechanism. The mechanism used in dry etching to achieve etch anisotropy is ion bombardment. Under the influence of an RF field, the highly energized ions impinge on the surface either to stimulate reaction in a direction perpendicular to the wafer surface or to prevent inhibitor species from coating the surface and hence re-enhance etching in the direction perpendicular to the wafer surface. Therefore, the vertical sidewalls, being parallel to the direction of ion bombardment, are little affected by the plasma.

Figure 3.16 is a schematic diagram of a planar etching system, which comprises vacuum chamber, two RF-powered electrodes, an etching gas inlet, and a pumping mechanism. The planar systems are also called parallel-plate systems or surface-loaded systems. These systems have been used in two distinct ways: (i) the wafers are mounted on a grounded surface opposite to the RF-powered electrode (cathode) or (ii) the wafers are mounted on the RF-powered electrode (cathode) directly. This latter approach has been called reactive ion etching (RIE). In this approach, ions are accelerated toward the wafer surface by a self-bias that develops

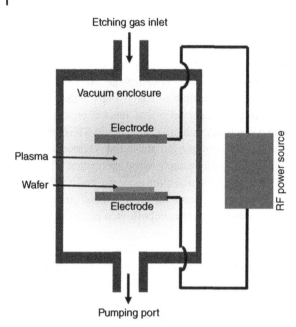

Etching gas inlet

Vacuum enclosure

Electrode

Plasma

Wafer

Electrode

RF power source

Pumping port

Figure 3.16 Schematic cross section of a plasma-etching system.

between the wafer surface and the plasma. This bias is such that positively charged ions are attracted to the wafer surface, resulting in surface bombardment. It has been demonstrated that a planar etching system, when operated in the RIE mode, is capable of highly directional and high-resolution etching.

3.6 Doping

When impurities are intentionally added to a semiconductor, the semiconductor is said to be "doped." Figure 3.17a shows a hypothetical 2D silicon crystal in which one silicon atom is replaced (or substituted) by an atom – in this example, a Group V element in the periodic table, namely, phosphorus. Phosphorus has five valence electrons, whereas silicon has only four. The phosphorus atom shares four of its electrons with four neighboring silicon atoms in covalent bonds. The remaining fifth valence electron in phosphorus is loosely bound to the phosphorus nucleus.

The ionization energy of an impurity atom of mass m in a semiconductor crystal can be estimated from a one-electron model. If this ionization energy is denoted by the symbol electron pair (bound electrons) in a covalent bond (Figure 3.17a) E_d, then,

(a)

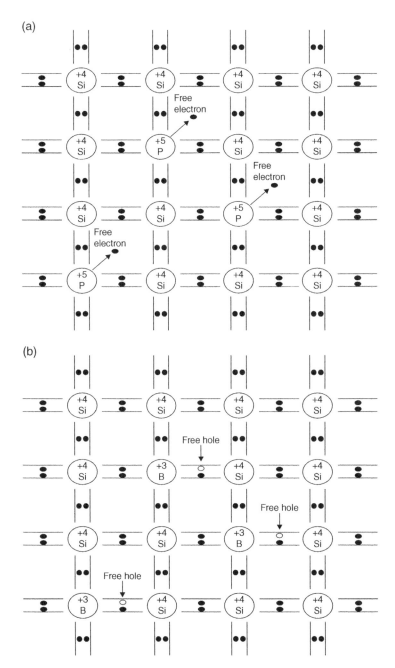

Figure 3.17 Hypothetical 2D silicon crystal doped with (a) phosphorus (n-type semiconductor) and (b) boron (p-type semiconductor).

$$E_d = \left(\frac{\varepsilon_0}{\varepsilon_r}\right)\left(\frac{m^*}{m}\right)E_n \tag{3.11}$$

where ε_0 is the permittivity of free space, ε_r is that of the semiconductor, and m^* is the effective electron mass in the semiconductor crystal. When the phosphorus atom in silicon is ionized, the released electron becomes a free electron that is available for conduction. The phosphorus atom is, hence, called a donor atom because it donates a free electron to the crystal. All atoms with five valence electrons, i.e. Group V elements, can behave in a similar manner to phosphorus in silicon, i.e. donate a free electron to the semiconductor crystal. Consequently, Group V elements, such as phosphorus or arsenic, are called donor atoms or simply donors, and the doped semiconductor is now referred to as an extrinsic semiconductor. When a single-crystal silicon is doped with phosphorus atoms of concentration ~10^{15} cm^{-3}, the free electron concentration in the conduction band increases from 10^{10} cm^{-3} to 10^{15} cm^{-3} at room temperature.

Now consider the situation in which a Group IV semiconductor is doped with atoms from an element in Group III of the periodic table, i.e. atoms that have only three valence electrons. To be more specific, let us take silicon doped with boron as an example, as is shown in the hypothetical 2D silicon lattice in Figure 3.17b. As can be seen from Figure 3.17b, the net effect of having a boron atom that substitute for silicon is the creation of a free hole (an electron deficiency in a covalent bond). This hole is generated as follows: because boron has three valence electrons, three neighboring silicon atoms will be bonded covalently with boron. However, the fourth nearest neighbor silicon atom has one of its four valence electrons sitting in a dangling bond; i.e. the whole system of the boron atom and the four neighboring silicon atoms has one electron missing. An electron from a neighboring Si-Si covalent bond may replace the missing electron, thereby creating an electron deficiency (a hole) at the neighboring bond. The net effect is, hence, the generation of a free hole in the silicon crystal. Therefore, this type of extrinsic semiconductor, silicon in this particular example, is called a p-type semiconductor or p-type Si. It is p-type because electrical conduction is carried out by positively charged free holes. The doping processes are of two types – (i) diffusion and (ii) ion implantation.

3.6.1 Diffusion

In a diffusion process, the dopant atoms are placed on the surface of the semiconductor by deposition from the gas phase of the dopant or by using doped oxide sources. Diffusion of dopants is typically done by placing the semiconductor wafers in a furnace and passing an inert gas that contains the desired dopant through it. Doping temperatures range from 800 to 1200 °C for silicon. The diffusion process is ideally described in terms of Pick's diffusion equation,

$$\frac{\partial C}{\partial t} = D\left(\frac{\partial^2 C}{\partial x^2}\right) \tag{3.12}$$

where C is the dopant concentration, D is the diffusion coefficient, t is time, and x is measured from the wafer surface in a direction perpendicular to the surface (Figure 3.18a). The initial conditions of the concentration $C(x,\ 0) = 0$ at time $t = 0$ and the boundary conditions are that surface concentration $C(0,\ t) = C_s$ at surface and that a semi-infinite medium has $C(\infty,\ t) = 0$. The solution of Eq. (3.12) that satisfies the initial and boundary conditions is given by,

$$C(x,t) = C_s \mathrm{erfc}\left(\frac{x}{2\sqrt{Dt}}\right) \tag{3.13}$$

where erfc is the complementary error function and the diffusion coefficient D is a function of temperature T expressed as,

$$D = D_0 \exp\left(\frac{-E_a}{kT}\right) \tag{3.14}$$

where E_a is the activation energy of the thermally driven diffusion process, k is Boltzmann's constant, and D_O is a diffusion constant. The diffusion coefficient

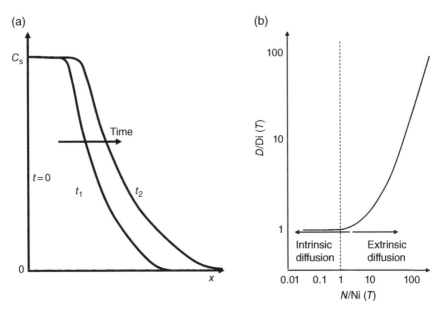

Figure 3.18 (a) Theoretical diffusion profile of dopant atoms within a silicon wafer and (b) relationship between the diffusion coefficient and doping concentration of a charge carrier in a semiconducting material.

is independent of dopant concentration when the doping concentration is low. However, when the doping concentration exceeds some temperature-dependent characteristic value, called the intrinsic carrier concentration [Ni(T)], the diffusion coefficient becomes dependent on dopant concentration. When D is independent of dopant concentration, the diffusion process is called intrinsic diffusion, whereas when D is dependent on the doping concentration, the diffusion process is called extrinsic diffusion (Figure 3.18b). In intrinsic diffusion, the dopant diffusion profiles are complementary error functions as given by Eq. (3.13); however, extrinsic diffusion profiles are somewhat complex and deviate from the basic linear theory. Instead, more complex models or empirical lookup tables are used to predict the diffusion depth. The diffusion coefficients of commonly used dopants are considerably smaller in silicon dioxide than in silicon. Hence, while doping silicon, silicon dioxide can be used as an effective diffusion barrier or mask.

3.6.2 Ion Implantation

Ion implantation is induced by the impact of high-energy ions on a semiconductor substrate. Typical ion energies used in ion implantations are in the range of 20–200 keV and ion densities could be between 10^{11} and 10^{16} ions/cm^2 incident on the wafer surface. Figure 3.19 shows the schematics of a medium-current ion implanter. It consists of an ion source, a magnet analyzer, resolving aperture and lenses, acceleration tube, x- and y-scan plates, beam mask, and Faraday cup. After ions are generated in the ion source, the magnetic field in the analyzer magnet is set to the appropriate value, depending on charge-to-mass ratio of the ion, so that desired ions are deflected toward the resolving aperture where the ion beam is collimated. These ions are then accelerated to the required energy by an electric field in the acceleration tube. The beam is then scanned in the x–y plane using the x- and y-deflection plates before hitting the wafer that is placed in the Faraday cup.

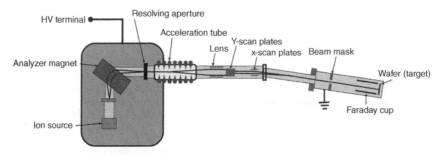

Figure 3.19 Schematic arrangement of an ion implanter for precise implantation of a dopant into a silicon wafer.

Commonly implanted elements are boron, phosphorus, and arsenic for doping elemental semiconductors, n- or p-type. After implantations, wafers are given a rapid thermal anneal to activate electrically the dopants. Oxygen is also implanted in silicon wafers to form buried oxide layers. The implanted ion distribution is normally Gaussian in shape and the average projected range of ions is related to the implantation energy.

3.7 Solution Processed Methods

Solution process-based device fabrication provides convenient way of materials integration compared to conventional growth techniques. System integration involving assembly of nanosized sensing materials in micron sized devices is crucial to (and still is a challenge in) the development and large-scale production of miniaturized sensors.

Solution casting approaches such as tape casting [39, 40], screen printing [40, 41], spray deposition [40, 42], inkjet printing [43–46], and drop dispensing [47, 48] are relatively easy and economical and thus are widely used in materials assembly. The first three techniques not only require temperatures in excess of 80°C for curing causing permanent change in electronic properties of biomodified materials, but also form thick films of sensing material in the order of micrometers that result in deviation in transport properties.

3.7.1 Inkjet Printing

Inkjet printing is presently one of the cheapest direct write techniques widely used in research and manufacturing industries especially in the electronic industry where conductive interconnects and other features are required to be printed easily and in cost-effective way on circuit boards for convenient assembly and packaging. Prior to the discovery of inkjet printing technique, the deposition of conductive ink on circuit boards and functional soluble materials was performed by screen-printing technique and spin coating, respectively. However, with the advent of inkjet printing solution processed inks including conductive solutions, functionalized inorganic, organic, and biomaterials have been successfully patterned on both rigid and flexible substrates. Printed circuit board (PCB) manufacturing, radiofrequency identification (RFID) tags, solar cells, batteries, and displays are some of the celebrated applications of inkjet printing.

The foundation for inkjet printing was laid in 1833 by Felix Savart [49] who showed that viscous liquid jets can be fragmented into a series of repeatable droplets and this principle was explained using laws of fluid dynamics in 1931 by Weber. Based on this phenomenon, Dr Sweet of Stanford University designed the first Continuous InkJet (CIJ) printer in 1965 [50]. In CIJ printing, a pressure was applied to a liquid jet to produce a series of droplets of uniform size and spacing which during their passage through an electric field gets charged to facilitate their deposition on their desired location on the substrate to form a pattern. The uncharged ink droplets are rendered into a gutter which was recirculated subsequently. The technique was commercialized by IBM in 1976 by the virtue of which the company launched IBM 4640 printer in commercial market. However, CIJ inkjet printers were not suitable to print functional materials as the recirculated ink may suffer degradation on continuous exposure to ambience. To combat this problem, drop-on-demand printers (DOD) were introduced where droplets were ejected only when it is required.

The thermal inkjet printer, invented by Sperry Rand Corporation in 1965, was the first DOD-type printer which was awarded a patent. In thermal inkjet printing, current is passed through a resistive coil, placed in the ink chamber near the nozzle, which superheats the ink to form nucleated bubbles. The enveloping vapor around the bubble insulated the ink from the heater, thereby preventing further heating of the ink in the chamber. The bubble formed in the ink chamber near the heater expands itself leading to the increase in pressure within the chamber. This increase in pressure in the enclosed ink chamber due to the expansion of the bubble forces the ink out of the nozzle in the form of a droplet. The heater is transiently switched off to facilitate the collapse of the bubble resulting in a transient pressure wave which separates the droplet from the nozzle and the subsequent refilling of the ink channel. The full cycle takes approximately 10 μs after which the bubble is recreated in the next cycle and the process continues to form a series of droplets ejected on the substrate. Functional materials are not printed using this technique as the thermal cycle involves increasing the heater to bubble nucleation temperature which would easily damage the material properties.

In 1972, Zoltan published a patent for piezoelectric-based DOD printer which consists of a hollow tube of piezoelectric material which contracts on the application of external voltage to squeeze the ink chamber for the ejection of ink droplets out of the nozzle. This technique was later termed as the squeeze-mode DOD printing. The bend-mode DOD printing, introduced by Kyser and Sears in 1972 [51], consists of a small ink chamber with an inlet tube at one end and outlet tube in the form of a nozzle at the other end. A piezoelectric disk which is mounted on one side of the chamber produces electric pulses that flexes the ink chamber inwardly, thereby squeezing the ink droplet out through the nozzle. The electric pulses of frequency of 700 Hz produce spot size as low as 100 μm. The push mode piezoelectric DOD print head, introduced in 1984, consists of a piezoelectric rod placed attached to a membrane on the ink chamber. The periodic excitation of the piezoelectric rod pushes the membrane inward, thereby decreasing the effective volume of the ink chamber. The periodic reduction in ink chamber volume causes ejection of ink droplets out from the nozzle. The shear-mode DOD printing technique was introduced in 1986 by Kenneth Fischbeck and Allen Wright of Xerox Corporation where the print head was designed such that shear deformation in the piezoelectric element forces the ink channel into a chevron shape with reduced volume and hence ejection of the ink droplet through the nozzle.

Figure 3.20 shows a schematic representation of piezoelectric based inkjet printer. Some of the earliest uses of inkjet printing of conductive materials are as described previously [50, 52]. The first demonstration of high-resolution piezo-electric DOD inkjet printing of a functional material was conducted by Sirringhaus et al. [53] with minimum feature sizes of 5 mm through the use of hydrophobic surface treatments. Inkjet printing of CNT films on substrates has proved to be an efficient and cost-effective technique used in a wide cross-section of applications. Okimoto et al. [54] designed high-performance thick-film SWNT transistors; the SWNT film being deposited 100 times at each position making the device fabrication process slow. The transistors showed mobility of 1.6–4.2 cm^2/V s and ON/OFF ratio of 10^4–10^5 and required a processing temperature of 80 °C. Kordas et al. [55] proposed a cost-effective method of generating conducting carboxyl group functionalized MWNT pattern on paper and polymer surface using commercial desktop printers. The FESEM image showed that the MWNTs were randomly

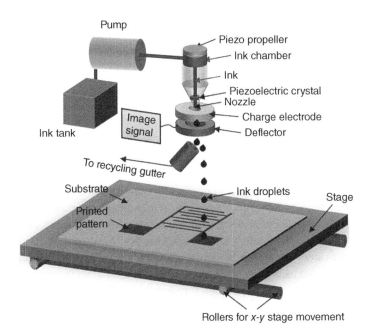

Figure 3.20 Schematic representation of the working of inkjet printing.

oriented and electrical characterization revealed that the conductance of the network increased with each print repetitions and percolation was only achieved in excess of 30 such repetitions. Song et al. [56] used inkjet printing to generate line patterns of SWNT that were 75 mm long and line width of 150 μm. However, it was observed that overwriting on the line pattern increased the line width. The electronic characterization of the pattern showed promise in high-frequency applications. Although inkjet printing has the ability to print locally on a substrate with minimum waste, it is relatively slow and the film thickness is larger than that in drop dispensing technique owing to the repetitive casting process required to achieve optimum percolation threshold.

3.7.2 Drop Dispensing

Shiraishi et al. [57] developed SWNT FET from solution processed SWNT aggregates, which showed a mobility of 10.9 cm^2/V s and showed promise of being fabricated on low heat-tolerant substrates. Li et al. [58] fabricated a chemical sensor by solution casting SWNT network on interdigitized electrode. The sensor showed high sensitivity at room temperature and good reproducibility for commercial applications. Since drop dispensing method suffers from drop-size control on different substrates, scientists have developed numerous solution dispensing technologies to solve the problem [59].

Drop dispensing technique also suffers from the problem of clogged orifice when the concentration of SWNTs in solution becomes high enough to form bundles. Covalent functionalization is extensively adopted to obtain highly dispersed solution of SWNT [60]. However, covalent functionalization of SWNT distorts the electronic property of the SWNTs [61] which brings about degradation in device performance. Noncovalent functionalization with ssDNA not only retains the excellent electronic property of the individual nanotubes, but also produces highly dispersed SWNT in solution [62].

Compared to the conventional syringe-based drop dispensing technique, the microcantilever-based drop dispensing method, first introduced by Paul et al. [63], showed excellent accuracy and control over position and diameter of the dispensed drop. The technique achieves a drop cast of nanometer thickness and desired diameter that not only enhances dynamic recovery of the sensor, but also prevents wastage of sensing material. The technique imparts consistency in film thickness and reduces stray signals which can evolve due to higher film thickness. This microcantilever-based technique of solution patterning is 30 times faster than inkjet printing [64] as the required percolation threshold is achieved by microcasting the as-prepared solution between Au-electrodes only once. However, this technique requires suspension of isolated SWNTs in solution to prevent aggregation leading to clogging of the dispensing orifice. Isolation of individual SWNT was achieved by passivating its surface by noncovalently attaching ssDNA to it. The negatively charged phosphate backbone of the DNA wrapped around the SWNT not only isolates individual nanotubes in solution due to mutual electrostatic repulsion, but also renders the SWNT network highly responsive so that any small physiochemical change in the neighborhood can be immediately sensed by change in conductivity of the network. The mutual repulsion among neighboring DFCs helps them to maintain a highly dispersed state in the solution.

Inspired by the work of Baba et al. [65], where they reported a strategy of micro-patterning of SWNT composites using microfluidic cantilevers, Paul et al. [63] extended this technique to the development of DFC-based chemical sensors. They used this microcantilever-based drop dispensing technique to cast DNA-functionalized carbon nanotube DFC solution between prefabricated electrodes and henceforth use it as a humidity sensor. The technique consists of a surface patterning tool which is mounted on a fixed support which can be lowered onto the substrate through z-control, while the substrate is kept on a stage capable of precise x–y movement in the order of micrometers. The SPT consists of a reservoir which stores the solution and dispenses it using capillary action through a narrow channel of 30-µm wide fabricated on the microcantilever as shown in Figure 3.21. This technique requires solution processed and uniformly dispersed active sensing material to prevent aggregation leading to clogging of the dispensing orifice. Micropatterning of solution processed active sensing material between Au-electrodes requires precise position and drop diameter control so that the DFC network just bridges the electrodes at a desired location thereby preventing wastage of sensing material. The diameter of the DFC drop dispensed by this technique

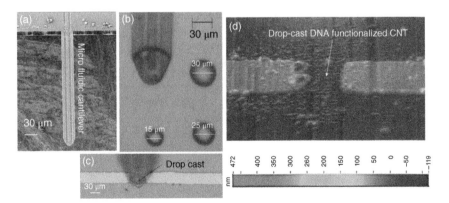

Figure 3.21 (a) SEM image of microfabricated Si-based microfluidic cantilever, commonly called the surface patterning tool (SPT) , Images of SPT dispensing (or drop-casting) DNA functionalized CNT (DFC) solution (b) of different drop cast diameters and (c) between Au electrodes, and (d) Digital image of drop cast DFC bridging the Au electrodes for use in sensor applications. *Source:* Adapted from Paul et al. [63].

depends on the RH, contact time between the cantilever tip, and the substrate *tc*, hydrophilicity of the substrate regulated by UV exposure time *t*UV, and viscosity of the solution. The UV/ozone exposure renders the substrate hydrophilic so that the DFC solution can be easily transferred to the substrate through capillary action. Higher contact duration provides ample time for the solution in the reservoir of the SPT to flow onto the substrate. The increased RH of the dispensing environment also favors higher drop diameter of the cast.

3.7.3 Spray Deposition

Spray deposition method is developed as a cost-efficient technique for fabrication of multilayer structures used in the organic-based devices (OBDs). The advantage of spray deposition technique lies in its ability for large area coverage [66] for multilayer deposition [67, 68]. The spray deposition technique produces smooth, defect-free, and uniform thin organic films [69] with cheap, vacuum-free process aimed for several sensor applications [70–72]. In the multilayer structures formed using the spray deposition technique, the film morphology is crucial for contact resistance reduction and increasing of the device current efficiency. The spray deposition technique consists of a spray gun bearing a spray nozzle which is connected to the reservoir containing the precursor solution to be spray deposited. The solution is ejected from the nozzle in the form of aerosol and deposited on the surface of the substrate kept on a heater for drying. The aerosol is formed due to the outflux of compressed gas through the nozzle and is controlled by a pressure regulator as shown in the Figure 3.22. The pressure of the compressed air flowing out

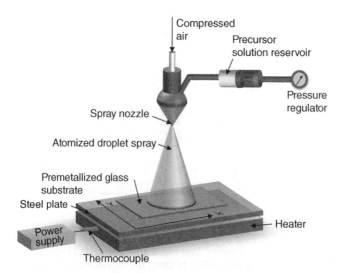

Figure 3.22 Schematic representation of the working of spray deposition.

through the nozzle determines the optimum distance between the nozzle and the substrate. The pressure of the compressed gas required for spray deposition depends on the material to be deposited and is typically around 4–6 atm. The thickness of each layer depends on the viscosity of the solution and the spraying time, whereas the surface roughness of the film is a function of volatility of the solvent used in the spray technique. The distance between the spray nozzle and the substrate also plays an important role in realizing a uniform film morphology. When the distance between the spray nozzle and the substrate is high, most of the solvent already evaporates during the flight stage, and dry powder hits the substrate forming an inhomogeneous and uneven surface morphology. At distances too low in the orders of a few centimeters, problems in surface wetting occurs which leads to streaming of solution off the substrate. Thus, optimum distance must be fixed between the substrate and the nozzle based on the heating and the choice of solvent [73]. Most efficient spray deposition is obtained with low concentration of solute. Low concentration solution allows the materials to coalesce through capillary effects, thereby forming a uniform film.

3.7.4 Screen Printing

The screen-printing process facilitates high-quality patterns to be repetitively produced at high rates and low cost and has been successfully adopted by electronics industries for the deposition of thick-film active sensing materials, patterning of conductive ink in the form of electrodes [74], and resistors on insulating substrates for the fabrication of complex hybrid electronic microcircuits [75]. The screen-printing technique overcomes the challenge of repetitive deposition of conductive ink onto substrate, and hence, allows controlled and precise printing of electrodes and active sensing materials.

In screen printing, the ink in the form viscous paste is deposited in a controlled way on the desired substrate through patterned apertures referred as a screen [76]. This process is accomplished by a flexible squeegee stroking the ink across the screen surface. There the screen is usually depressed to form a line contact with the surface thereby filling the apertures with ink when the line of contact advances with the squeegee stroke allowing the screen to peel away from the substrate behind the squeegee. The deposition of ink using screen printer is depicted in Figure 3.23.

Typically, the screen-printable inks comprise metal or metallic oxide particles [77], fiber-reinforced composites [78, 79], and polymer-conductive inks [80] with suitable binder having an organic carrier material to achieve desired ink viscosity which ensures ink flow through the apertures required for screen printing. Ideally the ink viscosity should be low enough to ensure free flow of ink through the apertures when encountered by shearing force exerted during screening. However, this viscosity must also be high enough so that the printed ink patterns retain their

(a)

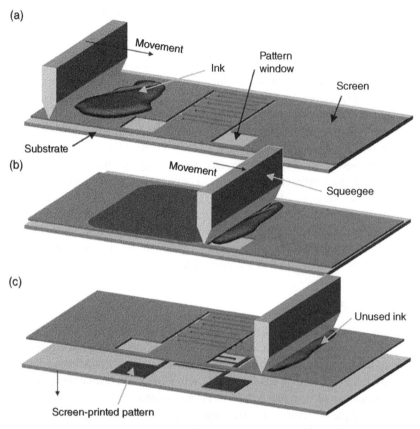

(b)

(c)

Figure 3.23 Schematic representation for the working of screen-printing technique. (a) The screen is placed atop a substrate and the ink is placed on top of the screen, and a squeegee pushes the ink through the holes in the mesh; (b) a flexible squeegee spreads the ink across the screen surface; (c) ink deposited on the substrate through the patterned apertures.

desired dimension when the shearing stress is withdrawn. Such viscous inks which flow through the apertures with moderate shear stress but acquire a highly viscous form after being printed on the substrate are termed as the pseudoplastic materials.

The screen-printed patterns on substrate are heat dried to prevent lateral flow of ink. The carrier material in the ink evaporates quickly to compensate for the lowering of ink viscosity with rise in temperature. In most applications such as the chemiresistors, it is desirable that the resistances of printed conductive electrodes and sensitive materials must be low to enhance sensor performance [81]. The thickness of the ink required to be deposited determines the resistance of the as-printed patterns and can only be ascertained by empirical means and through trials.

3.7.5 Tape Casting

Tape-casting technique is widely used to form thin films of sensory active materials for capacitors [82], chemiresistors, and CHEMFET. This technique is capable of addressing challenges in device miniaturization including the control of the film thickness. The underlying principle of tape casting is that a reservoir of slurry with a slit-shaped outlet at the base is moved across a surface, releasing a layer of slurry. Slurry characteristics and the speed of movement of the reservoir can be adjusted to influence the thickness of the tape. A doctor blade can also be drawn along the tape surface to modify tape thickness. The scraping (or doctor) blade is used to remove excess-deposited coating materials from a moving substrate. Film thicknesses from 5 µm to a few millimeters are currently obtained using tape casting. Additional postprocessing steps are required to control film composition and properties. One clear advantage of tape casting is the possibility of using the continuous process on flexible substrates. The schematic representation of tape casting is shown in Figure 3.24.

The preparation of film of desired thickness through tape casting is performed by the following steps [83]:

1) Slurry preparation
2) Tape casting
3) Drying

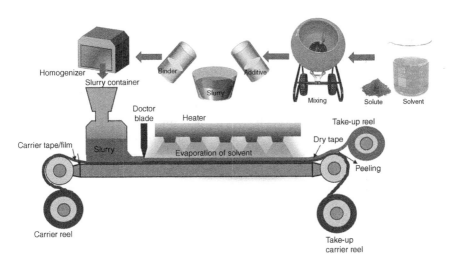

Figure 3.24 Schematic representation of the working of tape casting.

First, the active sensing material in powdered form is added to suitable liquid solution to form a slurry. The slurry generally includes the powdered material and a liquid consisting of a solvent, a deflocculant, a binder, and a plasticizer. The preparation of stable, concentrated powder dispersions in organic or aqueous solvents requires an effective stabilization of the particles in the medium. This was achieved by electrostatic steric or electrostatic stabilization of the powder dispersion in the desired solvent. In the case of electrostatic stabilization, the particle agglomeration and flocculation are suppressed by the repulsive coulombic forces between similarly charged species. In organic solvents, steric stabilization plays the dominant role with the effect being based on the repulsive forces increasing strongly when the adsorption layers of different particles start to overlap. In the case of low-molecular-weight surfactants, the geometrical range and thus the effective concentration window of the steric interaction is rather limited due to the maximum thickness of the adsorbed layer which can be reached.

During the second part of the process, the slurry is cast into a film under a special straight blade, called a "doctor blade," whose distance above the carrier determines the film thickness. The relative motion between the doctor's blade and the substrate cast the slurry in the form of thin film on the substrate. The tape-casting technique can be performed either on prefabricated electrodes or the metallization of the electrodes can be performed postcasting. In the formed case, it is required to cast the slurry consisting of active sensor material in a specified location between the electrodes to realize the device. Thus, tape casting at desired sites bridging the microelectrodes can be achieved by using hard mask with specified casting window over the electrodes. In the latter case, the electrodes can be fabricated on the processed slurry using either hard mask or process-compatible photolithography technique.

The third step involves drying of tape-casted slurry by evaporation of the solvent to form a lather-like tape. Once dried, the resulting film, called a green sheet, has the elastic flexibility of synthetic leather. The evaporation must be controlled to avoid differences in shrinkage of the green tape which causes cracking. The stresses result from a gradient in the pressure in the liquid in the pores of the drying body. With increasing drying rate and size of the body the stress increases, whereas the stress decreases with increasing permeability of the structure. In addition, binder migration caused by capillary forces must be minimized. Such migration processes would result in an inhomogeneous green tape with a low stability and problems with the waviness of the sintered substrate. The removal of the organic additives from the green substrate is a crucial step in the production of high-performance sensors. The tape-casting technique is used to form thin films of fiber-reinforced composites, metal oxide thin films, and polymers.

3.8 Conclusions

The new technological advancement of microfabrication has opened a platform for research and development to improve the performance and the characteristics of all types of microsensors for different applications in the realms of consumer and wearable electronics, health care, aviation, defense, security, and industry. A miniaturized solid-state sensor with a next-generation technology innovative architecture, smart choice of materials, efficient engineering capable of providing excellent sensor performance, reliable, and consumes low power is a gift to mankind. Microfabrication processes are more crucial for solid-state sensors used in precision operations such as defense, aviation, and health care, and are expected to have perfect temperature homogeneity for the long-term stability and high thermal and mechanical strength to intensify the robustness in different environments. Thus, appropriate use of fabrication tools and technologies are very important in dealing with complex designs. Again, the choice of such microfabrication tools is justifiably chosen so that the overall commercial cost of the product is optimized for best performance.

List of Abbreviations

APCVD Atmospheric Pressure Chemical Vapor Deposition
CIJ   Continuous Inkjet
CVD   Chemical Vapor Deposition
DFC   Deoxyribonucleic Acid Functionalized Carbon Nanotube
DOD   Drop-on-Demand Printers
EBL   Electron-beam Lithography
EBPVD Electron-beam Physical Vapor Deposition
EDTA   Ethylenediaminetetraacetic Acid
EUV   Extreme Ultraviolet Radiation
FESEM Field-effect Scanning Electron Microscope
FIBL   Focused Ion-beam Lithography
GelMA Gelatin Methacrylate
IBL   Ion-beam Lithography
LFP   Laminar Flow Patterning
LPCVD Low-pressure Chemical Vapor Deposition
MWNT   Multiwall Carbon Nanotube
NASA   National Aeronautics and Space Administration
OBD   Organic-based Devices
OLED   Organic Light-emitting Diode

PDMS   Polydimethylsiloxane
PECVD   Plasma-enhanced Chemical Vapor Deposition
PEG   Poly(ethylene Glycol)
PEGDA   Poly(ethylene glycol)diacrylate
PET   Polyethylene Terephthalate
PI   Polyimide
PLGA   Poly(lactic-*co*-glycolic Acid)
PLLA   Poly(lactic Acid)
PMMA   Polymethyl Methacrylate
PRINT   Particle Replication in Nonwetting Templates
PVA   Polyvinyl Alcohol
PVD   Physical Vapor Deposition
RF   Radiofrequency
RH   Relative Humidity
SFIL   Step and Flash Imprint Lithography
SPT   Surface Patterning Tool
ssDNA   Single-stranded Deoxyribonucleic Acid
SWNT   Single-walled Carbon Nanotube
XRL   X-ray Lithography

References

1 Xu, J., Wong, D.H.C., Byrne, J.D. et al. (2013). Future of the particle replication in nonwetting templates (PRINT) technology. *Angewandte Chemie (International Ed. in English)* 52 (26): 6580–6589.

2 Jeong, W., Napier Mary, E., and DeSimone Joseph, M. (2010). Challenging nature's monopoly on the creation of well-defined nanoparticles. *Nanomedicine* 5 (4): 633–639.

3 Rolland, J.P., Maynor, B.W., Euliss, L.E. et al. (2005). Direct fabrication and harvesting of monodisperse, shape-specific nanobiomaterials. *Journal of the American Chemical Society* 127 (28): 10096–10100.

4 Parrott, M.C., Luft, J.C., Byrne, J.D. et al. (2016). Tunable bifunctional silyl ether cross-linkers for the design of acid-sensitive biomaterials. *Journal of the American Chemical Society* 132 (50): 17928–17932.

5 Enlow, E.M., Luft, J.C., Napier, M.E., and DeSimone, J.M. (2011). Potent engineered PLGA nanoparticles by virtue of exceptionally high chemotherapeutic loadings. *Nano Letters* 11 (2): 808–813.

6 Kelly, J.Y. and DeSimone, J.M. (2008). Shape-specific, monodisperse nano-molding of protein particles. *Journal of the American Chemical Society* 130 (16): 5438–5439.

7 Zhang, H., Nunes, J.K., Gratton, S.E.A. et al. (2009). Fabrication of multiphasic and regio-specifically functionalized PRINT® particles of controlled size and shape. *New Journal of Physics* 11 (7): 075018.

8 Wheeldon, I., Farhadi, A., Bick, A.G. et al. (2011). Nanoscale tissue engineering: spatial control over cell-materials interactions. *Nanotechnology* 22 (21): 212001.

9 Thery, M. and Piel, M. (2009). Adhesive micropatterns for cells: a microcontact printing protocol. *Cold Spring Harbor Protocols* 2009: pdb.prot5255.

10 Otsuka, H., Ohshima, H., and Makino, K. (2014). Chapter 11 – Micropatterning of cell aggregate in three dimension for in vivo mimicking cell culture. In: *Colloid and Interface Science in Pharmaceutical Research and Development* (ed. H. Ohshima and K. Makino), 223–241. Amsterdam: Elsevier.

11 Rhee, S.W., Taylor, A.M., Tu, C.H. et al. (2005). Patterned cell culture inside microfluidic devices. *Lab on a Chip* 5 (1): 102–107.

12 Didar, T.F., Foudeh, A.M., and Tabrizian, M. (2012). Patterning multiplex protein microarrays in a single microfluidic channel. *Analytical Chemistry* 84 (2): 1012–1018.

13 Tan, W. and Desai, T.A. (2003). Microfluidic patterning of cells in extracellular matrix biopolymers: effects of channel size, cell type, and matrix composition on pattern integrity. *Tissue Engineering* 9 (2): 255–267.

14 Launiere, C., Gaskill, M., Czaplewski, G. et al. (2012). Channel surface patterning of alternating biomimetic protein combinations for enhanced microfluidic tumor cell isolation. *Analytical Chemistry* 84 (9): 4022–4028.

15 Geissler, M., Roy, E., Diaz-Quijada, G.A. et al. (2009). Microfluidic patterning of miniaturized DNA arrays on plastic substrates. *ACS Applied Materials & Interfaces* 1 (7): 1387–1395.

16 Khademhosseini, A., Eng, G., Yeh, J. et al. (2007). Microfluidic patterning for fabrication of contractile cardiac organoids. *Biomedical Microdevices* 9 (2): 149–157.

17 Demko, M.T., Cheng, J.C., and Pisano, A.P. (2010). High-resolution direct patterning of gold nanoparticles by the microfluidic molding process. *Langmuir* 26 (22): 16710–16714.

18 Cosson, S., Lutolf, M.P., Piel, M., and Théry, M. (2014). Chapter 7 – Microfluidic patterning of protein gradients on biomimetic hydrogel substrates. In: *Methods in Cell Biology*, vol. 121 (ed. M. Piel and M. Thery), 91–102. Academic Press.

19 He, X., Dandy, D.S., and Henry, C.S. (2008). Microfluidic protein patterning on silicon nitride using solvent extracted poly(dimethylsiloxane) channels. *Sensors and Actuators B: Chemical* 129 (2): 811–817.

20 Takayama, S., McDonald, J.C., Ostuni, E. et al. (1999). Patterning cells and their environments using multiple laminar fluid flows in capillary networks. *Proceedings* of the National Academy of Sciences of the United States of America 96 (10): 5545–5548.

21 Squires, T.M. and Quake, S.R. (2005). Microfluidics: fluid physics at the nanoliter scale. *Reviews of Modern Physics* 77 (3): 977–1026.

22 Whitesides, G. and Stroock, A. (2001). Flexible methods for microfluidics. *Physics Today* 54: 42–48.

23 Stone, H.A. and Duprat, C. (2016). Chapter 2 Low-Reynolds-number flows. In: *Fluid-Structure Interactions in Low-Reynolds-Number Flows*, 25–77. The Royal Society of Chemistry.

24 Rosenbluth, M.N., Berk, H.L., Doxas, I., and Horton, W. (1987). Effective diffusion in laminar convective flows. *Physics of Fluids* 30 (9): 2636.

25 Kane, R.S., Takayama, S., Ostuni, E. et al. (1999). Patterning proteins and cells using soft lithography. *Biomaterials* 20 (23): 2363–2376.

26 Berthier, E., Warrick, J., Casavant, B., and Beebe, D.J. (2011). Pipette-friendly laminar flow patterning for cell-based assays. *Lab on a Chip* 11 (12): 2060–2065.

27 Willson, C.G. and Colburn, M.E. (1999). Step and flash imprint lithography. US Patent Office, US6719915B2, University of Texas System.

28 Resnick, D.J., Sreenivasan, S.V., and Willson, C.G. (2005). Step & flash imprint lithography. *Materials Today* 8 (2): 34–42.

29 Xia, Y. and Whitesides, G.M. (1998). Soft lithography. *Angewandte Chemie International Edition* 37 (5): 550–575.

30 Stephen, Y.C., Peter, R.K., and Preston, J.R. (1996). Nanoimprint lithography. *Journal of Vacuum Science & Technology, B: Microelectronics and Nanometer Structures--Processing, Measurement, and Phenomena* 14 (6): 4129–4133.

31 Stewart, M., Johnson, S., Sreenivasan, S.V. et al. (2005). Nanofabrication with step and flash imprint lithography. *Journal of Micro/Nanolithography, MEMS and MOEMS* 4: 011002.

32 Gerard, M.S., Mike, M., Cynthia, B. et al. (2009). Step and flash imprint lithography for manufacturing patterned media. *Journal of Vacuum Science & Technology, B: Microelectronics and Nanometer Structures--Processing, Measurement, and Phenomena* 27 (2): 573–580.

33 Chung, Y.C., Chiu, Y.H., Liu, H.J. et al. (2006). Ultraviolet curing imprint lithography on flexible indium tin oxide substrates. *Journal of Vacuum Science & Technology, B: Microelectronics and Nanometer Structures--Processing, Measurement, and Phenomena* 24 (3): 1377–1383.

34 Lu, Y., Sturek, M., and Park, K. (2014). Microparticles produced by the hydrogel template method for sustained drug delivery. *International Journal of Pharmaceutics* 461 (1): 258–269.

35 Acharya, G., Shin, C.S., McDermott, M. et al. (2010). The hydrogel template method for fabrication of homogeneous nano/microparticles. *Journal of Controlled Release* 141 (3): 314–319.

36 Wunnemann, P., Noyong, M., Kreuels, K. et al. (2016). Microstructured hydrogel templates for the formation of conductive gold nanowire arrays. *Macromolecular Rapid Communications* 37 (17): 1446–1452.

37 Tamayol, A., Najafabadi, A.H., Aliakbarian, B. et al. (2015). Hydrogel templates for rapid manufacturing of bioactive fibers and 3D constructs. *Advanced Healthcare Materials* 4 (14): 2146–2153.

38 Chen, L., An, H.Z., and Doyle, P.S. (2015). Synthesis of nonspherical microcapsules through controlled polyelectrolyte coating of hydrogel templates. *Langmuir* 31 (33): 9228–9235.

39 Hotza, D. and Greil, P. (1995). Review: aqueous tape casting of ceramic powders. *Materials Science and Engineering A* 202 (1–2): 206–217.

40 Bensebaa, F. (2012). *Interface Science and Technology, Nanoparticle Technologies: From Lab to Market*, 1e, vol. 19, 560. Academic Press.

41 Ito, S., Chen, P., Comte, P. et al. (2007). Fabrication of screen-printing pastes from TiO_2 powders for dye-sensitised solar cells. *Progress in Photovoltaics: Research and Applications* 15 (7): 603–612.

42 Lavernia, E.J. and Grant, N.J. (1988). Spray deposition of metals: a review. *Materials Science and Engineering* 98 (February): 381–394.

43 Mabrook, M.F., Pearson, C., and Petty, M.C. (2006). Inkjet-printed polypyrrole thin films for vapour sensing. *Sensors and Actuators B: Chemical* 115 (1): 547–551.

44 Kamamichi, N., Maeba, T., Yamakita, M., and Mukai, T. (2008). Fabrication of bucky gel actuator/sensor devices based on printing method. *Presented at IEEE/RSJ International Conference on Intelligent Robots and Systems*, Acropolis Convention Center, Nice, France (22–26 September 2008). IEEE.

45 Wongchoosuk, C., Jangtawee, P., Lokavee, S. et al. (2012). Novel flexible NH_3 gas sensor prepared by ink-jet printing technique. *Advanced Materials Research* 506: 39–42.

46 Montilla, V., Ramon, E., and Carrabina, J. (2010). Frequency scan technique for inkjet-printed chipless sensor tag reading. *Presented at 17th IEEE International Conference on Electronics, Circuits, and Systems (ICECS)*, Athens, Greece (12–15 December 2010). IEEE.

47 Spannhake, J., Helwig, A., Schulz, O., and Muller, G. (2009). Micro-fabrication of gas sensors. In: *Solid State Gas Sensing* (ed. E. Comini, G. Faglia, and G. Sberveglieri), 1–46. Springer US.

48 Choi, S.J., Wang, C., Lo, C.C. et al. (2012). Comparative study of solution-processed carbon nanotube network transistors. *Applied Physics Letters* 101 (11): 112104–112104.

49 Perelaer, J., Kröber, P., Delaney, J.T., and Schubert, U.S. (2009). Fabrication of two and three-dimensional structures by using inkjet printing. *Presented at NIP25 and Digital Fabrication 2009: Technical Program and Proceedings*, Louisville, KY (20–25 September 2009). Society for Imaging Science and Technology.

50 Cummins, G. and Desmulliez Marc, P.Y. (2012). Inkjet printing of conductive materials: a review. *Circuit World* 38 (4): 193–213.

51 Kyser, E.L. and Sears, S.B. (1976). Method and apparatus for recording with writing fluids and drop projection means therefore. US Patent Office, U.S. Patent 3946398, Silonic Inc.

52 Prudenziati, M., Hormadaly, J., Prudenziati, M., and Hormadaly, J. (2012). 1 – Technologies for printed films. In: *Printed Films-Materials Science and Applications in Sensors, Electronics and Photonics* (ed. M. Prudenziat and J. Hormadalyi), 3–29. Woodhead Publishing.

53 Sirringhaus, H., Kawase, T., Friend, R.H. et al. (2000). High-resolution inkjet printing of all-polymer transistor circuits. *Science* 290 (5499): 2123–2126.

54 Okimoto, H., Takenobu, T., Yanagi, K. et al. (2010). Tunable carbon nanotube thin-film transistors produced exclusively via inkjet printing. *Advanced Materials* 22: 3981–3986.

55 Kordas, K., Mustonen, T., Toth, G. et al. (2006). Inkjet printing of electrically conductive patterns of carbon nanotubes. *Small* 2 (8–9): 1021–1025.

56 Song, J.-W., Kim, J., Yoon, Y.-H. et al. (2008). Inkjet printing of single-walled carbon nanotubes and electrical characterization of the line pattern. *Nanotechnology* 19: 095702–095708.

57 Shiraishi, M., Takenobu, T., Iwai, T. et al. (2004). Single-walled carbon nanotube aggregates for solution-processed field effect transistors. *Chemical Physics Letters* 394: 110–113.

58 Li, J., Lu, Y., Ye, Q. et al. (2003). Carbon nanotube sensors for gas and organic vapor detection. *Nano Letters* 3 (7): 929–933.

59 Ferraro, P., Coppola, S., Grilli, S. et al. (2010). Dispensing nano–pico droplets and liquid patterning by pyroelectrodynamic shooting. *Nature Nanotechnology* 5: 429–435.

60 Wang, Y., Iqbal, Z., and Mitra, S. (2006). Rapidly functionalized, water-dispersed carbon nanotubes at high concentration. *Journal of the American Chemical Society* 128 (1): 95–99.

61 Zhao, J., Park, H., Han, J., and Lu, J.P. (2004). Electronic properties of carbon nanotubes with covalent sidewall functionalization. *The Journal of Physical Chemistry B* 108: 4227–4230.

62 Zheng, M., Jagota, A., Semke, E.D. et al. (2003). DNA-assisted dispersion and separation of carbon nanotubes. *Nature Materials* 2 (May): 338–342.

63 Paul, A., Bhattacharya, B., and Bhattacharyya, T.K. (2014). Fabrication and performance of solution-based micropatterned DNA functionalized carbon nanotube network as humidity sensors. *IEEE Transactions on Nanotechnology* 13 (2): 335–342.

64 Teerapanich, P., Myint, M.T.Z., Joseph, C.M. et al. (2013). Development and improvement of carbon nanotube-based ammonia gas sensors using ink-jet printed interdigitated electrodes. *IEEE Transactions on Nanotechnology* 12 (2): 255–262.

65 Baba, A., Sato, F., Fukuda, N. et al. (2009). Micro/nanopatterning of single-walled carbon nanotube-organic semiconductor composites. *Nanotechnology* 20: 085301–085306.

66 Falco, A., Petrelli, M., Bezzeccheri, E. et al. (2016). Towards 3D-printed organic electronics: planarization and spray-deposition of functional layers onto 3D-printed objects. *Organic Electronics* 39: 340–347.

67 Aleksandrova, M. (2009). Spray deposition of multilayer polymer structures for optoelectronic applications. *e-Journal of Surface Science and Nanotechnology* 7: 859–862.

68 Maeda, M. and Horikawa, T. (2013). Photocatalytic properties of $TiO_2/WO_3/FTO$ multi-layer structures prepared by spray pyrolysis deposition. *MRS Proceedings* 1492: 84–89.

69 Abdellah, A., Virdi, K.S., Meier, R. et al. (2012). Successive spray deposition of P3HT/PCBM organic photoactive layers: material composition and device characteristics. *Advanced Functional Materials* 22 (19): 4078–4086.

70 Carey, T., Jones, C., Le Moal, F. et al. (2018). Spray-coating thin films on three-dimensional surfaces for a semitransparent capacitive-touch device. *ACS Applied Materials & Interfaces* 10 (23): 19948–19956.

71 Doojin, V., van Joel, E., Wallace, W.H.W., and Scott, W. (2015). Optically monitored spray coating system for the controlled deposition of the photoactive layer in organic solar cells. *Applied Physics Letters* 106 (3): 033302.

72 Friedel, B., Keivanidis, P.E., Brenner, T.J.K. et al. (2009). Effects of layer thickness and annealing of PEDOT:PSS layers in organic photodetectors. *Macromolecules* 42 (17): 6741–6747.

73 Chen, H., Ding, X., Pan, X. et al. (2018). Comprehensive studies of air-brush spray deposition used in fabricating high-efficiency CH3NH3PbI3 perovskite solar cells: combining theories with practices. *Journal of Power Sources* 402: 82–90.

74 He, P., Cao, J., Ding, H. et al. (2019). Screen-printing of a highly conductive graphene ink for flexible printed electronics. *ACS Applied Materials & Interfaces* 11 (35): 32225–32234.

75 Liu, L., Shen, Z., Zhang, X., and Ma, H. (2021). Highly conductive graphene/carbon black screen printing inks for flexible electronics. *Journal of Colloid and Interface Science* 582: 12–21.

76 Arapov, K., Rubingh, E., Abbel, R. et al. (2016). Conductive screen printing inks by gelation of graphene dispersions. *Advanced Functional Materials* 26 (4): 586–593.

77 McGhee, J.R., Sagu, J.S., Southee, D.J. et al. (2020). Printed, fully metal oxide, capacitive humidity sensors using conductive indium tin oxide inks. *ACS Applied Electronic Materials* 2 (11): 3593–3600.

78 Grabowski, K., Zbyrad, P., and Uhl, T. (2013). Development of the strain sensors based on CNT/epoxy using screen printing. *Key Engineering Materials* 588: 84–90.

79 Lu, Q., Liu, L., Yang, S. et al. (2017). Facile synthesis of amorphous FeOOH/MnO$_2$ composites as screen-printed electrode materials for all-printed solid-state flexible supercapacitors. *Journal of Power Sources* 361: 31–38.

80 Hong, H., Hu, J., and Yan, X. (2019). UV curable conductive ink for the fabrication of textile-based conductive circuits and wearable UHF RFID tags. *ACS Applied Materials & Interfaces* 11 (30): 27318–27326.

81 Kraft, U., Molina-Lopez, F., Son, D. et al. (2020). Ink development and printing of conducting polymers for intrinsically stretchable interconnects and circuits. *Advanced Electronic Materials* 6 (1): 1900681.

82 Jabbari, M., Bulatova, R., Tok, A.I.Y. et al. (2016). Ceramic tape casting: a review of current methods and trends with emphasis on rheological behaviour and flow analysis. *Materials Science and Engineering B* 212: 39–61.

83 Lin, Y., Kang, K., Chen, F. et al. (2018). 4.13 Gradient metal matrix composites. In: *Comprehensive Composite Materials II* (ed. P.W.R. Beaumont and C.H. Zweben), 331–346. Oxford: Elsevier.

4

Piezoelectric Sensors

4.1 Overview

In 1880, brothers Pierre Curie and Jacques Curie (Figure 4.1a) discovered that applying mechanical stress to crystals such as quartz, tourmaline, and Rochelle salt generates electrical charges on the surface of these materials [1]. Later, Hankel in 1881 termed this phenomenon as "piezoelectricity" to distinguish it from other sources of generation of electricity such as contact electricity (friction generated static electricity) and pyroelectricity (electricity generated from crystals by heating) [2]. Piezo comes from the Greek word "piezein," which means "squeeze" or "apply some pressure." The converse effect, implying the internal generation of mechanical strain on the application of an electrical field to these was predicted by Gabriel Lippman (Figure 4.1b) in 1881, via mathematical deduction from fundamental thermodynamic principles. This *inverse piezoelectric effect* was quickly demonstrated by the Curie brothers via experimentation.

They confirmed the existence of the converse effect and attributed this complete reversibility of piezoelectric phenomenon to the electroelastic mechanical deformations in the piezoelectric crystals. From the year 1882 onward the discovery of piezoelectricity generated significant interest in the European scientific community, where the researchers engaged themselves in the identification of piezoelectric crystals based on the asymmetric crystal structure and its reversibility in the electrical and mechanical forms of energy. As piezoelectricity developed as a new field of research in the last quarter of the nineteenth century, the thermodynamics in quantifying complex relationship among mechanical, thermal, and electrical variables using appropriate tensorial analysis were established. This work was published in 1910 in the form of Woldemar Voigt's (Figure 4.1c) Lehrbuch der Kristallphysik (Figure 4.1d) (Textbook on Crystal Physics) describing the 20 natural crystal classes in which piezoelectric effects occur. While Voigt was working on the identification and quantification of piezoelectric crystals in terms of

Solid-State Sensors, First Edition. Ambarish Paul, Mitradip Bhattacharjee, and Ravinder Dahiya.
© 2024 The Institute of Electrical and Electronics Engineers, Inc.
Published 2024 by John Wiley & Sons, Inc.

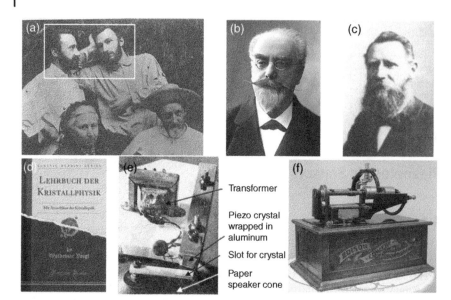

Figure 4.1 (a) Curie family portrait. Standing are Jacques and Pierre Curie (inset); Seated, their mother, Mme Cure, and Father, Dr Eugene Curie. Pierre shared the 1903 Nobel Prize in physics with his wife, Maria Sklodowska Curie (Marie Curie), and Henri Becquerel. (b) Gabriel Lippman in 1881 predicted inverse piezoelectric effect. (c) Woldemar Voigt's (d) Lehrbuch der Kristallphysik (Textbook on Crystal Physics) was published in 1910. (e) A Speaker in 1960's used Rochelle salt as the piezoelectric crystal, often termed as the 'singing crystals' then. (f) Phonograph used in WWI and WWII. *Sources:* (a) Unknown source/Wikipedia Commons/public domain. (b) Nobel Foundation/Wikipedia Commons/ public domain. (c) Woldemar Voigt/Wikipedia Commons/public domain. (d) Woldemar Voigt/ Forgottenbooks. (f) Flickr/Jalal gerald Aro.

measurable quantities, the world was maturing by transforming the science of piezoelectricity into application-oriented technology producing efficient machines. In 1917, Paul Langevin and his coworkers developed an ultrasonic submarine detector in France during World War I using natural piezoelectric crystals. Many new applications for piezoelectric crystals were developed in the years between World War I and World War II, such as speakers (Figure 4.1e) microphones, accelerometers, phonograph pick-ups (Figure 4.1f), and signal filters. However, due to the restricted availability of naturally occurring piezo crystals led to limited device performance and hence low commercial exploitation. During World War II, isolated research in the United States, Japan, and Soviet Union led to the emergence of a new class of piezoelectric crystal in the form of perovskites which showed improved piezo performance. These ceramic materials prepared in the laboratory by sintering metal oxides can be tailor made with specific characteristics to suit desired applications. In spite of this scientific progress in laboratory synthesis of piezo crystals, the technological advancements in application followed by commercial success could not be achieved; much attributed to the atmosphere of secrecy during the World War II and resisted the growth of industry and a new market for piezoelectricity-based technology. Around 1965, constant Japanese efforts in materials research led to the creation of new piezoelectric families that were free of patent restrictions. The emergence of new piezoelectric materials was used by the Japanese to manufacturer signal filters for the television and radio markets, and piezo-ceramic igniters for natural gas/butane appliances. As time progressed, the markets for piezoelectric materials grew and opened up avenues for new applications. A few such applications were audio buzzers, air ultrasonic transducers, and surface acoustic wave (SAW) filter devices. The research activities on piezoelectricity were stepped up at the end of the twentieth century as evident from the increased publications from US, UK, Russia, India, and China. The piezo materials emerged as the most important frontier material for the development of solid-state sensors by the stroke of the twenty-first century because of its usefulness and reasonably priced actuators which are low in power consumption and high in reliability and frequently used in electrostatic muscles.

4.2 Theory of Piezoelectricity

4.2.1 Direct Piezoelectric Effect

The origin of piezoelectricity can be explained using the molecular model. In the absence of any stress, the centers of the negative and positive charges of each molecule coincide – resulting into an electrically neutral molecule as schematically illustrated in Figure 4.2a. However, in the presence of an external mechanical

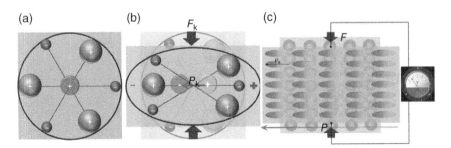

Figure 4.2 (a) Schematic representation of arrangement of negative and positive charges in a neutral molecule. (b) Separation of negative and positive charge centers due to application of force, and (c) external force causes polarization due to alignment of dipoles generating piezoelectricity under transverse compression and tension mode (Section 4.8.1).

stress the unit crystal cell of the material is deformed, thereby resulting in the separation of the positive and negative centers in the unit cell forming respective dipoles as represented in Figure 4.2b. As a result, the opposite facing dipoles inside the material cancel each other and fixed charges appear on the surface as shown in Figure 4.2c. Thus, due to the applied mechanical stress the material is polarized and this effect is termed as the *direct piezoelectric effect*. This polarization generates an electric field that can be used to transform the mechanical energy, used in the material's deformation, into electrical energy [3]. The piezoelectric effect is exhibited by 20 of 32 crystal classes and is always associated with *noncentrosymmetric* crystals. Naturally occurring materials, such as quartz, exhibit this effect as a result of their crystalline structure. Engineered materials, like lead zirconate titanate (PZT) for instance, are subjected to a process called *poling* to impart the piezoelectric behavior.

4.2.2 Poling

In a macroscopic crystalline structure that comprises several such unit cells, the dipoles are by default found to be randomly oriented as shown in Figure 4.3a. When the material is subjected to a mechanical stress, each dipole rotates from its original orientation toward a direction that minimizes the overall electrical and mechanical energy stored in the dipole. If all the dipoles are initially randomly oriented (i.e. a net polarization of zero), their rotation may not significantly change the macroscopic net polarization of the material, hence the piezoelectric effect exhibited will be negligible. Therefore, it is important to create an initial state

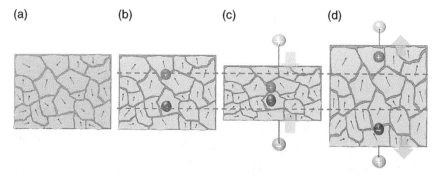

(a)   (b)   (c)   (d)

Figure 4.3 Macroscopic crystalline structure showing (a) the random orientation of dipoles producing net zero charge, (b) partial orientation of dipoles when the poling field is switched off, (c) generation of piezoelectricity when the material is compressed, and (d) generation of piezoelectric voltage of reverse polarity when the material in strained.

in the material such that most dipoles will be more or less oriented in the same direction. Such an initial state can be imparted to the material by poling it as shown in Figure 4.3b. The direction along which the dipoles align is known as the *poling direction*.

During poling, the material is subjected to a very high electric field that orients all the dipoles in the direction of the field. Upon switching off the electric field, most dipoles do not return back to their original orientation as a result of the pinning effect produced by microscopic defects in the crystalline lattice. This gives us a material comprising numerous microscopic dipoles that are roughly oriented in the same direction. It is noteworthy that the material can be depoled if it is subjected to a very high electric field oriented opposite to the poling direction or is exposed to a temperature higher than the Curie temperature of the material.

Using them in various sensing or actuating applications requires a systematic tabulation of their properties – for which, a standardized means for identifying directions is very important. That is to say, once the piezoelectric material is chosen for a particular application, it is important to set the mechanical and electrical axes of operation. Wherever crystals are concerned, the orthogonal axes originally assigned by crystallographers are used for this purpose. A general practice to identify the axes is to assign them the numerals, e.g. (i) corresponds to x-axis, (ii) corresponds to y-axis, and (iii) corresponds to z-axis. These axes are set during "poling"; the process that induces piezoelectric properties in the piezoelectric material. Conventionally, the poling direction is considered to be the third axis, except in quartz where the polarity is considered to be along the first axis. Hence, we need to interpret the material properties in terms of these principal directions. For example, the coupling coefficient d_{31} indicates how much the material will strain along its first principal direction when an electric field is applied across the third principal direction. This also means that the material properties can be used without any alteration only if the crystal's principal directions are aligned with the coordinate system used to describe the material's position.

The orientation of the DC poling field determines the orientation of the mechanical and electrical axes. The direction of the poling field is generally identified as one of the axes. The poling field can be applied in such a way that the material exhibits piezoelectric responses in various directions or combination of directions. The poling process permanently changes the dimensions of a piezoelectric material, as illustrated in Figure 4.3b. The dimension between the poling electrodes increases and the dimensions parallel to the electrodes decrease. In some materials, the poling step is also needed for the introduction of piezoelectric effect. For example, in virgin state the piezoelectric materials such as PVDF, P(VDF-TrFE), and ceramics are isotropic and are not piezoelectric before poling. Once they are polarized, however, they become anisotropic.

4.2.3 Static Piezoelectricity

Let us consider a poled piezoelectric material with a net polarization and having two metal electrodes deposited on opposite surfaces depending on the mode of operation of the piezoelectric crystal (Section 4.8.1). If the electrodes are externally short circuited, with a galvanometer connected to the short-circuiting wire, and compressive force is applied on the surface of piezoelectric material, a fixed charge density appears on the surfaces of the crystal in contact with the electrodes as shown in Figure 4.3c. This polarization generates an electric field which in turn causes the flow of the free charges existing in the conductor. Similarly, a tension force produces a polarization in the opposite direction (Figure 4.3d) causing the flow of free charges in the reverse direction in the conductor. Thus, depending on their sign, the free charges will move toward the ends where the fixed charges generated by polarization are of opposite sign. This flow of free charge continues until the free charge neutralizes the polarization effect. This implies that no charge flows in the steady state or in the unperturbed state – irrespective of the presence of external force. When the force on the material is removed, the polarization too disappears, the flow of free charges reverses, and finally the material comes back to its original standstill state. This process would be displayed in the galvanometer, which would have marked two opposites sign current peaks. If short-circuiting wire is replaced with a resistance/load, the current would flow through it and hence mechanical energy would be transformed into electrical energy. This scheme is fundamental to various energy-harvesting techniques that tap ambient mechanical energy such as vibrations and convert it into usable electrical form. Some materials also exhibit the reverse piezoelectric effect, i.e. a mechanical deformation or strain is produced in the material when a voltage is applied across the electrodes. The strain generated in this way could be used, for example, to displace a coupled mechanical load. This way of transforming the electrical electric energy into usable mechanical energy is fundamental to the applications such as nanopositioning devices.

4.2.4 Anisotropic Crystals

In some materials such as in the perovskites atomic structure (Section 4.5.2.2), the piezoelectric effect arises inherently out of anisotropy in the crystal structure and does not require poling unlike polymers (Section 4.5.2.1). This also means that piezoelectric material properties such as the stiffness (or compliance) matrix, coupling matrix, and permittivity matrix are defined in a certain crystal coordinate system that is typically denoted by the first, second, and third axes. *Anisotropy* can be defined as the ability of a material to be directionally dependent, i.e. showing different properties in different directions. The properties of an *anisotropic*

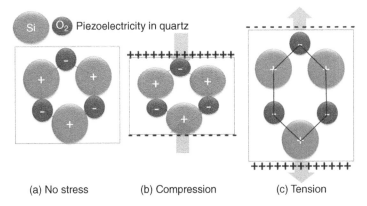

(a) No stress (b) Compression (c) Tension

Figure 4.4 Schematic representation of piezoelectric material at (a) unstressed, (b) under compression, and (c) tension showing the generation of charges of piezoelectric origin at the surface of the material.

material changes along the axis. Single crystals are anisotropic in nature. They exhibit directional dependent physical properties. The anisotropic nature comes from the fact that the linear density of atoms in a given particular crystallographic direction is not always the same. The interatomic spacing between the atoms also varies with the given crystallographic direction. Thus, when the anisotropic crystal is subjected to mechanical stress along a particular direction, the separation of positive and negative charge centers in the unit cell are separated from each other due to its *noncentrosymmetric* geometry leading to the generation of piezoelectricity (Figure 4.4a–c).

4.3 Basic Mathematical Formulation

It was discussed in the earlier section that piezoelectric material is electrically polarized when mechanically strained. When the piezoelectric material is polarized, fixed and opposite electric charges are developed at the opposite surface of the material. According to the linear theory of piezoelectricity [4], the density of as-generated fixed charges in a piezoelectric material is proportional to the applied stress and can be mathematically written as:

$$\vec{P}_{pe} = d \cdot \vec{T} \tag{4.1}$$

where P is the piezoelectric polarization vector, whose magnitude is equal to surface charge density σ_{pe} generated due to piezoelectric effect, d is the piezoelectric strain coefficient, and T is the stress applied on the piezo material. Similarly, the reverse piezoelectric effect is expressed as:

$$\vec{S}_{pe} = d \cdot \vec{E} \tag{4.2}$$

where S is the mechanical strain produced when the piezoelectric material is exposed to external electric field E. Thus, direct and reverse piezoelectric effects can be formulated as:

$$\vec{P}_{pe} = d \cdot \vec{T} = d \cdot \mathbb{C} \cdot \vec{S} = \phi \cdot \vec{S} \tag{4.3}$$

where $\mathbb{C} = (T/S) = (\phi/d)$ is the elastic constant of the material relating the magnitude of applied linear stress T and the corresponding mechanical strain S produced due to it and ϕ is the piezoelectric stress constant.

Again, when the piezo material is subjected to an externally applied electric field E, the negative and positive charge centers of the material are separated resulting in the generation of piezoelectric stress $\vec{T}_{pe}$ along specified direction, thereby producing piezoelectric strain $\vec{S}_{pe}$ in the same or different direction depending on the material. Thus, the reverse piezoelectric effect can be mathematically formulated using the equation:

$$\vec{T}_{pe} = \mathbb{C} \cdot S_{pe} = \mathbb{C} \cdot d \cdot \vec{E} = \phi \cdot \vec{E} \tag{4.4}$$

where $\mathbb{C}$ is the elastic constant of the material relating the generated stress T and the applied strain S as $T = \mathbb{C}S$.

4.3.1 Contribution of Piezoelectric Effect to Elastic Constant $\mathbb{C}$

The piezoelectric phenomenon causes an increase in the material's stiffness. Let us consider a piezoelectric material under strain S producing the following effects:

1) This strain S will generate elastic stress $\vec{T}_{elastic}$ proportional to the mechanical strain $\vec{S}$ such that $\vec{T}_e = \mathbb{C}\vec{S}$ and

2) The strain S will generate a piezoelectric polarization $\vec{P}_{pe} = \phi\vec{S}$ which creates an internal electric field $\vec{E}_{pe}$ in the material given by

$$\vec{E}_{pe} = \frac{\vec{P}_{pe}}{\varepsilon} = \frac{\phi S}{\varepsilon} \tag{4.5}$$

where ε is the dielectric constant of the material.

When piezoelectric material is poled with a specified polarity the material undergoes polarization due to the poling field. Under an externally applied compressive stress $\vec{T}$, an elastic stress $\vec{T}_e$ is generated in the material that restricts mechanical deformation of the material. Again due to the piezoelectric property of the material, the compressive stress on the material produces an electric field $\vec{E}_{pe}$

in the direction same as that of poling field. This electric field $\vec{E}_{pe}$ generated due to piezoelectric effect produces a stress $\vec{T}_{pe}$ which opposes the applied external stress $\vec{T}$.

Again for reverse piezoelectric effect, when an external electric field E is applied on a poled piezoelectric material in the direction of the poling field, the elastic stress $\vec{T}_e$, which is generated in the material, restricts material deformation. Again the generation of piezoelectric electric field $\vec{E}_{pe}$ produced in the direction of the poling field direction produces piezoelectric stress $\vec{T}_{pe}$. Thus, the presence of an applied electric field with polarity same as that of poling field results in positive strain S and hence the expansion of the material. Consequently, the stress T generated by the strain S can be calculated using the relations $T_e = \mathbb{C}S$ and $T_{pe} = \phi E_{pe}$ as:

$$T = T_e + T_{pe} = \left\{ (\mathbb{C}S) + \left(\frac{\phi^2}{\varepsilon} \times S \right) \right. = \left(\mathbb{C} + \frac{\phi^2}{\varepsilon} \right) \times S = \overline{\mathbb{C}} \times S, \qquad (4.6)$$

where $\overline{\mathbb{C}} = \mathbb{C} + \dfrac{\phi^2}{\varepsilon}$

Thus, piezoelectric effect results in an increased elastic constant $\mathbb{C}$, i.e. the material gets stiffened in the presence of piezoelectric effect. The constant $\overline{\mathbb{C}}$ is called the piezoelectrically stiffened constant.

4.3.2 Contribution of Piezoelectric Effect to Dielectric Constant ε

When an external electric field E is applied across two electrodes separated by a material of dielectric constant ε, an electric displacement field $\vec{D}$ is created between two electrodes and is expressed by the relation $\vec{D} = \varepsilon \vec{E}$. This is associated with polarization of the dielectric material resulting in the separation of positive and negative charges at the opposite sides of the material. For a material with piezoelectric property, the electric field E produces a strain $S_{pe} = d \times E$, which can be positive or negative depending on the direction of the external electric field with respect to the poling field. If the direction of external field is in the poling field direction, the strain is positive and material undergoes expansion along the direction of poling field. In the case of a direct piezoelectric effect, the expansion of material along poling field produces $\vec{P}_{pe} = \phi \vec{S}$, opposite to that of poling field or opposite to the external applied field. Thus, the surface charge density increases when the direction of applied external field is same as that of poling field. Again it can be shown that the surface charge density increases even if the direction of applied external field is opposite to that of the poling field. If the electric field is maintained constant, the additional polarization due to piezoelectric effect increases the electric displacement of free charges toward the electrodes by the

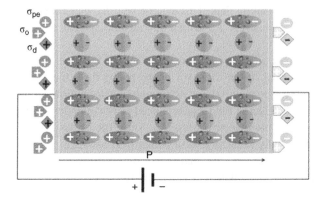

Figure 4.5 Schematic diagram indicating different electrical displacements due to dielectric polarization σ_d, orientational polarization σ_o, and piezoelectric polarization σ_{pe} associated with a piezoelectric and dielectric material.

same magnitude. Therefore, the total electrical displacement due to different types of polarizations as illustrated in Figure 4.5 is given by:

$$D = \left\{ \left(\varepsilon \vec{E} \right) + \vec{P}_{pe} \right\} = \left\{ \varepsilon \vec{E} + \phi S_{pe} \right\} = \varepsilon E + \phi d \vec{E} = (\varepsilon + \phi d) \vec{E} = \bar{\varepsilon} \vec{E}$$

(4.7)

where $\bar{\varepsilon} = \varepsilon + \phi d$

and $\bar{\varepsilon}$ is the effective dielectric constant. Thus, the piezoelectric behavior increases the dielectric constant of the material by ϕd.

4.4 Constitutive Equations

4.4.1 Piezoelectric

Linear piezoelectricity is the combined effect of the (i) linear electrical behavior and (ii) elastic behavior of the material. In order to construct the constitutive relations for piezoelectric effect let us consider the following equations which describe the elastic and the electrical behaviors.

The electrical behavior is defined by the relation $\vec{D} = \varepsilon \vec{E}$, (4.8)

where $\vec{D}$ is the electric displacement vector which is proportional to the accumulated charge density σ and the polarization $\vec{P}$, ε is the permittivity of the material, and $\vec{E}$ is the electric field generated in the material due to polarization.

The elastic behavior of the material is defined using Hooke's law as:

$$\vec{S} = \frac{1}{\mathbb{C}} \vec{T}, \quad \vec{S} = s \vec{T}$$

(4.9)

where s is the compliance of the material and is interpreted as the ability of the material to resists deformation.

These may be combined into coupled equations and represented as:

$$\vec{D} = d\vec{T} + \varepsilon^T \vec{E}$$
$$\vec{S} = s\vec{T} + d^t \vec{E}$$,

Or, $S_{ij} = s^E_{ijkl}T_{kl} + d_{kij}E_k$ and $D_i = d_{ikl}T_{kl} + \varepsilon^T_{ik}E_k$ (4.10, 4.11)

where S_{ij} is the mechanical strain applied in the i direction and observed in the j direction, D_i is the electric displacement in the i direction, T_{kl} is the mechanical stress applied in the k direction and observed in the l direction, E_k is the electric field, d_{kij} is the piezoelectric strain coefficient, s^E_{ij} is elastic compliance tensor applied in the i direction and observed in the j direction at constant electric field, and ε^T_{ik} is dielectric constant tensor under constant stress. The piezoelectric coupling coefficient d is a material property and depends on the material's mechanical properties (compliance or stiffness), its electrical properties (permittivity), and its piezoelectric coupling properties. The superscript E on the compliance s^E_{ijkl} indicates that the compliance was measured at constant E, preferably $E = 0$, while the superscript T on the permittivity matrix ε^T_{ik} means that the permittivity data were measured at constant stress field, preferably $T = 0$.

The equations can be expressed in matrix form as:

$$\{S\} = [s^E]\{T\} + [d^t]\{E\}$$
$$\{D\} = [d]\{T\} + [\varepsilon^T]\{E\}$$

or

$$\begin{bmatrix} S_1 \\ S_2 \\ S_3 \\ S_4 \\ S_5 \\ S_6 \end{bmatrix} = \begin{bmatrix} s^E_{11} & s^E_{12} & s^E_{13} & s^E_{14} & s^E_{15} & s^E_{16} \\ s^E_{21} & s^E_{22} & s^E_{23} & s^E_{24} & s^E_{25} & s^E_{26} \\ s^E_{31} & s^E_{32} & s^E_{33} & s^E_{34} & s^E_{35} & s^E_{36} \\ s^E_{41} & s^E_{42} & s^E_{43} & s^E_{44} & s^E_{45} & s^E_{46} \\ s^E_{51} & s^E_{52} & s^E_{53} & s^E_{54} & s^E_{55} & s^E_{56} \\ s^E_{61} & s^E_{62} & s^E_{63} & s^E_{64} & s^E_{65} & s^E_{66} \end{bmatrix} \begin{bmatrix} T_1 \\ T_2 \\ T_3 \\ T_4 \\ T_5 \\ T_6 \end{bmatrix} + \begin{bmatrix} d_{11} & d_{12} & d_{13} \\ d_{21} & d_{22} & d_{23} \\ d_{31} & d_{32} & d_{33} \\ d_{41} & d_{42} & d_{43} \\ d_{51} & d_{52} & d_{53} \\ d_{61} & d_{62} & d_{63} \end{bmatrix} \begin{bmatrix} E_1 \\ E_2 \\ E_3 \end{bmatrix}$$

(4.12)

$$\begin{bmatrix} D_1 \\ D_2 \\ D_3 \end{bmatrix} = \begin{bmatrix} d_{11} & d_{12} & d_{13} & d_{14} & d_{15} & d_{16} \\ d_{21} & d_{22} & d_{23} & d_{24} & d_{25} & d_{26} \\ d_{31} & d_{32} & d_{33} & d_{34} & d_{35} & d_{36} \end{bmatrix} \begin{bmatrix} T_1 \\ T_2 \\ T_3 \\ T_4 \\ T_5 \\ T_6 \end{bmatrix} + \begin{bmatrix} \varepsilon^T_{11} & \varepsilon^T_{12} & \varepsilon^T_{13} \\ \varepsilon^T_{21} & \varepsilon^T_{22} & \varepsilon^T_{23} \\ \varepsilon^T_{31} & \varepsilon^T_{32} & \varepsilon^T_{33} \end{bmatrix} \begin{bmatrix} E_1 \\ E_2 \\ E_3 \end{bmatrix}$$

(4.13)

Equation (4.13) shows that part of an electrical field applied to the material is converted into mechanical stress. Likewise, the second equation shows that part of a mechanical strain applied to the material is converted into electrical field. One can note that in the absence of electric field $\vec{E}$, the second equation is $\vec{S} = s\vec{T}$, which is Hooke's law. Likewise, in the absence of mechanical stress the first equation is $\vec{D} = \varepsilon^T \vec{E}$, only describing the electrical behavior of the material. Expressing the equations in an alternative form, we have:

$$\frac{D - dt}{\varepsilon^T} = \frac{S - s^E T}{T}$$

or $\vec{S} = s^E\left(1 - k^2\right)\vec{T} + (d/\varepsilon^T)\vec{D}$, where $k^2 = (d^2/(\varepsilon^T s^E))$

and k is known as the electromechanical coupling coefficient. It is an indicator of the effectiveness with which a piezoelectric material converts electrical energy into mechanical energy or converts mechanical energy into electrical energy. In the case where the electric displacement is equal to zero, the formula becomes:

$$\vec{S} = s^E\left(1 - k^2\right)\vec{T} \tag{4.14}$$

The strain is still proportional to the stress, but the compliance is multiplied by the term $(1 - k^2)$. When k is equal to zero, the equation is simply Hooke's law, which is logical as it means that all the energy in the material is strain energy. However, one must know that this expression of k has been obtained considering that the system is not connected to a circuit. A new expression of k in the case of a system linked to a circuit is now developed.

4.4.2 Sensor Equations for Electrical Circuits

Since the current and the voltages are measurable quantities in a circuit in contrast to electric field and electrical displacement, let us constitute the equations for the same. To replace the electric displacement and the electric field in the above equations, one must realize that a constant electric displacement results in zero current, hence the fact that these two entities are linked. Likewise, a zero electric field results in a zero voltage. The voltage and the current can be expressed in terms of electric field E and the electric displacement, respectively, as follows:

$$V = \int_0^x \vec{E} \cdot d\vec{x} \tag{4.15}$$

and

$$I = \frac{d}{dx}\int_A \vec{D} \cdot da, \tag{4.16}$$

where

I current

V voltage

x thickness of piezoelectric material

A surface area of piezoelectric material

$\vec{E}$ electric field

$\vec{D}$ electric displacement field

When the piezoelectric material is subjected to an externally applied stress, the electric field of piezoelectric origin is developed which gives rise to voltage and hence current in a closed electrical circuit. Assuming electric field E is uniform along the thickness of the material and the electric displacement D is uniform on the material's surface, and taking Laplace transform, we have:

$$V(\zeta) = \vec{E}(\zeta)\vec{x}$$
$$I = \zeta \cdot A \cdot \vec{D}\left(\vec{x}\right)$$

where ζ is the Laplace parameter. Expanding in the Laplacian space, we have:

$$I(\zeta) = \zeta C V(\zeta) + \zeta A d T(\zeta), \tag{4.17}$$

with $C = (A\varepsilon^T)/x$, where C is also known as the inherent open-circuit capacitance of the piezoelectric material and

$$S(\zeta) = \frac{d}{x}V(s) + s^E T(\zeta) \tag{4.18}$$

Suppressing $V(\zeta)$ and $T(\zeta)$, we have:

$$I(\zeta) = \zeta C\left(1 - k^2\right)V(\zeta) + \frac{\zeta A \varepsilon^T k^2}{d}S(\zeta) \tag{4.19}$$

$$S(\zeta) = \frac{k^2 s^E}{dA\zeta}I(\zeta) + s^E\left(1 - k^2\right)T(\zeta) \tag{4.20}$$

It is evident that even at $V(\zeta)$, the current will not be equal to zero, which means that there is a source of current in the circuit. The source of this current has a piezoelectric origin and is triggered once there is mechanical loading. Again, it is also to be noticed that the simple proportionality between stress $T(\zeta)$ and strain $S(\zeta)$ is lost as soon as both electrical and displacement fields are not equal to zero. Finally, when k is equal to zero, the second equation becomes Hooke's law, which is logical.

Expressing the Eqs. (4.19) and (4.20) in matrix form:

$$\begin{bmatrix} I \\ S \end{bmatrix} = \begin{bmatrix} \zeta C & \zeta A d \\ \dfrac{d}{x} & s^E \end{bmatrix} \begin{bmatrix} V \\ T \end{bmatrix} \tag{4.21}$$

The first term ζC in the matrix is the piezoelectric admittance. When a circuit of impedance Z_{ext} is connected, the total admittance is $\zeta C + (1/Z_{ext})$, and the matrix is expressed as:

$$\begin{bmatrix} I \\ S \end{bmatrix} = \begin{bmatrix} \zeta C + \dfrac{1}{Z_{ext}} & \zeta A d \\ \dfrac{d}{x} & s^E \end{bmatrix} \begin{bmatrix} V \\ T \end{bmatrix} \tag{4.22}$$

The modified k when the piezoelectric material is connected to the circuit can be obtained by using the ratio of the amount of electrical energy produced to the total energy of the system as:

$$k_{circuit}^2 = k^2 \frac{\zeta C Z_{ext}}{1 + \zeta C Z_{ext}} \tag{4.23}$$

The expression of strain is given by replacing k with $k_{circuit}$ as:

$$\vec{S} = s^E \left(1 - k_{circuit}^2 \right) \vec{T} \tag{4.24}$$

4.4.3 Piezoelectric Constants for a Material

When a piezoelectric material is electrically stressed by a voltage the material suffers change in material dimensions. On the other hand, when the piezoelectric material is stressed mechanically by an externally applied force an electrical voltage is generated in the closed circuit. Thus, these piezo elements capable of sensing or transmitting element are widely used in solid-state devices due to its rugged, compact, reliable, and efficient performance in devices. These piezoelectric devices are designed for suitable application by appropriate recognition of the piezoelectric properties of the material in terms of scientifically relevant parameters.

The electrical and mechanical responses of piezoelectric materials under external mechanical and electrical activation are highly directional and are represented and characterized by double subscripts. Piezoelectric coefficients with double subscripts link electrical and mechanical quantities. The first subscript gives the direction of the electric field associated with the voltage applied, or the charge produced. The second subscript gives the direction of the mechanical stress or strain. Several piezo ceramic material constants may be written with a "superscript" which specifies either a mechanical or electrical boundary condition. The superscripts are T, E, D, and S, signifying:

T = constant stress = mechanically free
E = constant electric field = short circuit
D = constant electrical displacement = open circuit
S = constant strain = mechanically clamped

Let us now discuss the piezoelectric constants to understand the behavior of different materials to be assigned with desired application.

4.4.3.1 Piezoelectric Strain Constant *d*

The piezoelectric constants relating the mechanical strain produced by an applied electric field are termed the strain constants, or the "d" coefficients. The units may then be expressed as m/V.

$$d = \frac{\text{Strain developed}}{\text{Applied electric field}} = \frac{\text{Short circuit charge density}}{\text{Applied mechanical stress}}$$

The d constants are calculated from the equation:

$$d = \frac{k}{\varepsilon_0 K_E^T s^E}$$

Large d_{ij} constants relate to large mechanical displacements suitable for motional transducer devices. Conversely, the d coefficient relates the charge collected on the electrodes, to the applied mechanical stress. For example, the coefficient d_{33} signify the parameter when the force is applied in the three directions (along the polarization axis) and the piezo response is felt on the same surface on which the charge is collected. Again d_{31} applies when the charge is collected on the same surface as before, but the force is applied at right angles to the polarization axis. The subscripts in d_{15} indicate that the charge is collected on electrodes which are at right angles to the original poling axis and that the applied mechanical stress is shear. The units for the d_{ij} coefficients are commonly expressed as coulombs/square meter per Newton/square meter.

4.4.3.2 Piezoelectric Voltage Coefficient *g*

The piezoelectric voltage coefficient "g" is the ratio of the electric field produced to the mechanical stress applied and is expressed as Vm/N. The g constants are obtained from the formula:

$$g = \frac{\text{Strain developed}}{\text{Applied charge density}} = \frac{\text{Field developed}}{\text{Applied mechanical stress}} = \frac{d}{\varepsilon}$$

The coefficient g_{ij} denotes that the applied stress or the piezoelectric induced strain is in the j direction and the electrodes are established normal to the i direction. Thus, a "31" subscript signifies that the pressure is applied at right angles to the polarization axis, but the voltage appears along the polarization axis. A "15" subscript implies that the applied stress is shear and that the resulting electric field is perpendicular to the polarization axis. Materials with high g_{ij} are most suitable for sensor applications.

4.4.3.3 Piezoelectric Coupling Coefficients *k*

The coupling coefficient (sometimes referred to as the electromechanical coupling coefficient) is defined as the ratio of the mechanical energy accumulated in response to an electrical input or vice versa. The piezoelectric coupling coefficient can be expressed in the following equation:

$$k = \sqrt{\frac{\text{Mechanical energy stored}}{\text{Electrical energy applied}}} = \sqrt{\frac{\text{Electrical energy stored}}{\text{Mechanical energy applied}}}$$

The subscripts in k_{ij} refer to the relative directions of electrical and mechanical quantities and the kind of motion involved. They can be associated with vibratory modes of certain simple transducer shapes; k_{33} is appropriate for a long thin bar, with electrodes on the ends, and polarized along the length, and vibrating in a simple length expansion and contraction. The k_{31} relates to a long thin bar, with electrodes on a pair of long faces, polarized in thickness, and vibrating in simple length expansion and contraction. The k_p signifies the coupling of electrical and mechanical energy in a thin round disk, polarized in thickness, and vibrating in radial expansion and contraction. The k_{15} describes the energy conversion in a thickness shear vibration. Since these coefficients are energy ratios, they are dimensionless.

4.4.3.4 Mechanical Quality Factor Q_M

The mechanical quality factor Q_M is defined as the ratio of the reactance to the resistance in the series equivalent circuit representing the piezoelectric resonator. The Q_M is also related to the sharpness of the resonance frequency. This parameter characterizes the sharpness of the electromechanical resonance system. The Q_M can be mathematically expressed as:

$$Q_M = \frac{1}{2\pi f_r Z_m C_0} \left(\frac{f_a^2}{f_a^2 f_r^2} \right)$$

where

f_r resonance frequency (Hz)

f_a anti resonance frequency (Hz)

Z_m Impedence at fa (ohm)

C_0 static capacitance (F)

4.4.3.5 Acoustic Impedance

When a voltage is applied across the piezoelectric crystals it produces a pressure field (a stress) on the atoms in their lattice with an accompanying overall contraction or expansion (a strain) in one or more dimensions of the material. The acoustic impedance Z_{acous} is the parameter which evaluates the acoustic energy transfer within the crystal and signifies the resistance in the propagation of acoustic energy. Acoustic impedance is the product of the density and speed of the wave in the piezoelectric crystal.

In solid materials, $Z_{acous} = \sqrt{\rho}\sqrt{k}$, where ρ is the density and k is the elastic stiffness of the material.

4.4.3.6 Aging Rate

The aging rate of a piezoelectric ceramic is a measure of how certain material parameters vary as a function of time (age). The most important parameters that age with time are the dielectric constant, frequency constants, and the resonant frequency. The aging of ceramics has a logarithmic function with time.

$$\text{Ageing rate} = \frac{1}{\log t_1 - \log t_2} \cdot \frac{P_2 - P_1}{P_1}$$

where t_1 and t_2 represent number of days after polarization.

P_1 and P_2 represent measured parameters.

4.4.3.7 Dielectric Constants K_{ij}^T

The relative dielectric constant K_{ij}^T is the ratio of the permittivity of the material ε to the permittivity of free space ε_0 in the unconstrained condition, i.e. well below the mechanical resonance of the part. The dielectric constant is derived from the static capacitance measurements at 1 kHz using a standard impedance bridge. The relations are illustrated in Figure 4.6 and expressed below:

$$K_{ij}^T = \frac{C_0 h}{\varepsilon_0 A}$$

K_{ij}^T: relative dielectric constant of material at constant stress T

(a)   (b)   (c)

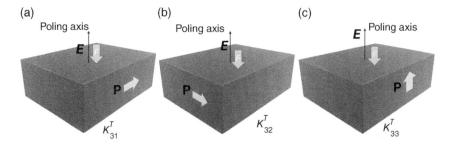

Figure 4.6 Schematic representation of relative dielectric constant K_{ij}^T showing the direction of mechanical stress and corresponding direction of induced polarization. Mechanical stress applied along the poling direction produces an induced polarization (a) perpendicular to it but in the direction along x-axis (referred as "1") (b) perpendicular to it but in the direction along y-axis (referred as "2") and (c) in the same direction along z-axis (referred as "3").

where electric field applied in j direction and stress generated along i

- C_0 measured capacitance
- h distance between electrodes
- ε_0 relative permitivity of free space
- A area of electrodes

4.5 Piezoelectric Materials

Piezoelectric materials can be classified into natural or man-made and their sub-divisions are illustrated in Figure 4.7. The naturally occurring piezoelectric materials are subdivided into single-crystal organic materials and biopiezoelectric materials. The single-crystal piezo materials include materials with anisotropic dielectrics with noncentrosymmetric crystal lattice, while the organic piezo materials involve hydrocarbon materials with piezoelectric properties. Some low structurally symmetric biomaterials which are associated with biological system show piezoelectric response to mechanical stress, are excessively used nowadays in medical devices due to its biocompatibility needs (separate mention in this section), and are termed as biopiezoelectric materials. However, it is the man-made piezoelectric materials including ceramics, polymers, and composites that dominate the piezoelectric sensor industry due to their enhanced properties as compared to natural piezoelectric material. Materials with *ferroelectric properties* are often used to prepare piezoelectric materials. Man-made piezoelectric materials

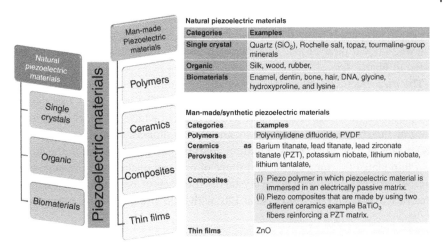

Figure 4.7 Classification of piezoelectric materials and their examples.

are subdivided into four main categories – polymers, ceramics, composites, and thin films. Different categories of the piezoelectric materials and their respective principles for the generation of piezoelectricity are described next.

4.5.1 Natural Piezoelectric Materials

4.5.1.1 Piezoelectric Single Crystals

Single crystal studies present a number of advantages for investigating piezoelectricity in any material, including biological structures. First, in terms of theoretical methodology, quantum mechanical modeling most often begins with the unit cell. Density functional theory (DFT) utilizes periodic boundary conditions to simulate bulk material behavior and can quantify material physical properties of crystals, including the dielectric, elastic, and piezoelectric constants. By studying biomolecular crystals in this way, the predicted physical properties can be directly related to single-crystal experiments, allowing effective screening of organic crystals for experimental investigation. A combination of modeling and characterization can provide much-needed insight into how piezoelectric properties are modulated by unit cell properties, such as dipole moments, molecular packing, and composition. Crystallizing biomolecules create a network of unit cell dipoles identical to the mechanisms of classical inorganic piezoelectrics, which allows for biological single crystals to easily fulfill the role of piezo ceramics, e.g. in stack actuation. Single crystals can often be grown quite easily from aqueous solutions, and the quality and shape of organic single crystals can be modulated using additives or pH buffers. This simple idea that the properties of single crystals can be modified chemically to enhance their properties suggests that this could be extended to piezoelectric properties using theoretical predictions to guide experiments. Some of the well-known piezoelectric single crystals along with their salient features are listed in Figure 4.8.

The piezoelectricity effect in the natural crystalline material is discussed by using a piezoelectric – a specially cut quartz crystal. The parameters that decide the piezoelectric nature of the crystal are: (i) angle at which the wafer is cut from natural quartz crystal, (ii) plate thickness, (iii) dimension of the plate, and (iv) means of mounting. Thus, a natural quartz crystal has to be cut in the shape of a thin plate of rectangular or oval shape of uniform thickness to investigate its piezoelectric characteristics. Each crystal has three sets of axes – optical axes, three electrical axes OX1, OX2, and OX3 at 120° with each other, and three mechanical

Quartz

• Quartz is the most popular single-crystal piezoelectric material. Single-crystal materials exhibit different material properties depending on the cut and direction of the bulk wave propagation. Quartz oscillator_operated in thickness shear mode of the AT-cut are used in computers, TVs, and VCRs.
• In SAW. devices ST-cut quartz with X-propagation is used. Quartz has extremely high mechanical quality factor **QM > 105**

Source: Parent Géry/Wikimedia Commons/Public Domain

Rochelle Salt

• Rochelle salt ($NaKC_4H_4O_6-4H_2O$) is one oldest material showing ferroelectricity between the two Curie temperatures $T_{c1} = 255K$ and $T_{c2} = 297K$ showing orthorhombic structure in the paraelectric phase and the monoclinic structure in the ferroelectric phase.
• It has a very low decomposition temperature equal to 55°C. The piezoelectric constants of Rochelle salt are $d_{21} = 7.0 \times 10^{-10}$C/N, $d_{22} = 2.2 \times 10^{-9}$C/N, $d_{23} = 2.1 \times 10^{-9}$C/N, and $d_{25} = 3.7 \times 10^{-11}$C/N

Source: dmishin

Topaz

• Topaz crystallizes in the orthorhombic structure and is mostly prismatic terminated by topaz pyramidal and other faces.
• Piezoelectricity in topaz can probably be attributed to ordering the (F,OH) in its lattice, which is otherwise centrosymmetric: orthorhombic bipyramidal (mmm). Topaz has anomalous optical properties which are attributed to such ordering.

Tourmaline

• Tourmaline [$(Na,Ca) (Mg, Li, Al, Fe^{2+})_3Al_6(BO_3)_3Si_6O_{18}(OH)_4$] is employed in pressure devices because of its piezoelectric properties – i.e. its ability to generate electric charge under mechanical stress or its change in shape when voltage is applied. It has been used in depth-sounding apparatus and other devices that detect and measure variations in pressure.

Figure 4.8 Different naturally occurring single-crystal piezoelectric materials.

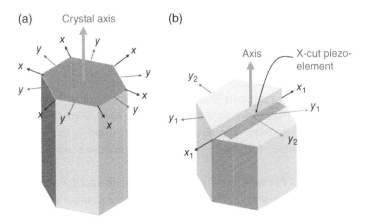

Figure 4.9 Display of (a) electrical (x) and mechanical (y) axes of a piezoelectric crystal and (b) cutting technique of a piezoelectric crystal, showing the cut axis that is perpendicular to the largest face.

axes OY1, OY2, and OY3 also at 120° with each other. The mechanical axes will be at right angles to the electrical axes as shown in Figure 4.9a, b. When a mechanical strain is applied along the y-axis, the atomic structure in the crystal is deformed which causes the shifting of the positive and negative charge centers of the atomic dipoles in the crystal, leading to the generation of electrical polarization. Due to this electrical polarization, a net charge is generated on the crystal faces perpendicular to the xy plane, thereby an external electrical field is achieved. On the other hand, if an electric stress is applied in the directions of an electric axis (x-axis), the atoms of the crystal are subjected to an electrical pressure, due to which the atoms are moved to rebalance themselves, leading to the deformation of the crystal and hence generation of mechanical strain in the direction of the y-axis.

4.5.1.2 Organic Materials

The organic piezoelectric effects are attributed to the lack of symmetry in most organic molecules. Thus, piezoelectricity is most prevalent in organic noncentrosymmetric structures where piezoelectric charges are generated when a mechanical stress is applied. The piezoelectric effect is enhanced in organic molecules when molecular dipoles within the bulk material are orienting in a particular direction through the application of a high electrical field termed as poling. Many organic material including the biomaterials like amino acids, peptides, DNA, and glycine form noncentrosymmetric crystals and exhibit piezoelectricity and are dwelled in the next section on piezoelectric biomaterials. Natural piezoelectric materials especially biomaterials usually exhibit weak piezoelectric property and uncontrollable shapes. Other organic piezoelectric materials like glycine, collagen, cellulose, keratin, and graphene have also attracted attention in recent years and will be discussed in this section.

Cellulose, the most abundant biopolymer on earth, is generally obtained from plant sources. Wood is the most common example of crystalline cellulose forming aligned fibers and oriented at a particular direction. The piezoelectricity of wood, i.e. the change of electrical polarization in a material in response to mechanical stress, has been known for decades. The piezoelectricity in wood is attributed to the cellulose crystal lattice, formed by unit cells of cellulose molecules ($[C_6H_{10}O_5]n$).

Let us consider a section of the wood is cut perpendicular to the grain direction and electrode in the z-plane as shown in Figure 4.10a. On application of a shearing force parallel to the direction of the fibers along the z-direction clockwise in the z-plane, it is found that positive charges are developed on the front side and negative charges of piezoelectric origin are developed on the backside of the wood section as shown in Figure 4.10b. A shear stress in one plane, including the grain direction, produced electrical polarization perpendicular to it. This origin of the piezoelectricity in wood originate from the crystalline structure of natural cellulose fiber consisting of micellar structure. The crystalline structure belongs to the monoclinic symmetry space group C2∥x3, and the possible piezoelectric constant d_{ij} for a single crystal of cellulose is as follows:

$$d_{ij} = \begin{bmatrix} 0 & 0 & 0 & d_{14} & d_{15} & 0 \\ 0 & 0 & 0 & d_{24} & d_{25} & 0 \\ d_{31} & d_{32} & d_{33} & 0 & 0 & d_{36} \end{bmatrix}$$

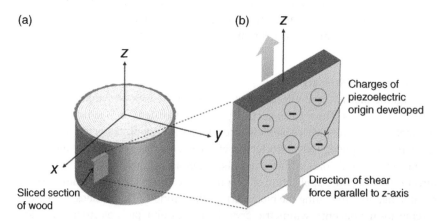

(a)

(b)

Charges of piezoelectric origin developed

Direction of shear force parallel to z-axis

Sliced section of wood

Figure 4.10 Schematic representation of wood section in Cartesian coordinates with (a) z-axis aligned along the direction of the fibers and (b) generation of piezo charges on application of shearing forces along the z-direction.

The piezoelectric coefficient d_{ij} describes the charge density generated under a certain applied stress. The components of third-rank tensor with piezoelectric coefficients d_{ij} can be expressed using a 3×6 matrix, where $m = 1, 2, 3$ refers to the electrical axis and $n = 1, 2, ..., 6$ to the mechanical axis [5]. The main axes 1, 2, and 3 correspond to length, width, and thickness, whereas the shear around these axes is expressed by indices 4, 5, and 6. Taking into consideration the parallel arrangement of cellulose fibers alone, the z-axis in wood the piezoelectric constant d matrix gets modified. The long axis of micelle are oriented parallel to the fiber axis and have a fair good orientation. These distribution of micellar fiber in the x–y plane are however random. Hence, the d constant of the piezoelectric system is given by d_{14} and d_{25}, where the relation $d_{25} = -d_{14}$ is also satisfied. Hence, the d constant of the system is given as:

$$d_{ij} = \begin{bmatrix} 0 & 0 & 0 & d_{14} & 0 & 0 \\ 0 & 0 & 0 & 0 & -d_{14} & 0 \\ 0 & 0 & 0 & 0 & 0 & 0 \end{bmatrix}$$

The chemical treatments which transform the lattice structure from cellulose I–II or III, increased the piezoelectric modulus. However, gamma ray irradiation up to a dose sufficiently high to decrease the molecular weight had only little influence on the piezoelectric modulus. The piezoelectric polarization in wood can be utilized in technical problems such as the measurement of shock velocity in timber.

Collagen, an organic polymer and one of the major components in biological tissue, possesses significant piezoelectric properties. Type I collagen (Figure 4.11a) accounts for more than 90% of the total collagen content in the body and constitutes the bones, tendons, and the ligaments, while type II collagen (Figure 4.11b) is found in cartilages. Collagen molecules are generally 1.5 nm in diameter and 300 nm long and exist as a spiral triple helix which self-assembles through extensive hydrogen bonding of amine and carbonyl functionalities and packs into a quasi-hexagonal (C6) lattice of crystalline fibrils of diameter 50–200 nm. The piezoelectric effect in collagen comes from polar and charged groups in the molecule. When a mechanical stress is applied to the collagen molecule the displacement of the hydrogen bonds reorients the dipole moments along the long axis of the collagen molecule, inducing permanent polarization in the molecule. The crystalline symmetry and structural stability of collagen is maintained by crystalline water molecules bridged through carbonyl groups of polypeptide chains. Together, these effects result in the overall piezoelectric effect in collagen. The shear piezoelectric constant of collagen is found to be around $d_{14} = 0.1$ pm/V. The piezoelectric properties can be improved by the addition of chitosan to collagen and adjusting the resultant mixture to neutral.

(a)

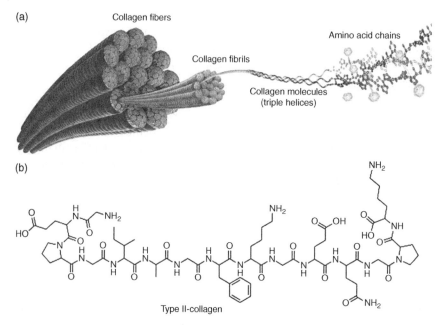

Figure 4.11 (a) Representation collagen fibers composed of aligned collagen fibrils which is made from triple helix collagen molecules and (b) molecular structure of type II collagen fiber.

Bone is a nanocomposite consisting of 90% type I collagen matrix and 10% hydroxyapatite filler. Here, collagen and the hydroxyapatite both possess piezoelectric properties and form the organic and the inorganic parts of the nanocomposite, respectively. Since collagen forms the major part of the bone, we discuss the piezoelectric contribution from the organic collagen fibers. The piezoelectricity in bone is largely attributed to the dense fibrillar packing of noncentrosymmetric collagenous matrix with embedded crystalline nanoapatite crystals of dimensions (size: $50 \times 25 \times 3$ nm). The highest shear mode piezoelectric coefficient d_{14} in femoral bone was recorded to be 0.7 pC/N. However, due to structural anisotropy, the bone displays anisotropic piezo response. The axial piezoelectricity and vertical or radial piezo resistivity of bone were suggested to be related to directional orientation of collagen fibrils as well as p–n junctions between collagen (n-type material) and apatite (p-type material). Anderson found the piezoelectric constants of dry bone to be:

$$d_{ij} = \begin{bmatrix} d_{11} & d_{12} & d_{13} \\ d_{21} & d_{22} & d_{23} \\ d_{31} & d_{32} & d_{33} \end{bmatrix} = \begin{bmatrix} 0.55 & 0.11 & -0.25 \\ -1.6 & 0.04 & 0.7 \\ 0.2 & 0.32 & 0.45 \end{bmatrix}$$

where d_{ij} is defined by the following relationship:

$$P_i = \sum_i d_{ij} T_i (i = 1, 2, 3 \text{ and } j = 1, 2, 3)$$

where P is the electric polarization; T, mechanical stress; Axis 1 is aligned anterior to posterior; Axis 2 is along the lateral direction; and Axis 3 comprises the longitudinal direction parallel to the long axis of the bone. The collagen fibers are oriented parallel to the long axis of the bone (axis 3). The piezoelectric constants, d_{21}, d_{23}, and d_{11}, showed greatest values, signifying that the generated piezoelectric charges are largest at the sides of the collagen fiber. Under physiological compression, polar hexagonal crystalline units of collagen in dry bone undergo dipolar reorientation resulting in a negatively charged surface. When the bone was immersed in the saline water the piezoelectric coefficient is modified to:

$$d_{ij}^{wet} = \begin{bmatrix} d_{11} & d_{12} & d_{13} \\ d_{21} & d_{22} & d_{23} \\ d_{31} & d_{32} & d_{33} \end{bmatrix} = \begin{bmatrix} 0.54 & 6.7 & 1.33 \\ 8.9 & 1.12 & 0.96 \\ 28 & 11.5 & 2.5 \end{bmatrix}$$

It was inferred that in the bone under wet state the applied stress drives the flow of fluid containing charged ions through the canalicular space, resulting in streaming potentials that provide electric signals to cells and hence the large increase in the piezoelectric constants d_{31} and d_{32}. The piezoelectric effect in bone promotes bone formation by the deposition of osteoblasts driven by the generated electrical dipoles. The electrical dipoles induce the deposition of minerals – primarily calcium – on the stressed side of the bone, and hence help in bone fracture healing.

Silk is highly disordered natural fiber obtained from moth or spider. Commercial silk fibers are obtained from *Bombyx mori* moth silk. The silk thread consists of longitudinal striations with two fibroin filaments of 10–14 μm each, and embedded in sericin. The sericin binder acts as an adhesive binder to hold the structure of the fiber. The silk fibroin is a natural fibroin protein with semicrystalline structure, providing fiber stiffness and strength. Silk fibroin is a high-molecular-weight block copolymer which consists of a heavy ($\approx$370 kDa) and light chain ($\approx$26 kDa) linked together by a single disulfide bond. Two crystalline polymorphs are reported for the *B. mori* silk fibroins. The first polymorph silk I represents the structure of the silk fibroin when stored in the silk gland of *B. mori* before spinning, while the silk II represents the structure of the fibroin after spinning. The polymorph silk I is mostly helical and less extended conformation of silk chains as compared to silk II. Under shearing forces, the metastable silk I structurally transforms into silk II. The silk II polymorph is composed of both amorphous and crystalline phases. The crystalline regions of the fibroin consist of alternating sequence of structurally pleated antiparallel β-sheet secondary structure and amorphous spacers. The crystalline heavy chain with antiparallel β-sheet structure is composed of highly repetitive alanine (Ala) and glycine (Gly) amino acids sequence motif. The β-sheet crystallite is the molecular network constructed by cross-linking β-sheet conformation of the molecular structures within several neighboring silk protein molecules. The Ala-Gly phase is hydrophobic, while the hydrophilic amorphous spacer impart polyelectrolytic properties to the heavy chain. The β-sheet secondary structure can be indexed as a monoclinic space group with a rectangular unit cell parameter of $a = 0.938$ nm, $b = 0.949$ nm, and $c = 0.698$ nm for *B. mori* silk.

Shear piezoelectricity is a characteristic property of the silk fibroin, the structural, self-assembling protein of silkworm fibers. Fukada et al. [6] reported that the piezoelectric response of silk fibroins was ~1 pC/N, which is comparable to that of the quartz crystal (~2 pC/N). Silk films with draw ratio = 2.7 exhibit shear piezoelectric coefficients of $d_{14} = -1.5$ pC/N. Silk's piezoelectricity is attributed cumulatively to high degree of silk II, β-sheet crystallinity, and crystalline orientation. Silk piezoelectricity is explained by net polarization of uniaxially aligned, silk II crystals with a noncentrosymmetric, monoclinic unit cell perpendicular to the plane of the applied shear force. The shear forces, when applied in the plane of the β-sheet, could result in intra- and interchain slide within pleated β-sheet domains and a subsequent rotation of amide dipoles, resulting in the formation of an internal polarization. While an isotropic distribution of β-sheet crystal domains would lead to cancellation of internal polarizations, uniaxial orientation of crystal domains could facilitate a net polarization and hence resulting in a piezoelectric effect. The interest in silk fibroin, a piezoelectric material for biotechnological and biomedical applications, is due to the highly controllable β-sheet content, biocompatibility, and controllable biodegradation rates. Piezoelectric silk could potentially enable development of exciting piezoelectric silk-based wearable sensors.

In addition to the organic materials discussed previously, many hard tissues like dentin and cementum together with soft tissues like cartilage, ligament, tendon, muscle, and hair have shown piezoelectric behavior. This property generally originates from the nanocrystalline or liquid crystalline-ordered nature of complex extracellular matrix components like collagen, keratin, elastin, glycosaminoglycans, and hydroxyapatite which compose these materials.

4.5.1.3 Biopiezoelectric Materials
Biological materials are low-symmetry, highly ordered structures and lack of inversion center has facilitated various biological materials to possess inherent piezoelectric properties with large linear electromechanical coupling coefficient. The piezoelectric behavior in biological materials is attributed to the complex dipolar

properties and dipole–dipole interactions mediated by intricate hydrogen bonding networks with different levels of self-assembly and hierarchy. Although the piezo-electricity derived from biological materials are not considered technologically advantageous for commercial success with d coefficients ranging from 0.1 to 2 pC/N2 as compared to 8 pC/N and 800 pC/N for ceramic-based piezo materials, new materials including viruses, prawn shells, and fish bladders are continued to be investigated for improved performances in the development of piezoelectric generators. It is expected that biological piezoelectric material will find applications in various sectors including high-performance sensors, actuators, and energy harvesters. Biological piezo materials show unique promise for applications in sensing and drug delivery and have the potential to be an alternative to Pb-based ceramic piezo materials. The difficulty in the crystallizing fibrous and transmembrane proteins is a major set-back in the investigation and understanding of biological piezo materials. However, breakthroughs in the crystallization techniques for amino acids and peptides have opened up fresh scope and interest in the investigation of biopiezo materials. The understanding of the piezoelectric properties of amino acids holds the major key to prediction of piezo behavior of biological material. Lysozyme, a well-known globular protein found in tears and saliva, and thymine, one of the nucleobases, were reported to exhibit piezoelectric properties in crystalline forms.

Generally amino acids are known to crystallize in low-symmetry, noncentro-symmetric, orthorhombic, and monoclinic space groups, thereby providing them with piezoelectric properties. Glycine, the simplest amino acid, can form crystals with α, β, and γ phases as shown in Figure 4.12a. Among them, both the β phase with the noncentral symmetric space group P_{21} and the γ phase with the noncentrosymmetric space group P_{32} exhibited piezoelectricity. The single-layer β glycine is metastable and converts easily into α-glycine in air. γ-Glycine crystals with a trigonal hemihedral symmetry are stable at room temperature, but γ crystals are tough to grow. The high shear piezoelectric constant d_{16} of up to 178 pm/V was achieved for β-glycine crystal. The γ glycine exhibited a low longitudinal piezo coefficient of 10 pm/V and piezoelectric constant d_{33} of 9.93 pm/V. The amino acid-based device enabled a maximum output voltage of ~0.45 V under a force of ~0.172 N. Doping of centrosymmetric α-glycine crystals with other ʟ amino acids also leads to crystals with piezoelectric properties. Piezoelectric pressure sensors using glycine as the active material have been reported [7].

The peptide that has attracted the most attention for its piezoelectric properties is diphenylalanine (FF) (Figure 4.12b), which is easily influenced by external electric fields in aqueous solution. FF peptide nanotubes possess piezoelectric and pyroelectric properties along with high stability and large Young's modulus, which make them promising alternative piezo materials for consumer electronics. FF was also found to generate voltages of 0.6–2.8 V, which makes them

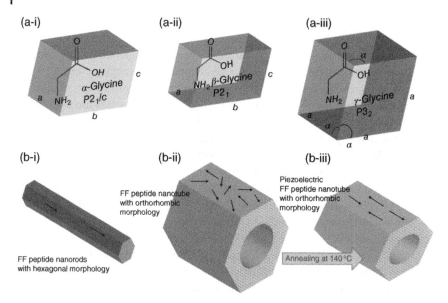

Figure 4.12 (a) (i) α Phase (ii) β phase, and (iii) γ phase of glycine crystal. (b) A schematic of (i) FF peptide nanorods with hexagonal symmetry, (ii) the change in the domain and crystallite size in orthorhombic FF peptide crystal, and (iii) as a function of the annealing temperature 140°C.

alternative beside glycine and alanine for self-powered brain implants and pace-makers. It was reported that the piezoelectric constant d_{33} of FF peptides ranged from 5 to 30 pm/V. Since as grown FF peptides are often distributed in random directions, electric field are applied during growth to promote self-assembly of the FF strands resulting in the alignment of the strands along the polarization direction, thereby generating an effective dipole moment to achieve a piezoelec-tric coefficient $d_{33} = 17.9$ pm/V. FF peptide nanorods were also synthesized where piezoelectric coefficient $d_{33} = 60$ pm/V was obtained. The piezoelectric coeffi-cients of FF peptide under shear orientation were obtained to be $d_{15} = 80$ pm/V, which is higher than the piezoelectric constant ($d_{31} = 4$ pm/V), ($d_{33} = 18$ pm/V), and ($d_{14} = 10$ pm/V). This is attributed to the FF hexagonal structures which produces strong dipole moments of six FF rings that point at the same orientation. The orthorhombic structure has antiparallel orientations in six FF rings and a zero total polarization. The FF peptide shows piezoelectric character only at tempera-ture above 140 °C due to the formation of an orthorhombic crystal structure above this temperature. The FF microrod possessed a piezoelectric constant d_{33} as high as 17.9 pm/V. Other piezoelectric peptides, cyclo-GW, with a monoclinic ($P21$) crystal structure, has an effective piezoelectric constant of $d_{33} = 14.1$ pC/N.

When the cyclo-GW peptide was fabricated into a nanogenerator, an output voltage of 1.2 V at a force of 65 N was generated. W-based aromatic dipeptides such as cyclo-FW peptides with an orthorhombic crystal structure used as voltage generators produced a high open-circuit voltage reaching 1.4 V. DNA's piezoelectric properties are attributed to the internal rotation of the dipoles created by phosphate groups. The piezoelectric properties in DNA were observed at lower water content when the bonds holding the DNA helix together gets weaker. This demonstrates the importance of bonding, structure, and experimental conditions when determining piezoelectric properties. The M13 bacteriophage's piezoelectric effect is caused by extruding proteins and it can be fabricated into thin films that exhibit strengths of 7.8 pm/V.

Piezoelectric biological materials have opened up new avenues in the design and construction of energy-harvesting devices in spite of multiple challenges relating the amount of water content in the biomaterial and the role of water in the modulation of piezoelectric properties. Furthermore, the softness of biological materials has restricted the measurement of piezoelectric parameters using traditional instruments. The challenge is even magnified due to the nanoscopic size of the biomolecules and that they cannot be sliced, polished, and electrode with relative ease.

4.5.2 Man-made/Synthetic Piezoelectric Material

4.5.2.1 Polymers

Polymer material is lightweight, low cost, and flexible, which is promising for various applications. There are mainly two kinds of piezoelectric polymer materials as mentioned earlier. First, the polymer materials (Table 4.1) intrinsically have the

Table 4.1 Mechanical and electric properties of different piezoelectric materials.

	PVDF [9, 10]	PVDF-TrFE	Parlyene-C	Polyimide
Young's modulus (GPa)	2.5–3.2	1.1–3	2.8	2–3
Dielectric constant	12	12	3.15	4
Dielectric loss	0.018	0.018		0.01
Mechanical loss	0.05	0.05	0.06	0.06
d_{33}	13–28	24–38	2	5.3–16.5
d_{31}	6–20	6–12		
k_{33}	0.27	0.37	0.02	0.048–0.15
k_{31}	0.12	0.07		
Maximum usability temperature	90	100	115	220

piezoelectric effect. These kinds of polymer materials mainly are PVDF and its copolymer of trifluoroethylene (PVDF-TrFE) nylon-11 and polyurea [8]. However, most polymer-based piezoelectric generators are fabricated from PVDF and its copolymer. The second kind is called an electret that can preserve the induced charges for a long time. An electret can be considered a piece of dielectric material with the presence of quasi-permanent real charges on the surface or in the bulk of the material, or frozen-in aligned dipoles in the bulk. An electret behaves like a battery or acts as an electrical counterpart of a permanent magnet. A piece of poled ferroelectric material can also be an electret. It can be used not only in the fabrication of piezoelectric generators but also in triboelectric generators based on electrostatic induction effect. Here, we mainly discuss on PVDF and its copolymers. Crystallographically, PVDF has very low symmetry because poling is achieved by uniaxial or biaxial stretching rather than the application of an electric field. This leads to an orthorhombic structure; class 2 mm and it therefore has 17 independent coefficients. The properties that make it attractive are its low acoustic impedance, $Z = 3.9$ MRayl, its low Q_m, and its high piezoelectric voltage constant, $g_{33} = 0.23$ Vm/N. Together, these make it an excellent broadband receiver for underwater sonar and biomedical imaging applications, where the ultrasonic medium has an acoustic impedance, $Z \approx 1.5$ MRayl. Its physical form as a thin plastic film also finds it applications and assists with its use as a broadband receiver because this places its fundamental resonance at a high frequency, allowing spectrally flat operation at lower frequencies.

The piezoelectric effect stems from hydrogen and fluorine atoms in the PVDF, which are positioned perpendicularly to the polymer as shown in Figure 4.13. The crystallinity of the PVDF polymer will be a major factor on the piezoelectric

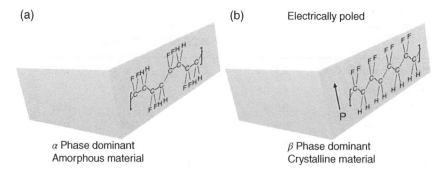

(a)

α Phase dominant
Amorphous material

(b) Electrically poled

β Phase dominant
Crystalline material

Figure 4.13 Molecular arrangements in (a) unpoled and (b) electrically poled PVDF showing the separation of positively charged H-ions and negatively charge F-ions contributing to the enhancement in piezoelectric behavior.

constant of polymers. Typically, PVDF has three crystalline phases, namely α, β, and γ, and it is the α phase that typically forms in most situations. The piezoelectric polymers have a crystalline region that has an internal dipole moment. These dipole moments are randomly oriented without any mechanical or electrical poling process, and the net dipole moment is zero in this condition. This type of structure is called α-phase PVDF film (Figure 4.13a) that has no piezoelectric response. The α-phase PVDF film is commonly used as insulating material because of its low thermal conductivity, low density, and high chemical and heat resistance. With post processes such as mechanical stretching and electrical poling under a high electric field, crystalline regions inside the bulk PVDF film will align in electric field direction. The PVDF structure with this morphology is called β-phase film (Figure 4.13b). As the polar β phase shows the strongest piezoelectric behavior, the material needs to be electrically poled using an electric field with the order of 100 MV/m or mechanically stretched. A higher β-phase crystalline can lead to a higher piezoelectric coefficient d_{33}. Typically, around 90–95% of β-phase portion shows a strong piezoelectric response for PVDF polymer. Copolymers of PVDF such as PVDF tetrafluoroethylene (PVDF-TrFE) P(VDF-TrFE) [(CH2-CF2)n-(CHF-CF2)m] crystallizes more easily into the β phase due to steric factors, resulting in better piezoelectric response. Upon application of post processes to the PVDF film, the β-phase PVDF film retains its morphology unless there are severe changes in temperature to the film. The maximum operating temperature for the β-phase PVDF film is 80 °C and 110 °C for the β-phase PVDF-TrFE film.

Piezoelectric polymers have certain common characteristics as

- Small piezoelectric d constant which makes them a good choice for the actuator.
- Large g constant which makes them a good choice as sensors.
- These materials have good acoustic impedance matching with water or human body due to their light weight and soft elasticity.
- Broad resonance bandwidth due to low QM.
- These materials are highly opted for *directional microphones* and *ultrasonic hydrophones*.

4.5.2.2 Ceramics

Piezo ceramics, a significant group of piezoelectric materials, are ferroelectric materials with polycrystalline structures (perovskite, tetragonal/rhombohedral crystals). Above the Curie temperature, these crystals exhibit simple cubic symmetry in structure, where there are no dipoles present and the positive and negative charge sites are coincident due to the centrosymmetric structure. However, this symmetry is broken below the Curie temperature, where the charge sites are no more coinciding as depicted in Figure 4.14. This results in built-in electric dipoles, (a-i)   (a-ii)   (b-i)   (b-ii)

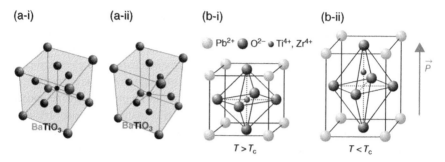

Figure 4.14 The transition of (a) BaTiO₃ from (i) tetragonal to (ii) cubic and (b) PbZrO₃ from (i) tetragonal to (ii) cubic beyond the Curie temperature T_c, $\vec{P}$ denotes the direction of polarization.

which are reversible. Neighboring dipoles realign locally to form Weiss domains. Lead zirconate titanate (PZT), barium titanate (BaTiO₃), lead titanate (PbTiO₃), potassium niobate (KNbO₃), lithium niobate (LiNbO₃), lithium titanate (LiTaO₃), sodium tungstate (Na₂WO₃), and zinc oxide (ZnO) are some of the most typical piezo ceramics. Barium titanate (BaTiO₃, BT) ceramics were discovered independently by US, Japan, and Russia during the WWII, which marked the dawn of the "glory of piezoelectric perovskites." Based on the widely used "Tita-Con" (titania condenser) composed of TiO₂–MgO, researchers doped various ternary oxides to enhance the permittivity of the materials. When the researchers investigated three dopants, CaO, SrO, and BaO, in a wide stoichiometric proportions, they found that the permittivity was maximum when the oxides were in compositions of ABO3 type, e.g. CaTiO₃, SrTiO₃, and BaTiO₃, where such structures were termed as perovskites. Initially BaTiO₃ was not recognized as a piezoelectric material but an insulator due to its high capacitance. Later Gray at Erie Resister asserted that electrically poled BaTiO₃ exhibited piezoelectricity owing to the domain realignment. BaTiO₃ has a noncentrosymmetric crystal structure with a net non-zero charge in each unit cell of the crystal. However, as a result of the titanium ion sitting slightly off-center inside the unit cell, an electrical polarity develops, thereby turning the unit cell effectively into an electric dipole. A mechanical stress on the crystal further shifts the position of the titanium ion, thus changing the polarization strength of the crystal. This is the source of the direct effect. When the crystal is subjected to an electric field, it also results in a relative shift in the position of the titanium ion, leading to the distortion of the unit cell and making it more (or less) tetragonal. This is the source of the inverse effect. Due to its high dielectric constant and low loss characteristics, BaTiO₃ has been used in applications such as capacitors and multilayer capacitors (MLCs). While doped BaTiO₃ has found wide application in semiconducting devices, PTC thermistors, and piezoelectric devices, and has become one of the most important ferroelectric ceramics. Piezoelectric $BaTiO_3$ (Figure 4.14a-i,ii) ceramics had a reasonably high coupling coefficient and nonwater solubility, but the bottlenecks were (i) a large temperature coefficient of electromechanical parameters because of the second-phase transition (from tetragonal to rhombohedral) around room temperature or operating temperature, and (ii) the aging effect due to the low Curie temperature (phase transition from tetragonal to cubic) around only 120 °C. In order to increase the Curie temperature higher than 120 °C, to decrease the second transition temperature below −20 °C, various ion replacements, such as Pb, Zr, and Ti, were studied. From these investigation and using the solid solutions of $PbTiO_3$, $PbZrO_3$, and $SrTiO_3$, a new system, lead zirconate titanate $(Pb[Zr(x)Ti(1-x)]O_3)$ (PZT), was discovered (Figure 4.14b).

PZT has a perovskite crystal structure, with each unit consisting of a small tetravalent metal ion in a lattice of large divalent metal ions. In the case of PZT, the small tetravalent metal ion is Ti or Zr, while the large divalent metal ion is usually Pb. Under conditions of tetragonal or rhombohedral symmetry on the PZT crystals, each crystal has a dipole moment. As compared to $BaTiO_3$, the PZT materials exhibit greater sensitivity and have a higher operating temperature. Piezoelectric thin films of PZT are widely used in the field of sensors and actuators, because of their ability to generate large displacements, the higher sensitivity and higher energy densities with wide dynamic range, and low power requirements. Due to its excellent polarization values, dielectric constant, easy integration to various devices, and piezoelectric constant, PZT thin film is often used as the actuating/ sensing component. Some examples where PZT thin films are used as active layer/component are force sensors, ultrasonic sensors, thermal sensors, micropumps and valves, probes for medical imaging and nondestructive testing, accelerometers, and new range of electronic components.

Due to excellent piezoelectric and dielectric properties of PZT, it has been on the forefront in sensor and actuator technologies for more than 60 years. Later in the 1990s, single-crystal piezoelectric material such as PMN-PT and PZN-PT were made, which along with PZT have been well commercialized and found many applications in the sensor and actuator fields because of their very high electromechanical coupling coefficient and large piezoelectric coefficient. However, the biggest challenge which is being faced by Pb-based materials is its environmental compatibility, because of toxicity of Pb. Recent global restrictions on manufacture, use, and disposal of Pb-based material requires pressing need to develop Pb-free and environment friendly material with piezoelectric properties similar to that of Pb-based. Among the Pb-free piezoelectric material, BNT $((Bi_{1/2}Na_{1/2})TiO_3)$-based, KNN $((K_{1/2}Na_{1/2})NbO_3)$-based, or BKT $((Bi_{1/2}K_{1/2})TiO_3)$-based, and $BaTiO_3$-based materials are widely investigated. Of these materials, $BaTiO_3$, because of its ease of fabrication and better electromechanical properties, provides a good model to understand the physics of Pb-free material.

Barium Titanate

• These materials with **dopants** such as Pb or Ca ions can stabilize the **tetragonal phase** over a wider temperature range.
• These are initially used for **Langevin** -type piezoelectric vibrators.

PZT

• Doping PZT with donor ions such as Nb^{5+} or Tr^{5+} provides soft PZT's like PZT–5.
• Doping PZT with acceptor ions such as Fe^{3+} or Sc^{3+} provides hard PZT's like PZT-8

Lead Titanate Ceramic

• These can produce clear ultrasonic imaging because of there extremely low planar coupling.
• Recently, for ultrasonic transducers and electromechanical actuators single crystal relaxer ferroelectrics with morphotropic phase boundary (MPB) are being developed.

Lithium Niobate and Lithium Tantalate

• These materials are composed of oxygen octahedron.
• Curies temperature of these materials is **1210** and **6600 °C**, respectively.
• These materials have a high electromechanical coupling coefficient for surface acoustic wave.

Figure 4.15 Salient features of various piezoelectric ceramics.

Lithium niobate and tantalate have the same chemical formula ABO_3 as $BaTiO_3$ and $Pb(Zr,Ti)O_3$. However, the crystal structure is not perovskite, but ilmenite. Since the Curie temperatures in these materials are high (1140 and 600 °C for LN and LT, respectively), perfect linear characteristics can be observed in electro-optic, piezoelectric, and other effects at room temperature. Although fundamental studies had been conducted, particularly in electro-optic and piezoelectric properties, the commercialization was not accelerated initially because the figure of merit was not very attractive in comparison with perovskite ceramic competitors. Some of the piezo ceramics along with their salient features are listed in Figure 4.15.

4.5.2.3 Piezoelectric Composites

To modulate and enhance the properties of different piezoelectric materials, relevant composites/nanocomposites have been developed. Piezo composites composed of a matrix phase and the filler phase where one or both may possess piezoelectric properties. The piezo composites are promising materials because of their excellent tailorable properties. The advantages of this composite are high coupling factors, low acoustic impedance, good matching to water or human tissue, mechanical flexibility, broad bandwidth in combination with a low mechanical quality factor, and the possibility of making undiced arrays by simply patterning the electrodes. Piezoelectric composite materials are especially useful for underwater sonar and medical diagnostic ultrasonic transducer applications.

Newnham et al. [11] introduced the concept of "connectivity" for classifying the various diphasic piezoelectric composites. A diphasic composite is identified with this notation with two numbers "m–n," where m stands for the connectivity of a filler phase (such as PZT) and n for a matrix phase (such as a polymer). A 0–0 composite is depicted as two isolated filler and matrix phases, while a 1–0 composite has phase 1 (filler) connected along the z-direction. A 1–3 composite has a one-dimensionally connected filler rods arranged in a three-dimensionally connected polymer matrix, and in a 3–1 composite, a honeycomb-shaped filler contained in one-dimensionally connected polymer phase. A 3–3 is composed of a network of filler embedded in a 3D connecting polymer. Figure 4.16 depicts the concept of connectivity in piezoelectric composites.

PVDF and its copolymers are widely used as piezoelectric material and emerged as an alternative to the ceramic-based piezoelectric due to its bendability. Moreover, its ease of processability, availability, and high piezoelectric coefficients in PVDF has encouraged researchers to use it as a matrix in piezocomposites.

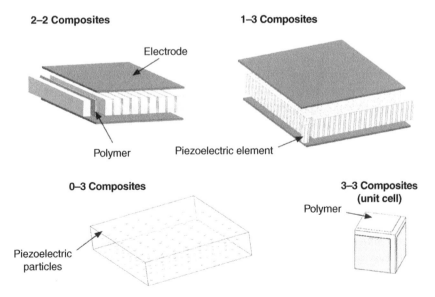

Figure 4.16 Schematic representation of the concept of connectivity in piezoelectric composites.

The β phase of PVDF which provides piezoelectric character to the polymer may be enhanced by the inclusion of different fillers, the addition of which promotes the formation of the β phase. Thus, fillers of different classes, including ceramics, metal- and carbon based, hybrid material, biomaterials, in its macro, micro, and nano sizes, are being widely used to achieve improved piezo performance and tailor ability. Here are some of the frequently used fillers that are widely used in polymer-PVDF matrix.

Ceramic fillers: Ceramic fillers include nonmetallic solids like an inorganic compound of a metal, nonmetal, or metalloid atoms. Some of the piezo ceramic fillers with highest piezoelectric coefficients and frequently used with PVDF matrix are lead-zirconate-titanate (PZT), barium titanate (BaTiO$_3$), sodium potassium niobate (KNN), and bismuth sodium titanate (BNT). These ceramic materials show excellent piezoelectric and dielectric properties where their inherent properties, such as brittleness, nonductility, and poor shape ability, are minimized due to its 0–3, 1–3 connectivity in the matrix.

Carbon-based fillers: Carbon-based fillers such as carbon nanotube (CNT), graphene oxide (GO), and reduced graphene oxide (rGO) have lured researchers for the development of polymer nanocomposites due to their high mechanical strength and thermal conductivity. These carbon fillers have also found to improve the piezoelectric and dielectric characteristics of the piezoelectric matrix material. Surface functionalized MWCNTs with –CF2 promotes the formation of β phase in PVDF polymer matrix and was attributed to the strong interaction between the functionalized MWNT and the PVDF where the strong electronegativity of fluorine atom in dipole reacted with the π electron-rich surface of MWCNT, thus forming PVDF chains with all *trans–trans* zigzag conformation. Again, Alamusi et al. [4] reported that incorporation of rGO in PVDF induced the formation of ß phase, where the rGO–PVDF nanocomposite film with 0.05 wt% rGO loading possesses the highest output voltage (3.28 V) at a frequency of 30 Hz, which is around 293% of that of the pure PVDF film.

Metal-Based Fillers: It is well established that the piezoelectric behavior of PVDF-based materials increase significantly when reinforced with metallic fillers such as zinc oxide (ZnO), titanium dioxide (TiO$_2$), ferrites, iron oxide (Fe$_3$O$_4$). In 2012, Dodds et al. [12] found that the remnant polarization of PVDF-TrFE/ZnO nanoparticle-composite-based thin films increased from 5.8 to 15.2 mC/m^2 under the applied electric field of 60 MV/m, at 20% ZnO nanoparticles concentration.

Biobased Materials: Biobased filler materials have attracted much interest due to its easy availability, biocompatibility, and multifunctional behavior. Many research groups have developed several PVDF-based biopolymer composites for improving the device performance and reliability toward the next generation.

Hybrid Fillers: The hybrid fillers comprise two components – an inorganic compound and an organic material – which may be nano or micro sized. Hybrid fillers not only reduce the filler loading, and therefore, the cost and processing difficulties, but also help in the achievement of superior properties through synergism. Qiu and coworkers synthesized a nanocomposite by adding TiO_2 nanoparticles coated on MWCNTs as hybrid fillers in PVDF matrix for enhancing the dielectric and ferroelectric behavior. The prepared $PVDF/TiO_2@MWCNT$ composite films showed enhanced dielectric constant, breakdown strength of $210\,V/\mu m$ at 0.3 wt% of $TiO_2@MWCNTs$, and lower dielectric loss compared to the pristine MWCNT/PVDF composite film.

As discussed earlier, the improvement in the piezoelectric properties of the nanocomposite depends on the reinforcing filler dispersion geometry. Piezoelectric composites with 0–3 geometry (fillers in the form of particles in polymer matrix) shows poor piezoelectric properties, while composites with 1–3 (fillers in the form of fiber) or 2–2 (fillers in the form of laminates) geometry exhibits superior piezoelectric properties, when the longitudinal direction of the fillers remains parallel to the direction of poling.

Let us consider a 1–3 composite where longitudinally aligned one-dimensional piezo electric fibers forms phase 1 and the polymer matrix forms the phase 2. The effective piezoelectric coefficients d^* and g^* of the composite with piezo ceramic fillers in polymer can be determined as follows. During reverse piezoelectric effect, when an electric field E is applied to this composite in the longitudinal poling direction, the piezo ceramic fibers extend easily because the polymer is elastically very soft (assuming the electrode plates which are bonded to its top and bottom are rigid enough). Thus, d_{33}^* is almost the same as d_{33}^1 of the PZT itself.

$$d_{33}^* = d_{33}^1.$$

Similarly, $d_{33}^* = {}^1V \cdot d_{33}^1$, where ${}^1V$ is the volume fraction of phase 1 (piezoelectric).

Again, in case of direct piezoelectric effect when an external stress is applied to the composite in the longitudinal direction along the aligned fibers, the elastically stiff piezo ceramic fibers will support the applied stress and is enhanced and inversely proportional to the volume fraction. Thus, larger induced electric fields and larger g^* constants are expected:

$$g_{33}^* = \frac{d_{33}^*}{\varepsilon_0 \varepsilon_3^*} = \frac{d_{33}^1}{{}^1V \varepsilon_0 {}^1\varepsilon_3} = \frac{{}^1g_{33}}{{}^1V}$$

In conclusion, the piezoelectric g^* coefficient of the composite can be enhanced by two orders of magnitude with decreasing volume fraction ${}^1V$ of the piezo fiber fillers, while the d^* coefficient of the composite remains constant.

The salient features of piezo composites are as follows:

- Piezoelectric composites made up of piezoelectric ceramic and polymer phases form excellent piezoelectric materials.
- High coupling factor, low acoustic impedance, and mechanical flexibility characterize these materials.
- These materials are especially used for underwater sonar and medical diagnostic ultrasonic transducer applications.

4.5.2.4 Thin Film

Piezoelectric polymer film sensors have become one of the fastest growing technologies in the global sensor market. With the advent of piezoelectric polymer films in the last two decades there has been a significant rise in the technological activity which have revolutionized sensor applications to combat various problems. Piezoelectric polymer films find wide use in sensor electronics due to its large tensile strength flexibility and castable in a wide range of thicknesses and areas. As a transducer, the piezoelectric film can be molded into unique designs and can also be glued with commercial adhesives. The major advantages of piezoelectric polymer over piezo ceramics that encourages usage in bioelectronic devices are low acoustic impedance which is closer to that of human tissue, water, and other organic materials. Such a close impedance match allows efficient transduction of acoustic signals in tissue and water. Piezoelectric polymer films also exhibit excellent sensitivity and low density. When piezoelectric polymers are extruded into thin film, they can be joined to a structure without affecting its mechanical motion. Piezoelectric films are ideal for strain sensing applications, which demand high sensitivity and wide bandwidth. Moreover, piezoelectric film offer more benefits in the form of wide dynamic range, and high sensitivity, large displacement, and low power consumption make it suitable for the development of actuators. However, on the other side when compared with piezoelectric ceramic film, the piezoelectric polymer film makes a relatively weak electromechanical transmitter, especially in low frequency and resonance applications. Piezoelectric thin films of lead zirconate titanate (PZT), lead-free piezoelectric, piezo polymer films, and cellulose-based electroactive paper (EAPap) used for electromechanical transduction. Enhanced piezoelectric activity in a thin film leads to improved performances in devices. PZT films are widely used in energy-harvesting devices due to its high piezoelectric coefficients, high power output density, and relatively low permittivity. Thin films of $BaTiO_3$ deposited on a flexible substrate are used to convert mechanical energy to electrical energy. $BaTiO_3$–metal-structured ribbons when transferred on flexible substrate and periodically deformed by a nanogenerator can generate an output voltage of up to 1.0 V. Piezoelectric polymers like PVDF having low permittivity, low thermal conductivity, and flexibility with low acoustic loss are used extensively in shock sensors, vibration control, and tactile sensors. Thin films of PVDF possess enhanced piezoelectric constant of 6–7 pC/N, which makes them suitable for transducer application in ultrasonic imaging with operating frequencies of 60–85 kHz. Aluminum nitride piezoelectric film does not need poling to exhibit piezo properties and can be used in electroacoustic device due to its enhanced electroacoustic properties such as insertion loss, coupling coefficient, quality factor, bandwidth, and dielectric constant.

In a nut shell, lightweight, high dielectric strength, high mechanical strength, and excellent stability and impact resistance together with high elastic compliance, wide dynamic range, and low acoustic impedance of piezoelectric films make them suitable for wide range of applications.

4.5.2.5 Choice of Piezoelectric Material for Desired Applications

Piezoelectric materials are chosen based on the requirement of our applications. The material that possesses the desired electromechanical properties would meet the desired requirement and hence can be considered for suitable applications. Here are the salient features of the materials that makes them suitable for different applications.

1) The parameters to be considered for applications working under mechanical resonance are the mechanical quality factor, electromechanical coupling factor, and dielectric constant. Higher the magnitude of these parameters, the more suitable the materials are for the application.
2) Materials with large piezoelectric strain coefficient, large nonhysteretic strain levels are best for an actuator.
3) Materials with high electromechanical coupling factor and high dielectric permittivity are best as transducers.
4) Low dielectric loss is important for materials used in off-resonance frequency applications accounting for low heat generation.
5) Materials like PVDF with large piezoelectric voltage coefficient are suitable for sensor application.
6) Polymers have low piezoelectric constant d as compared to ceramics, but are bendable and suitable for wearable applications.
7) Shape change of ceramic-based materials is more than that of polymer-based materials when the same amount of voltage is applied.

4.6 Uses of Piezoelectric Materials

The uses of piezoelectric material can be classified into three major ways – transducers, actuators, and generators or energy harvesters as described next.

4.6.1 Piezoelectric Transducer

Transducers are very important part of the sensor circuit as it converts the input energy received from the sensor element into output energy in the form of electrical signals as shown in Figure 4.17. A piezoelectric transducer is electroacoustic transducer that transforms mechanical energy into electrical energy by some forms of solid-state materials. The piezoelectric transducers utilize direct piezoelectric effect where the piezoelectric materials generate electrical charges proportional to the applied stress. The mechanical compression and elongation on the piezo material determines the polarity of the generated potential in close sensor circuit.

Not all piezoelectric materials can be used in piezoelectric transducers. The piezoelectric transducer materials are so chosen that the electrical signal output generated by the transducer are measurable. The material properties required in piezoelectric transducers are frequency stability, high output values, and insensitive to the extreme temperature and humidity conditions. The materials must be available in various shapes or should be flexible to be manufactured into various shapes without disturbing their properties. Unfortunately, there is no piezoelectric material that meets all these properties. Thus, the piezoelectric materials for transducer applications must be chosen carefully.

The advantages of piezoelectric transducers are listed as follows:

- *Self-generating*: Piezo materials have the ability to produce voltage and do not require any external power source.
- *Accessible*: Piezoelectric transducer circuits are portable due to their small dimensions and large measuring range.
- *High-frequency response*: The high-frequency response of these transducers makes a good choice for various applications.
- *Flexibility*: Piezoelectric polymers like PVDF are bendable and can be castable into different forms.

However, the piezoelectric transducers also carry some limitations:

- Small electric charge: While they might be self-generating, you will require a high impedance cable to establish a connection with an electrical interface.

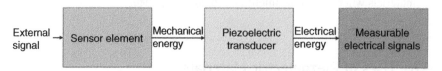

Figure 4.17 Block diagram representation showing the conversion of mechanical energy to electrical energy in piezoelectric sensor leading to output signal.

- Temperature and environmental conditions can affect the behavior of the transducer.
- They can only measure changing pressure, hence they are useless while measuring static parameters.

4.6.2 Piezoelectric Actuator

In a direct piezoelectric effect, a mechanical stress applied on the piezoelectric material produces electrical charges of different polarity at opposite surfaces of the material thereby generating piezoelectric voltage. However, in reverse piezoelectric effect, the electric field applied parallel to the polarization direction produces an elongation of the crystalline material. The electrical field generates a torque over the electrical dipoles found in the structure of the material, which will be aligned along the field, producing in turn a change in the length of monocrystalline partitions. Thus, in reverse piezoelectric effect, the electrical voltage applied across the piezo material along the poling axis produces mechanical strain and hence microscopic and precise displacements in relevant directions in the piezoelectric material. The piezoelectric actuator works on the principle of reverse piezoelectric effect where it converts an electrical signal into a precisely controlled physical displacement (stroke). If this small displacement or strain in the piezoelectric material is resisted, a force (blocking force) is developed which can be used in wide range of application and this arrangement is called the actuator. Thus, the actuators can produce a small displacement of the blocking object with a high force capability when voltage is applied. The performance of the piezoelectric actuators depends on the amount of displacement, and hence force generated, and the operating voltage of the actuator. Other factors to consider are stiffness, resonant frequency, and capacitance. Stiffness is a term used to describe the force needed to achieve a certain deformation of a structure. Resonance is the frequency at which the actuators respond with maximum output amplitude. The capacitance is a function of the excitation voltage frequency. The design and the configuration of the actuator depend on the desired applications. A disk actuator is a device in the shape of a planar disk which generated displacements along the axis perpendicular to the plane. Piezoelectric stack or multilayer actuators consists of piezoelectric disk or plate actuators that are stacked together where the axis of stacking is the axis of linear displacement when triggered by a voltage. Tube actuators are monolithic devices that contract laterally and longitudinally when a voltage is applied between the inner and outer electrodes. Ring actuators are annular disk actuators with a central hole, where the axis of displacement is made accessible. Block, disk, bender, and bimorph actuators are also available. The piezoelectric actuators are used in wide variety of applications where precise control on small mechanical displacement at high speed is required. This include adjusting fine machine tools, lenses, mirrors, or other equipment. A piezo actuator can be used to control hydraulic valves, act as a small-volume pump or special-purpose motor, and in other applications requiring movement or force. Although piezoelectric actuators provide plenty of advantages such as large generated force table displacement and ease of use, they suffer from small displacement and require large voltage. Moreover, piezo actuators undergo rate-independent hysteresis exhibited between the input voltage and the output displacement, which hampers positioning accuracy of the device. Undesirable oscillations and instabilities are some of the disadvantages frequently face that limits the usage of piezoelectric actuators.

The piezoelectric stick–slip actuators use a piezoelectric material, sandwiched between the distal and a proximal element PE (Figure 4.18a) to displace the distal element kept on the surface by a short distance through static friction. When a voltage is applied across the piezoelectric material, the mechanical strain developed within the material leads to stretching of the material by a small distance ~1 μm as shown in Figure 4.18b. On retracting the piezoelectric material rapidly, the inertia of the distal element causes slipping relative to the piezoelectric element, resulting in a net displacement of the distal element as shown in Figure 4.18c. The distal element undergoes displacement in small stochastic steps

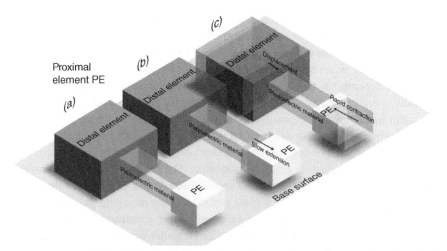

Figure 4.18 Schematic representation showing the working of piezoelectric stick-slip actuator (a) piezoelectric material is sandwiched between the distal and a proximal element PE (b) on application of voltage the mechanical strain in the piezoelectric material stretches the PE slowly and (c) on withdrawal of voltage, the piezoelectric material rapidly contracts producing a net displacement of the distal element.

size and by taking multiple successive steps, large net motions are possible. Piezo-electric stick–slip actuators behave much like admittance-type actuators during normal operation, in that they are very precise and they maintain their position when not commanded to move.

The salient features of the piezoelectric actuators are:

1) They have a theoretical unlimited resolution, often resolving to subnanometer values, i.e. smallest changes in supply voltage are detected and converted into linear motion without skipping or stepping.
2) High actuation forces can be attained without significant loss in precision.
3) Extremely high response times of under 1 ms.
4) There are no moving parts and thus no friction or free play. Elongation of a piezoelectric actuator is based only on deformation of the material and there is also no fatigue or aging involved. Endurance tests have proved that there is no altering in such an actuator's operation even after 500 million cycles.
5) Very low power consumption as the piezo effect converts energy directly into motion, the execution element absorbs energy only when elongating.
6) No magnetic fields are generated when the actuator is operating.

4.6.3 Piezoelectric Generator

The piezoelectric generator is a power generation device and works on the principle of conversion of energy from mechanical to electrical energy where vibrations in piezoelectric materials are transduced to electrical signals. The piezoelectric generator also includes an energy storage electronic circuitry for continuous supply of electrical energy to the load. Such energy-harvesting devices can be fabricated using all kinds of piezoelectric materials – monocrystals, ceramics, polymers, and composites. The architecture of the piezoelectric generator (energy harvester) is composed of a piezoelectric crystal sandwiched between two metal plates and an energy storing component (capacitor) and a full-wave rectifier circuit as shown in Figure 4.19a. Under unloading condition, the charges in the piezoelectric material are in structural equilibrium and polarization of the charges does not occur to generate a voltage $V_{out} = 0$. Under the influence of an external load on the piezoelectric material, the polarization of charges across the opposite faces of the material produce a voltage as a result of which electrical signal is generated. Since the piezoelectric transducer is unable to generate output voltage under static mechanical loading, the generator produces voltage signal only under change in pressure on the piezo element. Thus, in the reverse cycle when the load is withdrawn from piezoelectric element after the loaded condition produces a voltage output but with a opposite polarity. Considering the forward and reverse cycles

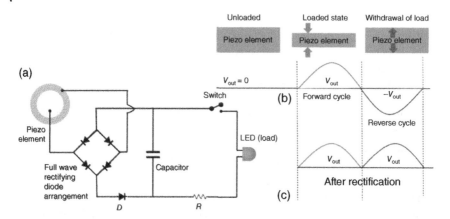

Figure 4.19 (a) Schematic diagram of piezoelectric generator and (b) mechanical to electrical energy-harvesting device. (c) Output waveform of the forward and reverse cycles before and after rectification using the diode arrangement.

of loading and unloading states, respectively, the piezoelectric element generates an alternating voltage as depicted in Figure 4.19b and needs to be converted to direct current for suitable applications. The full-wave rectifier circuit comprising four diodes connected to the piezo element converts the alternating current into direct current as shown in Figure 4.19c, and is subsequently stored in a capacitor for a short time as illustrated in the figure. Once enough energy is stored, the switch is flipped to light up an LED. To achieve an efficiency of the energy-harvesting device, the soft piezoelectric material with high flexibility and rough-houses to damage and environmental effects is to be used as this material has increased hysteresis property. The energy-harvesting efficiency (η) of the device is given as:

$$\eta = \left(\frac{P_{\text{out}}}{P_{\text{in}}} \times 100\right)\% = \frac{1}{m}\sum_{n=2}^{m}\frac{\{(V_n + V_{n-1})^2/R\}}{[\{(F_n + F_{n-1})(d_n - d_{n-1})\}/(t_n - t_{n-1})]}$$

where V is the voltage drop across resistance R, F is the force applied to the base of the plate, d is the displacement of the plate, t is the time increment between data points, n is the data point index, and m is the total number of measured data points. If flexibility of the material is high, the plates are able to return to their original positions after the release of the force, leading to the achievement of nearly equal displacement of the piezoelectric material for the data point index n and $(n-1)$. Thus, the $(d_n - d_{n-1})$ is reduced, which in turn leads the generation of increased hysteresis effect $(V_n + V_{n-1})$ in the material and hence, the achievement of high energy-harvesting efficiency of the device.

4.7   Piezoelectric Transducers as Sensors

4.7.1   Pressure Sensor

When pressure is applied on a piezoelectric crystal, it generates electric charges on the crystal surface, leading to generation of the piezoelectricity. Thus, the piezo-electric crystal can be used as a pressure transducer with a piezoelectric crystal inserted between a solid base and the diaphragm as shown in Figure 4.20. The charges generated in response to the applied pressure (or force) can be measured in terms of output voltage (V_{out}) across the output terminals and found to be pro-portional to the applied pressure (or force). However, the piezoelectric crystal will show no physical deformation when a fixed voltage is applied across the crystal, i.e. the piezoelectric material shows no inverse piezoelectric effect. The pressure trans-ducer is not suitable for measuring static applied pressure as the output signal gradually drops to zero, even in the presence of constant pressure. The piezoelec-tric pressure transducers are sensitive to dynamic changes in pressure across a wide range of frequencies and pressures. The piezoelectric transducers as pressure sensors are suitable only for measuring rapid small variations in pressure over a wide dynamic range of the sensor. Unlike piezo resistive and capacitive transdu-cers, piezoelectric sensor elements require no external voltage or current source as they are capable of generating an output signal directly from the applied strain. The disadvantages of the piezoelectric sensor lie in the easy neutralization of sur-face charges produced by the applied force by charges from the environment, leak-age current through the dielectric, and through the input resistance of the circuitry. This makes the sensor behave as a high-pass filter for input signals, pre-venting static measurements. Second, the piezoelectricity is affected by tempera-ture which may induce crystal deformation and even output voltage.

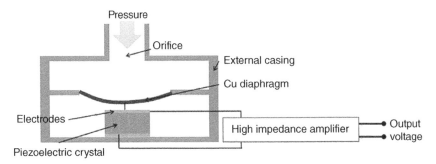

Figure 4.20   Schematic diagram of the piezoelectric pressure sensor demonstrating its working principle.

4.7.2 Accelerometer

In a piezoelectric pressure sensor, the pressure is applied to a thin metal diaphragm and is mechanically transferred to the crystal. Piezoelectric pressure
sensors are affected by acceleration (a), because the mass (m) of the object
housing the piezo pressure sensor produces an inertial force on the crystal
when accelerated. Such devices which are used to measure acceleration are
called accelerators. An accelerometer consists of one or more piezoelectric crystals and a *proof mass* (or *seismic mass*) which is fixed by a preload onto the
crystal. Figure 4.21a shows the architecture of the piezoelectric compression
mode accelerometer where the piezoelectric component is in the form of piezoelectric stack as shown in Figure 4.21b. The piezoelectric stacks increase the
sensitivity of the accelerometer by generating more output voltage in response
to small displacement (Section 4.8.2). When the accelerometer is subjected to
acceleration/vibration of amplitude $= a$, a force ($F = $ ma) is generated which
acts on the piezoelectric element. Due to the piezoelectric effect, a charge (q)
and voltage (V_{out}) output proportional to the applied force is generated and
expressed as:

$$Q = d_{33}F \text{ and } V_{out} = \frac{d_{33} \times d}{\varepsilon_{33} \times A}F$$

where d_{33} and ε_{33} are piezoelectric constants. d and A denote the thickness and
area of the piezo disc. Since the seismic mass is constant, the charge output signal
is proportional to the acceleration of the mass. Over a wide frequency range both
sensor base and seismic mass have the same acceleration magnitude and hence,
the sensor measures the acceleration of the test object. The charge (S_q) and voltage
($S_{V_{out}}$) sensitivity are given as, $S_q = q/a$ and $S_{V_{out}} = V_{out}/a$.

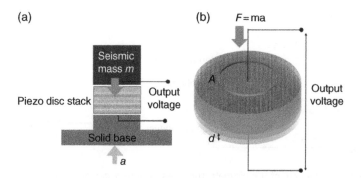

Figure 4.21 (a) Schematic representation of a piezoelectric accelerometer and
demonstration the working principle of the accelerometer. (b) Schematic representation
of Piezo disc stacking arrangement.

A piezoelectric accelerometer consists of a base plate, a seismic mass, and a (pre-loaded) crystal. The mechanical characteristics of this spring mass system follow from the equation of motion:

$$m\frac{d^2x}{dt^2} + \alpha\frac{dx}{dt} + kx = F(t)$$

where m is the seismic mass, α the damping constant, and k the spring constant between seismic mass and base. The natural frequency ω_o of the system equals:

$$\omega_0 = \sqrt{\frac{k}{m}}$$

Since the damping constant α of a piezoelectric accelerometer is usually very small, the natural frequency of the accelerometer is approximately equal to the resonance frequency ω_{res} under damped condition. A high resonance frequency requires a small mass m and a large stiffness k. The mass determines the sensitivity in the frequency range 50–60 Hz and is chosen for suitable applications. The resonance frequency of commercial accelerometers ranges from 1 to 250 kHz: the smaller the size, the higher the resonance frequency. The piezoelectric accelerometer can have (i) high and (ii) low impedance outputs. In the case of high impedance device, the output voltage generated in proportion to the acceleration is given as the input to other measuring instruments. The instrumentation process and signal conditioning of the output is considered high and thus a low impedance device cannot be of any use for this application. This device is used at temperatures greater than 120 °C. The low impedance accelerometer produces a current due to the output voltage generated and this charge is given to a FET transistor so as to convert the charge into a low impedance output. This type of accelerometers is most commonly used in industrial applications.

4.7.3 Acoustic Sensor

The scope of the acoustic sensors has widened over the years. The common application of acoustic sensors includes measuring displacement, concentration of compounds, stress, force, and temperature. An acoustic wave can propagate either on the surface of the material (SAW) or through the bulk material (BAW). The acoustic sensor has a pair of piezoelectric transducer at each end of the device. At the transmitting end the piezo transducer generates acoustic waves, while at the receiver end, the acoustic waves that are received from the transmitter end are converted into an electrical signal as shown in Figure 4.22. The electrical circuit is coupled to the mechanical structure using direct and inverse piezoelectric effect. A transmitter transducer end forces atom of the solid into vibratory motions about their equilibrium position. The neighboring atoms then produce a restoring force

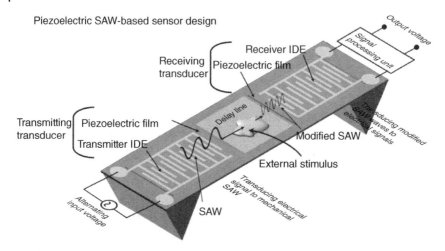

Figure 4.22 Three-dimensional representation of the surface acoustic wave (SAW) sensors.

tending to bring the displaced atoms back to their original positions. Any variation in the vibratory characteristics due to external stimulus affects its phase velocity and/or the attenuation coefficient of the propagating acoustic wave at the receiver end. In acoustic sensors, external stimuli, such as mechanical strain in the material, increase the propagating speed of sound, while in chemical sensors the adsorption of molecules, attachment of bacteria cause a reduction of acoustic wave velocity. Again, in viscosity sensors, viscous liquid comes in contact with the active region of an elastic wave sensor to decrease the amplitude of the wave. The changes in the velocity or amplitude produces variation in the natural frequency or phase characteristics of the sensor, which can then be correlated to the corresponding physical or chemical stimulus to measure the magnitude of the external agent. In chemical and biological sensors, the acoustic path, where mechanical waves propagate, may be coated with chemically selective compound which interact only with the stimulus of interest. Piezoelectric materials are generally used to generate and detect acoustic waves. Piezoelectric materials provide the transduction of electrical signal into mechanical acoustic waves in the generation of acoustic wave and vice versa in the detection of the same. Conventional piezoelectric materials which are used as acoustic sensor include quartz, ZnO, $LiNbO_3$, PZT ceramic, AlN, and $LiTaO_3$.

Based on the acoustic wave propagation modes the piezoelectric acoustic wave sensor can be classified into two categories-(i) SAW sensors and (ii) bulk acoustic wave (BAW) sensors. Since the waves must propagate through the cross-sectional thickness of the BAW sensors, the electrodes are positioned on the opposite sides of the propagating medium and spans over an large area. For the SAW sensors, the waves propagate on the surface of the material on which the interdigitated excitation electrodes are fabricated.

i) *SAW Sensors*:

When an electric signal with alternating polarity is applied on the electrodes an alternating region of tensile and compressive strain between two fingers of the electrodes are developed due to piezoelectric effect of material. The mechanical wave propagates on the surface of the material in both directions from the transmitting electrodes crosses the delay line to reach the receiver electrodes. For a chemical sensor the delay line is sensing area, where the sensor material absorbs the target analytics thereby inducing variations in the propagating waves.

The response of the piezoelectric SAW sensor is attributed to the changes in the surface properties of the propagation medium in the delay line. These changes in the properties at the delay line modulate the velocity and amplitude of the SAW.

The operation frequency of the SAW device depends on the interdigital transducer's design and the nature of piezoelectric material and expressed as, $f_{res} = (V_R/\lambda)$, where V_R is Rayleigh wave velocity determined by material properties and λ is the wavelength defined as the periodicity of the interdigitated electrodes (IDE). The surface wave velocity (v) can be perturbed by various factors, each of which represents a possible sensor response as:

$$\frac{\partial v}{v_0} = \frac{1}{v_0}\left(\frac{\partial v}{\partial m}\Delta m + \frac{\partial v}{\partial c}\Delta c + \frac{\partial v}{\partial T}\Delta T + ...\right)$$

where v_0 is unperturbed wave velocity, m is mass, T is temperature, and c is stiffness. Therefore, these kinds of devices can be used in mass, pressure, and temperature sensing applications. The change in wave velocity also causes the change in f_{res} of the SAW device.

ii) *BAW Sensors*:

The quartz crystal microbalance (QCM) is the oldest and the simplest acoustic wave device which utilizes BAW for the mass sensing applications. The device consists of a thin disk of AT-cut quartz with parallel circular electrodes patterned on both sides of the crystal as shown in Figure 4.23. An alternating applied voltage at the transmitting electrodes produces shear deformation of the piezoelectric crystal. Bulk adsorption of target analyte onto the coated crystal causes an increase in effective mass, which reduces the resonant frequency of the crystal, in direct proportion to the concentration of target analyte. For ideal sensing material, this sorption process is fully reversible with no long-term drift effect, giving a highly

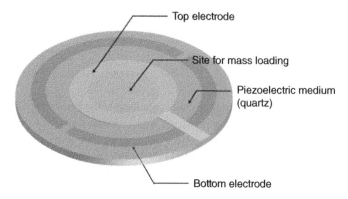

Top electrode

Site for mass loading

Piezoelectric medium (quartz)

Bottom electrode

Figure 4.23 Illustration of QCM-based piezoelectric bulk acoustic wave sensor, showing top and bottom circular electrodes.

reliable and repeatable measurement. The relation between the frequency shift and the mass loading is given as:

$$\Delta f = \frac{2f_0^2}{A\sqrt{\rho_q \mu_q}} \Delta m$$

where f_0 = resonant frequency depends on the wave velocity (v) and the piezoelectric material thickness = $f_0 = v/2d$, Δf = frequency change, Δm = mass change, A = active area, ρ_q = density of piezoelectric material, and μ_q = shear modulus of piezoelectric material.

The SAW and BAW sensors that are used to measure environmental properties such as temperature, pressure, stress, acceleration, field, and charge have been reported over recent years. Furthermore, through proper design and choice of piezo material, different chemical analytes in the form of atoms/molecules have also been detected. Such chemical sensing applications include analytes monitoring the deposition of monolayers, biochemical trace detections including monitoring DNA mutations, and commercial application like environmental monitoring. In addition to chemical and biological sensors, these materials have also been configured to measure physiochemical properties such as dew/melting point, curing, adsorption/desorption, and viscosity. The success of acoustic wave sensors lies in its capability to monitor a wide range of properties while using the measurement electronics.

Moreover, piezoelectric materials can be used as temperature sensor due to their pyroelectric property where the crystals generate electric energy in response to temperature. However, pyroelectricity and the application of piezoelectric materials as temperature sensors are not included in this chapter.

4.8   Design of Piezoelectric Devices

The piezoelectric sensor operated using the direct piezoelectric phenomenon generates an output voltage V_o at the electrodes when force F is applied along suitable direction. This phenomenon is frequently utilized in the development of various sensors as described in Sections 4.6.1 and 4.7. Similarly, with the same piezoelectric arrangement under inverse piezoelectric effect, the crystal undergoes deformation when an external voltage is applied at the terminals. This phenomenon is used for actuation and described in Section 4.6.2. Thus, depending on the desired application of the device the piezoelectric crystal may be used as direct or inverse piezoelectric effect.

4.8.1   Orientation of Piezo Crystals

To construct a device for a given target application, the device architecture plays an important role and determines the performance of the device. The piezoelectric devices are designed using suitable piezoelectric material (Section 4.5) considering its various material properties (Section 4.4.3) where the orientation of the material or crystal in the device occupies a pivotal importance. Since the piezoelectric properties are direction dependent, the orientation of piezoelectric materials in devices for a given application is decided by their piezoelectric strain coefficient d_{ij} (Section 4.4.3.1). The piezoelectric crystal/material are able to generate charges at the different surfaces based on the various modes of structural deformation experienced by the crystal under the influence of external forces F. The piezoelectric crystal may operate in the following modes such as: (i) parallel compression and tension mode, (ii) transverse compression and tension mode, (iii) transverse shear mode, and (iv) parallel shear mode. The piezoelectric crystal operates in parallel compression and tension mode when a pair aligned forces F producing compression and expansion of the crystal generates a polarization P in the direction of the force. Similarly when the pair aligned forces F producing compression and expansion in the crystal, polarizes it in the direction perpendicular to the direction of the applied force the piezoelectric crystal is said to be operating in transverse compression and tension mode. Piezo materials with high d_{ii} tensor elements are frequently used in parallel compression and tension mode, while that with high d_{ij}, $i, j = 1, 2, 3$, and $i \neq j$ tensor elements, are used in transverse compression and tension mode. The piezo crystal operates in the transverse shear mode when a pair of opposite unaligned forces produces shearing deformation in the crystal giving rise to resultant polarization in the direction perpendicular to the direction of force. Similarly when this applied force produces polarization in the crystal in the direction of the force, the crystal operates in parallel shear mode. The performance of the piezoelectric crystal or material used as sensor in shear modes involve dominant participation of d_{ij}, $i = 1, 2, 3$ and $j = 4, 5, 6$ tensor elements.

To understand the d_{ij} tensor elements of the piezoelectric crystals and the charges generated at different crystal faces due to various structural deformations due to applied forces on it, let us consider a piezoelectric crystal of length a, width b, and thickness c, with electrodes deposited on the opposite faces of crystal perpendicular to the z-axis and as shown in Figure 4.24a. Here the x-, y-, and z-axes are denoted as direction 1, direction 2, and direction 3. With reference to Figure 4.24b, when force F is applied along z-axis (direction 3), the dipoles are polarized P along the same direction ($\mathbf{P}$ along direction 3). Due to this deformation charge $Q = F$. The d_{33} is accumulated across the terminals along the z-axis and the piezoelectric crystal operates in parallel compression and tension mode. Similarly when the aligned F is applied on the crystal along the x-axis (direction 1) produces polarization along the z-direction (direction 3), then the accumulated charge at the electrodes is given by $Q = F$. The d_{31} and the crystal operate in the transverse compression and tension mode as shown in Figure 4.24c. The shear forces are in the form of opposite and unaligned forces acting on the faces of the crystal. The forces may be imagined as torque acting about an axis. The shear force acting on the crystal about the axes x, y, and z is denoted by directions 4, 5, and 6, respectively. In Figure 4.24d, the crystal operates in transverse shear mode where shear forces are applied on the crystal along the z-axis and about the y-axis (direction 5), producing a polarization in the x-direction (direction 1). The charge accumulated at the terminals is given by $Q = F.d_{15}$. The piezo crystal under parallel shear mode is shown in Figure 4.24e where shear force is applied along the x-axis and about the y-axis (direction 5), producing a polarization in the x-direction (direction 1). The charge accumulated at the terminals in this case is also given by $Q = F.d_{15}$.

4.8.2 Piezo Stacks

A single-layer piezoelectric actuator is made from a single piezoelectric element. A single piezoelectric element produces an electric charge or force which may not be sufficient to operate a device for useful work. To obtain greater accumulation of charges or generate higher forces, multiple piezo elements are put together on top of each other in a casing to produce enough electric charge to create significant displacement that can be harnessed as useful work. However, piezo stacks and multilayer piezo stacks must not be confused with each other. A piezo stack has multiple layers of piezo elements, while a multilayer piezo stack has multiple layers of piezo stacks.

As mentioned earlier, a multilevel piezoelectric actuator delivers a much wider range of forces and displacements for the same amount of applied voltage. This is attributed to the multiplicative effect resulting from the stacking-up of

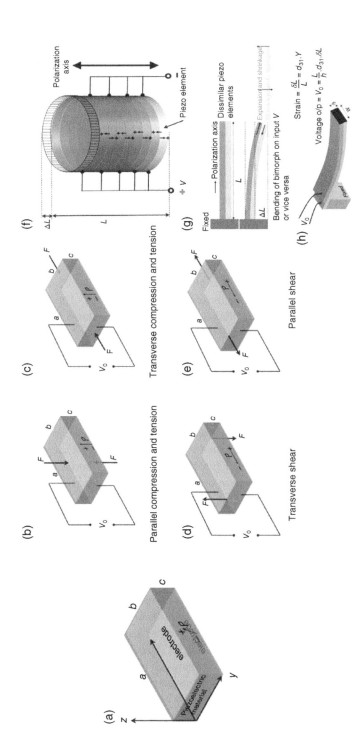

Figure 4.24 Illustration of a piezoelectric crystal in (a) equilibrium, under (b) parallel compression and tension, (c) transverse compression and tension, (d) transverse shear, and (e) parallel shear modes of operation. (f) Illustration of multiple piezoelectric elements stacked together to form a piezoelectric stack. (g) Schematic representation of the working of a piezoelectric bimorph through a cantilever architecture and (h) possible scheme for energy harvesting using the piezoelectric bimorph where to and fro bending of the cantilever produces output voltage.

piezoelectric elements. Suppose a single-layer piezoelectric actuator produces 1 μ m displacement at 1 kV. When such piezo elements are stacked together on top of each other through parallel connection, produces 100 μm displacement at 100 V. Therefore, a multilevel piezoelectric element produces 100 times more force at 10 times less voltage. Thus, performance of piezo stacks are significantly more pronounced than each piezo elements. Low operating voltages with high force magnitudes extend the capabilities of piezoelectric actuators and make them suitable for a wider range of control applications.

Figure 4.24f shows multilayer piezoelectric stack consisting of ceramic piezo element laminates. Within operational range of voltage, the piezo actuator reaction versus controlled voltage is linear. The mechanical displacement ΔL or stroke of a piezo stack, consisting of n piezo elements with piezoelectric stain coefficient d_{ij} under applied voltage V, depends on the number of layers and is given by $\Delta L = d_{ij} \cdot n \cdot V$. Thus, the more the number of layers a piezo stack has, the higher the mechanical displacement or stroke. The blocking force of the piezo stack is directly proportional to the cross-sectional area of the piezo stack. The wider the cross-sectional area the higher the blocking force of the piezo stack. In other words, any amount of pressure that can deform a piezo element is strong enough to create a piezo effect.

4.8.3 Bimorph Architecture

A piezoelectric bimorph consists of two active piezoelectric layers adhered together in the form of a cantilever architecture as shown in Figure 4.24g. In piezo actuator applications, if voltage is applied, one piezo layer expands and the other shrink, thereby bending the cantilever. Thus, the piezoelectric bimorph can be used to generate displacements. The piezo bimorphs greatly magnify the power and range of a piezoelectric actuator. While a single piezoelectric crystal can only shrink or expand by a few millimeters at most, bimorphs can bend well over ten or a hundred times that range at the same voltage. In piezo sensor applications, the piezo bimorph generates an electrical signal as a response to a mechanical input, such as force, pressure, or acceleration. A schematic diagram of a piezoelectric series bimorph is illustrated in Figure 4.24h, where the bimorph in the form of cantilever of length L, width W, and height h, bends in response to physical input, thereby generating a output voltage V_0. In piezo transducer applications, the piezoelectric bimorph can receive either electrical or mechanical inputs and produce either electrical or mechanical outputs depending on the desired results. The piezoelectric bimorph can also have a passive layer between the two active layers acting as a supportive substrate. The designs of piezoelectric bimorph are extremely common in sound production; one can often find a piezoelectric bimorph in fire alarms, buzzers, and other sound producing devices.

4.9 Application of Piezoelectric Sensors

4.9.1 Industrial Applications

4.9.1.1 Engine Knock Sensors

Knock sensors (KS) are vibration sensors that are used to detect structure-borne acoustic oscillations. The KS is a piezoelectric sensor that consists of a sensing crystal and a resister. The KS takes advantage of the unique property of piezoelectric crystal that generate a small voltage under vibration. Automobile engines are frequently subjected to undesirable problem called detonation (unintended combustion) where the air/fuel instead of burning smoothly, produces vibrations (also termed as knocking), which may destroy the pistons, rods, valves, gaskets, and plugs. To avoid this notorious problem, piezoelectric KS can be employed to sense the detonation before it becomes problematic. The KS attached to the engine senses the knocking sound (vibration) in the engine and will send a signal to the PCM (power train control module) of the car. The schematic diagram of the KS is shown in Figure 4.25. The KS is installed at a transduced reference voltage of 2.5 V produced by the piezoelectric crystal. When the engine produces knocking, the vibrations experienced by the sensor head are transmitted to the

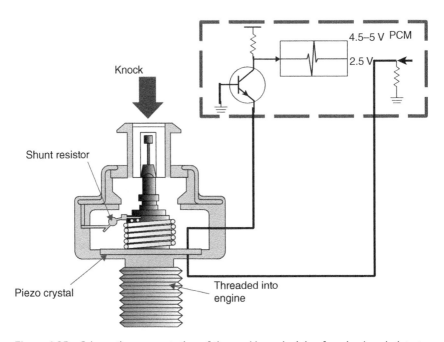

Figure 4.25 Schematic representation of the working principle of engine knock detector.

piezoelectric crystal which converts the vibrations to electrical signals. These AC signals arousing from the vibrations produce electrical spikes ~4.5 V in output signal and are communicated to the PCM. When the PCM receives the signal, it retards the engine ignition timing preventing harm to the engine. A knock sensor is important to a car's fuel efficiency and helps the car to adjust the fuel-to-air mix, which optimizes the amount of power to the engine.

4.9.1.2 Tactile Sensors

The use of piezoelectric polymers in the development of wearable sensors has seen a burst of research interests. A flexible PVDF film-based piezoelectric tactile sensor array was developed by Yu et al. [13] in 2016 to measure three-axis dynamic contact force distribution. The array consists of six tactile units arranged as a 3×2 matrix where the piezoelectric PVDF film is sandwiched between four square-shaped upper electrodes and one square-shaped lower electrode, forming four piezoelectric capacitors. A truncated pyramid bump on the four piezoelectric capacitors improved the sensing of forces. A three-axis contact force transmitted from the top of the bump will lead to the four piezoelectric capacitors underneath undergoing different charge changes, from which the normal and shear components of the force can be calculated. The sensor array can be easily integrated onto a curved surface, such as robotic and prosthetic hands, due to its excellent flexibility.

Chen et al. [14] in 2018 developed a flexible three-axial tactile sensor using piezoelectricity enhanced P(VDF-TrFE) micropillars. The three-axis force measurement was achieved using the vertically aligned P(VDF-TrFE) micropillars which are sandwiched between four square bottom electrodes and a common top electrode to form four symmetrically arranged piezoelectric sensing units. The arrangement uses high sensitivity and good flexibility of the imprinted P(VDF-TrFE) micropillars on four distributed piezoelectric units to provide highly sensitive detection of three-axis compressive and tensile stress. The relative generation of piezoelectric charges on the bottom electrodes gives the direction and the amplitude of the applied force. The flexible three-axial tactile sensor has high potential for use in advanced robots, wearable electronics, and a variety of human–machine interface implementations.

A multimodal temperature and force sensor on ultrathin, conformable, and flexible substrates was developed by Viola et al. [15] in 2018. The device involves coupling a organic charge modulated field-effect transistor (OCMFET) with a PVDF-based pyro/piezoelectric element. The device detects both pressure stimuli and temperature variations, opening avenues for low-cost, highly sensitive, and conformable multimodal sensors. The device is conformable on skin as the overall thickness of the device is 1.2 µm.

4.9.1.3 Piezoelectric Motors

The piezo motors have emerged as an excellent alternative to electric motors, opening new avenues for smart applications. A piezoelectric micro motor uses inverse piezoelectric effect, i.e. when an electric voltage is applied to a piezoelectric material, it deforms mechanically. These piezo motors have two parts – a stator and a rotor. Stator converts electrical energy of the piezoelectric element into oscillations. The movements of the stator are converted into the movement of a slider into frictional contact with the stator. This movement may either be rotational or linear, depending on the design of the motor structure.

There are three types of piezo motors – (i) stick and slip piezo motor, (ii) stepper piezo motor, and (iii) ultrasonic/resonant piezo motor. Figure 4.26a explains the working principle of the stick–slip piezo motor. The stick–slip piezo motor set up consist of piezoelectric stack which is fixed to a rigid support on one side, a moving slider, and bearings on which the slider moves. The piezo crystal is in contact with the slider through a rigid contact. The stick and slip piezo motor operates through distinct and alternating stick and slip phases [16]. In the stick phase, the voltage across the piezo crystal is increased gradually causing slow extension of the crystal allowing the slider to advance owing to the frictional force between the contact point and the slider. In the slip phase, the piezo actuator is rapidly retracted by applying a rapidly decreasing voltage, because of which the slider remains stationery due to inertia, while the contact point slips back to its original position. The alternating phases of stick and slip produce a net displacement ΔL of the slider in the desired direction. The phases are repeated to realize macroscopic movement of the slider.

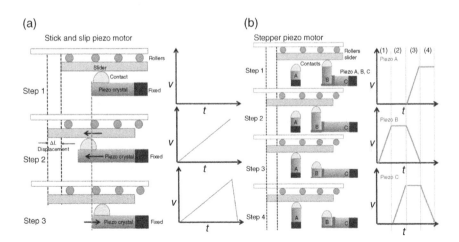

Figure 4.26 Schematic representation of the working of the stick and slip (a) and stepper (b) piezo motors through graphical representation of the applied voltage versus time plots.

The stepper type of piezo motors, also called walking piezo motors, are typically used for high-force applications. Generally PZT crystals are used in the construction of such piezoelectric motors [17]. It consists of three piezo actuators, of which some of the actuators are in contact with the slider (piezo A and piezo B) and serve as anchors to the slider, while the others (piezo C) are used to generate translational motion of the slider as shown in Figure 4.26b. Initially piezo A and piezo B are in contact with the slider when it is at rest. At the onset of translational motion of the slider, the piezo A is retracted, while piezo B is extended by application of suitable bias voltages to different piezo stacks. At this state only contact point of piezo B is in contact with the slider as shown in Figure 4.26b (Step 1). Now piezo C is extended to realize the translational motion of piezo B, and hence execute movement of the slider due to frictional force between the piezo B contact point and the slider as shown in Figure 4.26b (Step 2). Now, piezo A is extended and piezo B is retracted such that only piezo A is in contact with the slider as in Figure 4.26b (Step 3). Finally, piezo C is retracted to its original position as shown in Figure 4.26b (Step 4). The cycle steps 1–4 are repeated to realize movement of the slider.

The ultrasonic piezo motor arrangement consists of linear or circular arrangement of piezoelectric segments depending on the design of the motor and works using the generation of travelling or standing waves [5]. The piezo elements are capable of oscillating at one of its resonant frequencies, in the ultrasonic range under the influence of AC voltage. The oscillations of the piezo elements produce elliptical motion of the stator which creates movement of a slider due to frictional contact with the stator. All the piezoelectric segments are arranged circularly and have the same polarization direction in z-direction, perpendicular to the plane of the rotor. Thus, the piezo elements oscillate along the z-direction under applied voltages. The piezo segments are excited by applied voltages $+s$, $+c$, $-s$, and $-c$ with phase shift of $90°$ over one wavelength such that the superposition of multiple standing waves produces travelling waves which move along the piezo ring of the motor as illustrated in Figure 4.27a. As the surface travelling waves moves along the stator, the contact point(s) shift(s) along the motion direction, constantly pushing the slider in the direction opposite to the direction of movement of the travelling wave. Figure 4.27b shows a typical ultrasonic piezo motor showing the circularly arranged piezo segments generating a travelling wave. Patel et al. [18] through COMSOL simulation have depicted the travelling waves in the piezo ceramic ring using animations as shown in Figure 4.27c. Although travelling wave piezo motors achieve high forces, they are prone to wear and tear and have a limited lifetime. To reduce the wear between the piezo elements and the rotor, the piezo elements are coated with wear resistive material.

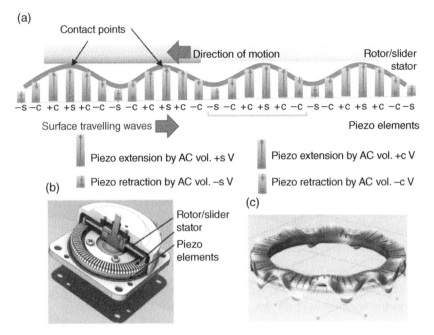

Figure 4.27 (a) Schematic representation of the working of the ultrasonic piezo motor, (b) typical ultrasonic piezo motor showing its components, and the (c) COMSOL illustration of the standing waves generated due to oscillations in the circularly arranged piezo elements.

4.9.1.4 SONAR

SONAR (Sound NAvigation and Ranging) is a technology that uses acoustical waves to detect the location of objects in the ocean as depicted in Figure 4.28. SONAR devices send out pulsating sound waves from a transducer, and then precisely measure the time it takes for the sound pulses to be reflected back to the receiver transducer. The distance to an object can be calculated using the time difference and the speed of sound in the water (~1500 m/s). SONAR was widely used during WWI to detect icebergs and enemy submarines. Lead indium niobate–lead magnesium niobate–lead titanate (PIN–PMN–PT) ternary single crystals are frequently used in SONAR applications due to its higher coercive field ($E_c \sim 5$ kV/cm) and higher Curie temperature ($T_c > 210\,°C$). PIN–PMN–PT single-crystal piezoelectric materials provide increased sensitivity and wider bandwidth for underwater transducers [19]. These piezoelectric crystals provide higher piezoelectric performance magnitude which helps to decrease the number of piezoelectric elements necessary. Thus, improved performance is achieved at lower power consumption levels. SONAR can be used for under water sea surface imaging, such as fish finder

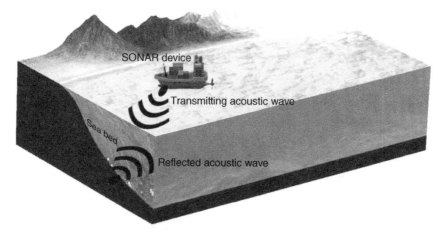

Figure 4.28 Schematic representation of the working of SONAR devices.

to locate objects like fish, wreck, vegetation, and sudden change in water temperature, in the search area of the sea, and under water communication.

4.9.2 Consumer Electronics

4.9.2.1 Piezoelectric Igniters

In a piezoelectric igniter, a piezoelectric rod generates a rapid rise in voltage when the piezo rod is suddenly hit by a spring release hammer. This mechanical shock generates a voltage which is high enough to ionize the air between sizeable spark gap to ignite the fuel. Piezoelectric igniters are commonly used for butane lighters, gas grills, gas stoves, blowtorches, and improvised potato cannons. Although lead zirconate titanate piezoelectric crystal is commonly used in kitchen gas oven igniters, other piezo crystals such as quartz, sodium potassium tartrate, and tourmaline are also used frequently for other ignition applications.

4.9.2.2 Drop on Demand Piezoelectric Printers

In a piezoelectric inkjet printer, piezoelectric actuators in the printer head serve as small diaphragms which deforms when external voltage is applied [20]. The deformation of the piezo actuator changes the internal volume of an ink well which forces the ink droplets onto paper through a nozzle as shown in Figure 4.29a, b. The advantages of piezo printer head include controlled discharge of ink droplets where high accuracy in printing can be regulated by controlling the applied voltage. This printing technique is highly durable and resistant to the influence of the operating environment as it does not apply heat

(a)   (b)

Drop on demand piezoelectric printers

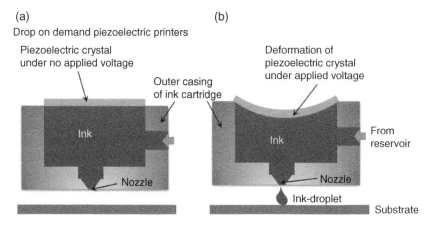

Figure 4.29 Schematic representation of the drop on demand piezoelectric printers under (a) no applied voltage and (b) presence of applied voltage showing the deformation in the piezoelectric crystal which oozes the ink out of the nozzle.

to the ink. The shortcomings of the piezo printing head involve clogging of nozzles when air bubbles are formed in the ink.

4.9.2.3 Speakers

A piezoelectric material produces a mechanical motion when a voltage is applied across it. The diaphragms and resonators convert the motion into audible sound. This is the working principle of the piezoelectric speaker. In ultrasound applications, piezoelectric speakers operate well in the range of 1–5 kHz and up to 100 kHz. Piezoelectric speakers generate sound in digital quartz watches and other electronic devices. The advantages of piezoelectric speakers over conventional loudspeakers lies in their resistance to overloads that would normally destroy most high-frequency drivers. However, piezoelectric speakers suffer disadvantage of distortion and damages to the amplifier when driving capacitive loads in piezoelectric. Piezoelectric buzzers are similar to piezoelectric speakers, but they are usually designed with lower fidelity to produce a louder volume over a narrower frequency range. Buzzers are used in a seemingly endless array of electronic devices.

4.9.2.4 Other Daily Use Products

Ultrasonic toothbrushes emit very high in frequency but low in amplitude vibrations that break up bacterial chains on the teeth. Many cool mist piezoelectric-based humidifiers use a piezoelectric transducer to transmit ultrasonic sound energy into a pool of water. The ultrasonic vibrations cause fine water droplets to break away and atomize from the surface of the pool where they become entrained in an air stream and enter the desired space.

4.9.3 Medical Applications

4.9.3.1 Ultrasound Imaging

Ultrasound imaging uses the principle of SONAR to detect changes in the appearance, size, shape, or contour of organs, tissues, and vessels, or used to detect tumors or other abnormal masses. An ultrasound technician uses a handheld piezoelectric transducer (Figure 4.30) to send acoustic waves into the body as well as to receive the echoing waves reflected by the soft tissue of organs. On pressing the transducer against the skin, the device sends tiny acoustic pulses of inaudible, high-frequency waves into the desired region of the patient's body. These transducers produce sound waves at frequencies well above the threshold of human hearing at 20 kHz and higher; most transducers used today operate at much higher frequencies in the megahertz (MHz) range. In ultrasound imaging, the piezoelectric material is used as acoustic resonators where the piezoelectric transducer resonates at $f_0 = c/2d$ when the waves reflected by the internal organs, tissues, and fluids return to the transducer to produce usual acoustic mismatch between the piezoelectric transducer and the organs. The transducer resonates at odd harmonics too ($3f_0$, $5f_0$, etc.) because these acoustic waves encounter softer materials and bounce back into the piezoelectric a number of times. The acoustic mismatch, i.e. the efficiency of piezoelectric crystal to transfer acoustic energy from piezoelectric crystal to soft tissue, is measured by the transmission factor (TF). From acoustic theory, the energy that is transferred from one acoustic material, Z_c, to an adjacent

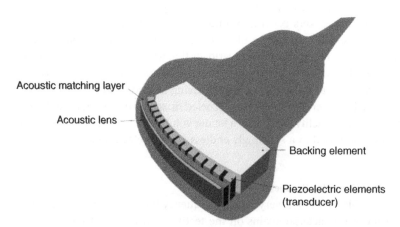

Figure 4.30 Schematic representation of the internal architecture of ultrasound imaging kit/probe.

material of impedance Z can be quantified as $TF = 4\,Z\,Z_c/(Z_c + Z)^2$. This factor can be used to compare different materials within the body and to detect position and size of other tissues. A computer measures and displays the signature waves to create a real-time picture on a monitor.

4.9.3.2 Surgery and Ultrasound Procedures

Piezoelectric substances are recently being used in medical surgery where piezo-electric instruments placed along the cutting edge are made to vibrate by applying a voltage across the crystal. These vibrations can be fine-tuned to only cut miner-alized tissue, making it much more precise and less invasive than normal surgery [21]. The advent of the harmonic scalpel has enabled surgeons to simultaneously incise and coagulate tissue during a surgical procedure without the need for cau-terization. This leads to less tissue damage, less blood loss, and faster healing times. Some noninvasive medical procedures rely on the use of focused ultrasonic waves to break up kidney stones (Lithotripsy) [22, 23] or destroy malignant tissue [24]. The piezo-based devices are also used to remove plaque and tartar deposits on the teeth with the aid of high-frequency sound waves (Figure 4.31). The waves gener-ated by the piezo material break up the hard deposits of plaque and tartar, and the attached water spray washes the deposits away.

4.9.3.3 Wound and Bone Fracture Healing

The application of piezoelectric crystal in shockwave therapy is relatively new, and is still being researched. It is being used extensively in developed countries for many years with successful clinical outcomes in treating various forms of

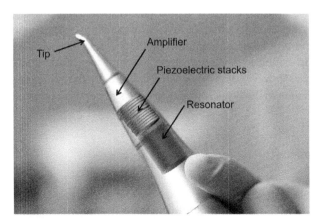

Figure 4.31 Representative image of piezoelectric dental scaler showing the piezoelectric stacks.

musculoskeletal conditions. The shockwave therapy also known as the piezoelectric impulse therapy is a nonsurgical and noninvasive procedure where an array of piezo crystals delivers gentle pulses of focused energy into the affected tissue in the region of treatment, typically "the painful area," to boost your body's natural healing and regenerative properties. The pulses promote tissue regeneration and remodeling to heal the scar tissues and calcific fibroblasts by initially breaking them down and then provoke metabolic activity and an inflammatory response which promotes and stimulates healing. Instead of air or radial impulse, the piezoelectric device uses electromagnetic impulses that encourage recovery without inflammation or side effects. The shockwave therapy is successfully used for the treatment of Achilles tendinopathy, plantar fasciitis, ankle and foot pain, ankle and foot muscle injuries, excessive scar tissues, and other conditions related to the ankle and feet. The advantages of the piezoelectric impulse therapy are nonsurgical and noninvasive procedures, reduces, mitigates, and eventually eliminates pain, reduces muscle tensions and sprain, encourages natural healing in the treatment area, improves microcirculation to promote tissue metabolism, and increases collagen production.

Induced polarization in cortical bone produces electric fields at the microfractures of the bones. The field is high enough at the tip of the microcrack to induce osteocyte apoptosis and thus initiate the crack-healing process. Piezoelectric materials are capable of stimulating the physiological electrical microenvironment, and can play a vital role to stimulate regeneration and repair [25]. Jacob et al. [26] had listed various piezoelectric biomaterials that can be used for tissue engineering especially for bone and cartilage repair. Zhang et al. [27] fabricated patient-specific bioactive three-dimensional scaffolds that possess controlled microarchitectures for bridging bone defects and promote bone repair. Damaraju et al. [28] fabricated flexible, three-dimensional fibrous scaffolds which can be used to stimulate human mesenchymal stem cell differentiation and corresponding extracellular matrix/tissue formation in physiological loading conditions. These piezoelectric scaffolds were observed to exhibit both chondrogenic and osteogenic differentiation leading to bone repair. However, it is to be noted that self-repair and regeneration in bones are attributed to the flexoelectricity of bone's mineral component – hydroxyapatite [29].

4.9.4 Defense Applications

4.9.4.1 Micro Robotics
Microsized and power-efficient piezoelectric actuators and sensors are often required to design robots capable of flying and crawling for gathering information from sites which are not reachable by humans. Such microrobots can be used in defense for spying, public security, law enforcements, and other public works

(a) (b)

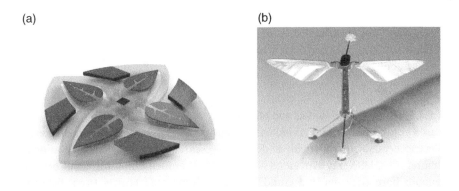

Figure 4.32 (a) 3D printed floating water lilies and (b) a biologically inspired insect-sized robot RoboBee. *Sources:* (a) Subramanian Sundaram/Massachusetts Institute of Technology. (b) Pakpong Chirarattananon/Wyss Institute.

sectors. Researchers at MIT developed 3D printed floating water lilies (Figure 4.32a) with petals equipped with arrays of actuators and hinges that fold up in response to magnetic fields run through conductive fluids. Chopra et al. [30] in 2019 designed and developed a strain sensing piezoelectric actuator with integrated position sensing for millimeter scale mobile robotics and was used to detect wing collisions when flapping near an obstacle and wing degradation when integrated in flying microrobotic system. Electrically isolated strain-sensing regions of the piezoelectric material undergo identical motion as the actuation layers and thus directly sense tip deflection through the piezoelectric effect. This development is considered a step toward closed loop control capabilities of microrobots using on-board sensors. Ma et al. [31] in 2013 demonstrated the first controlled flight of an biologically inspired insect-sized robot. The robotic insect called the RoboBee (Figure 4.32b), with sub-millimeter scale anatomy and two wafer-thin wings that flap at 120 times per second, is capable of vertical takeoff, hovering, and steering. The tiny robots flap their wings using ceramic piezoelectric actuators that expand and contract when an electric field is applied. Applications of RoboBee could include distributed environmental defense monitoring, search-and-rescue operations, and assistance with crop pollination. Later in 2017, the group reported the development of a hybrid aerial-aquatic microrobot capable of locomotion in aerial and aquatic environments. This 175 mg robot uses a flapping wing design, and the motion is driven by a pair of piezoelectric actuators [32]. The underwater robotic finger was developed in 2019 by Yu et al. [33] in which the three piezoelectric actuators were bonded to three phalanges of the finger and helps to push the finger forward using two joints by friction. This robotic finger exhibits potential to be employed as underwater manipulators. Recently Niu et al. [34] proposed a novel piezoelectric crawling robot with multiple degrees of freedom. The researchers demonstrated four modes of working where the superposition of different modes was performed to produce the linear and rotational motion of the robot.

4.9.4.2 Laser-Guided Bullets and Missiles

Laser guided bullets and missiles are the most exciting application of piezo-electric crystals in the field of defense. Researchers at the Sandia National laboratories, USA have invented a dart-like self-guided bullet for small caliber smooth bore firearms that could hit laser illuminated targets from distance of one mile. Sandia's 4-inch-long bullet consists of a optical sensor at the nose of the bullet to detect the laser spot illuminating the target. The identification of the target is performed using the laser spot on the moving vehicle as shown in Figure 4.33. The sensor sends the information to the guidance and control electronics that use an eight-bit CPU (housed within the bullet) to communicate with the electromechanical actuators. The electromechanical actuators steer the bullet's fins that guide the bullet to strike the target. An onboard tracking chip calculates the course corrections, carried out by four actuator-controlled fins on the bullet's body.

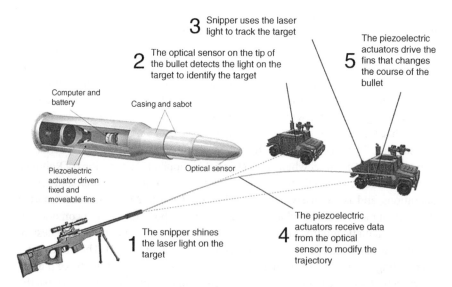

Figure 4.33 Schematic representation of the working of course changing bullets.

4.9.5 Musical Applications

4.9.5.1 Piezoelectric Pickups for Instruments

Many acoustic–electric-stringed instruments utilize piezoelectric pickups to convert acoustic vibrations to electric signals which are either recorded or amplified [35]. To electronically capture the voltage signals from the instrument, a piezoelectric strip is placed between the instrument body and a structure that supports the strings. As the strings vibrate, the piezo strip oscillates to generate an electric signal.

4.9.5.2 Microphones and Ear Pieces

The ability to convert vibrations into electrical output and vice versa (piezo transducer) makes piezoelectric substances extremely useful in the world of communications. Microphones and earphones exploit piezoelectric crystals as transducers by utilizing their piezoelectric and inverse piezoelectric effect, respectively. Microphones are used to amplify audio signal or record them to store as digital data. The main components of microphone are movable diaphragm which vibrates in response to sound waves, and a piezoelectric crystal which oscillates with the vibrating diaphragm to generate an output voltage across the terminal as shown in Figure 4.34a. In a microphone, the acoustic waves produce displacement of the diaphragm which causes twisting or bending of a piezoelectric crystal leading to variation in the voltage output. Rochelle salt is frequently used as transducer in piezoelectric microphones where the crystal is coupled to a diaphragm. When

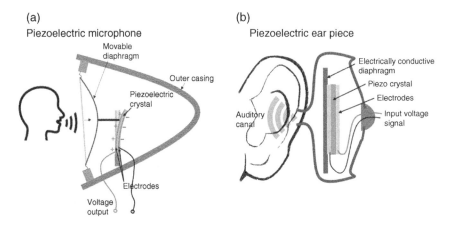

Figure 4.34 The schematic representation of the (a) operation of piezoelectric microphone and (b) design of piezoelectric earphone.

the crystal is strained by the displacement of the diaphragm, the ions of the crystal are displaced asymmetrically to generate electrical charges on its surface. This polarization of the charges generates an equivalent potential difference across the transducer resulting in a voltage output. Due to the market demand for large-scale use of piezoelectric microphones, synthetic piezoelectric crystals like barium titanate are widely used for frequencies up to hundreds of kHz. Piezoelectric microphones are preferred over other microphones as they are durable and cheap, generates relatively large electrical output, and can be easily bonded to a solid and even used in immersed state under nonconducting liquid. However, the piezoelectric transducers lack good linearity and so are inadequate for quality sound recording. Lee et al. [36] fabricated a piezoelectric microphone on a circular diaphragm, using ZnO as the piezoelectric material and low-stress Si_xN_y as the diaphragm.

Piezoelectric earphones convert input voltage signals into mechanical vibrations which produces sound waves. These earphones consist of a piezoelectric crystal attached to a diaphragm as shown in Figure 4.34b. The fluctuating input voltage is fed into piezoelectric crystal causing it to vibrate which moves the diaphragm producing sound waves in air. Usually this is done in an enclosure with a hole in it to capture the sound waves and release them in one direction. The hole usually has a tube stuck to it so that you can further direct those sound waves into your ear canal. Riding on newer advancements in miniaturization and fabrication, Shin et al. [37] designed a new piezoelectric transducer for round window (RW)-driven middle ear implants, suitable for compensation of sensorineural hearing loss. The transducer operates at 6 V and generates an average displacement of 219.6 nm in the flat band (0.1–1 kHz).

4.9.6 Other Applications

4.9.6.1 Energy Harvesters

Energy harvesters are considered as promising independent power sources that are believed to drive next-generation low-power electronic devices such as wearable sensors, medical implants, and various other IoT devices. Out of different energy-harvesting technologies, piezoelectric mechanical energy harvesting has emerged as a convenient method for powering small- and large-scale devices as they can be designed to handle a wide range of input frequencies and forces allowing for energy harvesting to occur. The piezoelectric effect has been widely adopted to convert mechanical energy to electricity due to its high energy conversion efficiency, ease of implementation, and miniaturization. We have discussed the piezoelectric generator in Section 4.6.3 where we have described the generation of electrical energy from mechanical vibrations, kinetic energy, or deformation energy. The energy harvesters, because of its high potential to achieve long lifespan self-powered operations, have witnessed a rapid growth in global market and were valued at \$511.6 million in 2020, and projected to reach \$1,057.7 million by 2030, growing at a CAGR of 7.5% from 2021 to 2030. Mechanical energy can be extracted and converted to electrical energy from wide variety of available sources including human wearables, human organs, and joints, roads, vehicles, and industrial machinery, mega structures – buildings and bridges, and water flow. The factors which play an important role in the efficiency of energy harvesters are property of the piezoelectric material, geometry/design of the piezo crystal/composites, architecture of the piezoelectric generator, type of excitation, mode of operation, and design of circuit. The power output of the piezoelectric energy harvester depends on intrinsic and extrinsic factors. The intrinsic factors include the frequency constant of the piezoelectric element, piezoelectric and mechanical properties of the material, and the temperature and stress dependence of the physical properties. Extrinsic factors comprise the input vibration frequency, acceleration of the base/host structure, and the amplitude of the excitation. Structurally multilayer piezo generator produces much lower output voltage as compared to the single-layer piezo generator, but the current produced by the former is significantly higher than its counterpart. Due to the fact that multilayer piezo generators do not create electromagnetic interference, they are excellent solid-state batteries for electronic circuits. As the power requirement in smaller devices is low, the solid-state multilayer piezoelectric generators can easily replace cell batteries as promising power source in some applications. Researchers have designed and fabricated piezoelectric energy harvesters to generate electrical energy from wearables and medical implants, tires of vehicles, walkways roads, and floors. The available sources of piezoelectric energy harvesting are listed below:

1) *Footwears:* Piezoelectric shoes were developed which utilized the vibrations produced by human stepping during walking, running, and dancing. Kymissis et al. [38] developed an insole with eight-layer stacks of 28-μm PVDF sheets with a central 2-mm flexible plastic substrate, which harnessed the parasitic energy in shoes where the average power measured was 1.1 mW at 1 Hz. Zhao et al. [39] designed a piezoelectric energy-harvesting shoe which converts mechanical energy of human motion into electrical energy, yielding an average output power of 1 mW during a walk at a frequency of roughly 1 Hz. The DC current is built through an integrated circuit where an average of 50 mW power is generated by two to three steps of 5 ms each. Zhang et al. [40] developed a spring energy storage device to harvest energy from human motion using a crank rocker mechanism and a compliant spring which was integrated with a piezoelectric beam to form a mechanical energy-harvesting system. This arrangement was placed in a shoe near the shoe heel. The electrical energy is generated when the heel touches the ground, thereby bending the

piezoelectric beam. When the heel lifted off the ground, the elastic potential energy stored in the spring when the heel touches the ground, was converted to kinetic energy, thereby bending the piezoelectric beam in the opposite direction. A 60-kg person footfall generated a power of 235.2 mJ per step. The system can store and release instantaneous footstep energy to provide uninterrupted piezoelectric electrical energy even when human motion is absent.

2) *Medical Implants:* Piezoelectric energy harvesters, due to their compactness in volume and restriction in weight, occupies less space and regarded as the most suitable energy-harvesting sources for medical implants [41]. There are mainly three excitation sources that are available for the design and development of energy harvesters for pacemakers – heart beats, blood pressure gradients, and arterial wall deformation. Ansari and Karami [42] in 2017 proposed a fan-folded structure that consists of bimorph beams folding on top of each other, when excited by a normal heartbeat waveform from a feedback controlled shaker, generated an average power of 16 mW. In 2014, Dagdeviren et al. [43] proposed a strain-based harvester for conformal energy harvesting from the natural contractile and relaxation motions of the heart, lung, and diaphragm of several animal models. The piezoelectric harvester consists of 500 nm thin PZT ribbons sandwiched between two metal electrodes that are attached on spin-cast PI substrate layer, and encapsulated with biocompatible materials. The harvester was integrated with SEH circuit and a energy storage component for simultaneous power generation and storage. This flexible piezoelectric patch was attached on the surface of the ventricles of bovine and bovine hearts and generated a maximum open-circuit voltage of 4–5 V, and maximum power density of 1.2 mW/cm^2 using multilayer piezo stack arrangement. Hwang et al. [44] proposed a flexible energy harvester using the high-efficiency single-crystal PMNPT which generated maximum open-circuit voltage of 8.2 V and short-circuit current of 145 mA. Later, the same group [45] advanced the findings using single-crystalline PMN-PZT-Mn thin film on porcine heart and obtained a short-circuit current of 1.75 mA and open-circuit voltage of 17.8 V. For harvesting energy from blood pressure gradients, piezoelectric diaphragm is the most widely used structure in this application. In 2014, Deterre et al. [46] proposed a microspiral piezoelectric transducer associated with a novel microfabricated packaging that captures energy from blood pressure variations in the cardiac environment for use as a leadless pacemaker. The device achieved a maximum power density of 3 mJ/cm^3 heartbeat under an excitation of 1.5 Hz and 180 mN (corresponding to the maximal amplitude pressure variations in the left ventricle on a surface of diameter of 6 mm). Harvesting energy from blood vessels may hamper free flow of blood in the tissue. Thus, the researchers developed soft and flexible devices using bendable PVDF films for harvesting energy from arterial deformation. In 2015, Zhang et al. [47]

developed a PVDF-based energy harvester on the ascending part of the aorta of a pig heart. The researchers studied the performance of the device on latex tubing and obtained maximum output power of 681 nW and a maximum voltage of 10.3 V, respectively, under a pressure variation of arterial pressure of 80 mmHg. When wrapped around the aorta of a porcine heart, the device offers a maximum voltage of 1.5 V under the heart rate of 120 beats/min and blood pressure of 160/105 mmHg. The implanted harvester is able to charge a 1-mF capacitor to 1.0 V within 40 s. It is roughly estimated that the average power output is about 12.5 nW. In spite of the advancements in piezoelectric energy-harvesting techniques, the energy harvesters are facing numerous hurdles and require solutions on various issues before they can be assimilated into commercial market as self-powered medical implants. The common issue faced by the researchers are miniaturization of piezoelectric energy harvesters, limitations for *in vitro* testing, and biocompatibility and reliability of the devices that still remains to be addressed.

3) *Roads, walkways, and floors:* Researchers have explored various sources for harnessing electrical power from natural vibrational phenomena from pedestrian footfalls. They utilized the vibrational energy from footfalls by placing piezoelectric tiles on walkways, stairways, roads, pavements, and floors to generate electrical energy to power appliances. Edlund et al. [48] compared vibrational energy harvesting using piezoelectric tiles in walkways and stairways and concluded that the power generated from piezoelectric tiles placed in a stairway perform better than the ones placed in walkways due to the increased pressure during pedestrian footfall while traversing the stairs. Liu et al. [49] developed a footstep energy harvester with a self-supported power conditioning circuit. The energy generator consists of 300 monolithic layers of multilayer piezoelectric stack with a force amplification frame to extract electricity from human walking locomotion. Using the synchronized switch harvesting on inductance (SSHI) technology, the power conditioning circuit was designed to optimize the power flow from the piezoelectric stack to the energy storage device under real-time human walking excitation. The researchers considered the force excitation of 114 N with human walking, and the amplified force resulted being 846 N. This excitation was applied on three separate circuits – standard energy-harvesting (SEH) circuit, series SSHI circuit, and the parallel SSHI circuit and obtained an output power of 1.35, 1.33, and 2.35 mW, respectively. The harvesting efficiency of parallel SSHI circuit was 74% higher than that of the SEH circuit. Hobeck et al. [50] developed lightweight, highly robust, energy harvester design in the form of piezoelectric grass which generated energy through turbulence-induced vibration (TIV) suitable for use in low-velocity, highly turbulent fluid flow environments, i.e. streams or ventilation systems. The biomimetic design

consisted of an array of six cantilevers made of piezoelectric PVDF (type 1) or an array of four cantilevers made of piezoelectric PZT (type 2). These cantilevers experience vigorous vibrations when exposed to proper turbulent flow conditions. The type 2 harvester achieved a power output of 1.0 mW per cantilever with a mean airspeed of 11.5 m/s, whereas identically sized type 1 harvester yielded an output of 1.2 μW per cantilever at 7 m/s. Despite 1000 times higher power generation capability f type 2 harvester than type 1 harvester, the latter is more suitable for long-term deployment.

A burst of research progress in energy-harvesting technology in the last decade has lured the researchers to undertake promising applications for self-powered autonomous operations of wearable electronics, medical devices, automotive sensors, and wireless sensor monitoring systems.

4.9.6.2 Sports-Tennis Racquets

When a player hits a tennis ball with a conventional racket, it undergoes a considerable deformation, i.e. bending resulting in the loss of control over the shot. To provide enhanced controllability in shot making, new racquets designed by HEAD's Intellifiber™ system (Figure 4.35a) incorporate bundles of piezoelectric zirconate titanate (PZT) fibers (Figure 4.35b) embedded into the body of the tennis racket frame (in its body and the throat region). When the ball comes in contact with the racquet it produces structural deformation (Figure 4.35c-i). The piezoelectric fibers generate an electrical current as a direct response to the deformation (Figure 4.35c-ii). The electric current is fed into a microchip located in the top of the handle. The microchip generates and amplifies the electrical response caused due to piezoelectric effect and supplies the voltage to the piezoelectric fibers, producing a counterforce which increases the stiffness and damped structural vibration (Figure 4.35c-iii). This energy transformation takes less than a millisecond while the ball is still on the strings, and the increased racquet stiffness also provides more control and power.

4.10 Conclusions

Developed through more than 50 years of extensive research and development, the field of piezoelectric sensors is still prominently active because of their striking benefits of harsh environment operation and their scope for integration on flexible substrates. A piezoelectric device has established themselves as the class of sensors with the highest commercial demand due to the linear behavior between the input force and the generated output voltage (in sensors) or vice versa (in actuators).

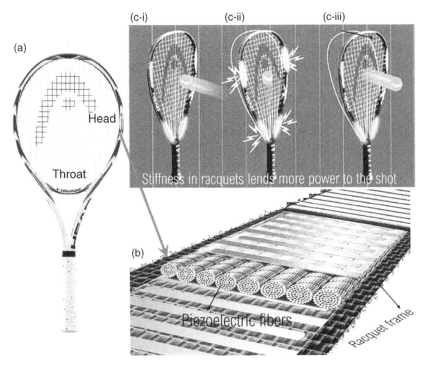

Figure 4.35 (a) Newly designed tennis racquet with piezoelectric boost, designed by HEAD's Intellifiber™ system. (b) Schematic view of the internal architecture of the racquet frame showing the bundles of piezoelectric zirconate titanate (PZT) fibers embedded into the body of the tennis racket frame and (c) illustration of the working of the racquet – (i) structural deformation when the ball hits the racquet, (ii) piezoelectric fibers generate an electrical current in response to deformation, and (iii) increase in stiffness and dampening of structural vibration. *Source:* (a) HEAD.

These sensors are widely used in variety of applications in the field of medical, consumer electronics, industry, sports, defense, and self-power energy-harvesting sensors. The researchers have explored various piezoelectric materials with attractive properties and developed next-generation piezoelectric sensors that are not only architecturally unique but also delivers enhanced sensor performance in terms of sensitivity and dynamic range as compared to other class of sensors. The piezoelectric sensors have opened new avenues for the development of highly durable, efficient, lightweight, portable, and self-powered sensors. The piezoelectric materials in a piezoelectric sensor operate as a transducer that converts mechanical energy (either force, pressure, strain, or vibration) into an electrical signal or vice versa. Although piezoelectric sensors are conventionally used as force, pressure, and acceleration sensors, they have a potential for use as immunosensors, gas sensors, ultrasonic sensors, and other wearable physiological sensors. Since the scope of piezoelectric sensors are widening its horizons in all possible fields of applications with each passing day, the commercial demand for newer piezoelectric materials with enhanced properties is in high demand in the global market.

The sensing materials used in piezoelectric devices are classified into natural and man-made or synthetic piezoelectric materials. The natural piezoelectric materials include single crystal, organic, and biomaterials, whereas the man-made materials incorporate polymer and ceramics in the form of composites and thin film. The piezoelectric ceramics are the most widely used piezoelectric material due to numerous benefits and are suitable for multifunctional applications, especially in high energy-harvesting devices. Piezoelectric ceramic material PZT has a perovskite atomic structure and possesses high dielectric constant, piezoelectric constant, and highly capable of producing high piezoelectric voltage corresponding to force pressure or vibration. However, the use of piezoelectric ceramics are restricted due to the brittleness of the materials, complex and cost-intensive processing method, and the presence of toxic materials – lead as in PZT. Lead-free perovskite-based piezoelectric ceramics like the $BaTiO_3$, $KNbO_3$, $LiNbO_3$, $LiTaO_3$, Na_2WO_3, and ZnO can be used. Since piezoelectric ceramics are brittle they cannot be used in the fabrication of flexible piezoelectric sensor. Piezoelectric polymers are promising in diverse multifunctional applications because of their superior advantages in mechanical flexibility, easy processing, cost-effectiveness, chemical resistiveness, low density, and biocompatibility over ceramics. Alternatively, bio-piezoelectric materials show spontaneous piezoelectricity, but they suffer from the drawbacks of low piezoelectric coefficient and dielectric constant. Piezoelectric composites are fabricated by reinforcing nonpolarized piezoelectric ceramics or polymers into a passive polymer matrix, such as polymers and metals. The materials properties can be altered and modified to suit device requirement by regulating the spatial concentration of the components and architecture of the composites.

The piezoelectric materials can be suitably used as a transducer where mechanical energy can be converted to electrical energy and vice versa. This transducer action of the piezoelectric material to convert mechanical energy to electrical output signals are utilized in the development of piezoelectric sensors. The performance of the piezoelectric sensors depends on the piezoelectric material, orientation of the polarization vector in response external stimuli, and piezoelectric device architecture. Since the piezoelectric response in the form of output voltage is direction dependent, the piezoelectric devices must be designed suitably for a given target application. When the sensor is designed for large output force and low displacement, piezoelectric stacked architecture is preferred, whereas the bimorph architecture is suitable when the applied force is low, but the displacement is large. The use of piezoelectricity in the development of medical imaging tools and SONAR has created global stir where modern research will explore more on the piezoelectric sensors. In a piezoelectric actuator, electrical energy is converted to mechanical energy in the device and finds application in piezoelectric motors, ink cartridges, etc. Piezoelectric energy harvesters derive energy from mechanical vibrations using a full-wave rectifier circuit and store the energy in electronic components and can be later used to power other devices. In spite of many shortcoming and restriction of piezoelectric materials and devices including temperature sensitivity and neutralization of surface charges by environmental agents or current leakage, the piezoelectric sensor still holds a pivotal edge in the field of solid-state sensors.

List of Abbreviations

BAW	Bulk Acoustic Wave
BNT	Bismuth Sodium Titanate
CNT	Carbon Nanotube
cyclo-FW	*Cyclo*-phenylalanine-tryptophan
cyclo-GW	*Cyclo*-glycine-tryptophan
DC	Direct Current
DFT	Density Functional Theory
EAPap	Cellulose-based Electroactive Paper
FF	Di-phenylalanine
IDE	Interdigitated Electrodes
KNT	Sodium Potassium Niobate
KS	Knock Sensors
LN	Lithium Niobate
LT	Lithium Tantalate
MLC	Multilayer Capacitors
OCMFET	Organic Charge-modulated Field-effect Transistor
P(VDF-TrFE)	Copolymer of Trifluoroethylene
PCM	Powertrain Control Module
PIN-PMN-PT	Lead Indium Niobate – Lead Magnesium Niobate – Lead Titanate
PVDF	PolyVinyliDene Fluoride
PZT	Lead Zirconate Titanate
QCM	Quartz Crystal Microbalance
QM	Mechanical Quality Factor
SAW	Surface Acoustic Wave

SEH circuit   Standard Energy Harvesting Circuit
SONAR   Sound NAvigation and Ranging
SSHI   Synchronized Switch Harvesting on Inductance
TF   Transmission Factor
TIV   Turbulence-induced Vibration (TIV)

References

1 Curie, J. and Curie, P. (1880). *Development by pressure of polar electricity in hemihedral crystals with inclined faces. Bulletin de la Société Botanique de France* 3: 90.

2 Hankel, W.G. (1882). *Ueber die actino-und piezoelectrischen Eigenschaften des Bergkrystalles und ihre Beziehung zu den thermoelectrischen. Annalen der Physik* 253 (9): 163–175.

3 Ikeda, T. (1996). *Fundamentals of Piezoelectricity*. Oxford University Press.

4 Alamusi et al. (2012). *Evaluation of piezoelectric property of reduced graphene oxide (rGO)-poly(vinylidene fluoride) nanocomposites. Nanoscale* 4: 7250–7255.

5 Uchino, K. (1998). Piezoelectric ultrasonic motors: overview. *Smart Materials and Structures* 7 (3): 273–285.

6 Fukada, E. (1956). *On the piezoelectric effect of silk fibers. Journal of the Physical Society of Japan* 11 (12): 1301A–1301A.

7 Hosseini, E.S. et al. (2020). *Glycine-chitosan-based flexible biodegradable piezoelectric pressure sensor. ACS Applied Materials & Interfaces* 12 (8): 9008–9016.

8 Peng, H. et al. (2017). *Energy harvesting based on polymer*. In: *Polymer Materials for Energy and Electronic Applications* (ed. S. Miyashita), 151–196. Academic Press.

9 Jean-Mistral, C., Basrour, S., and Chaillout, J. (2010). *Comparison of electroactive polymers for energy scavenging applications. Smart Materials and Structures* 19 (8): 085012.

10 Laput, G., Chen, X.A., and Harrison, C. (2016). SweepSense: ad hoc configuration sensing using reflected swept-frequency ultrasonics. In: *Proceedings of the 21st International Conference on Intelligent User Interfaces (7–10 March)*. Sonoma, CA: Association for Computing Machinery.

11 Newnham, R.E., Skinner, D.P., and Cross, L.E. (1978). *Connectivity and piezoelectric-pyroelectric composites. Materials Research Bulletin* 13 (5): 525–536.

12 Dodds, J., Meyers, F., and Loh, K. (2013). *Piezoelectric nanocomposite sensors assembled using zinc oxide nanoparticles and poly(vinylidene fluoride). Smart Structures and Systems* 12 (1): 055-071.

13 Yu, P., Liu, W., Gu, C. et al. (2016). Flexible piezoelectric tactile sensor array for dynamic three-axis force measurement. *Sensors (Basel, Switzerland)* 16 (6): 819.

14 Chen, X., Shao, J., Tian, H. et al. (2018). Flexible three-axial tactile sensors with microstructure-enhanced piezoelectric effect and specially-arranged piezoelectric arrays. *Smart Materials and Structures* 27 (2): 025018.

15 Viola, F.A. et al. (2018). Ultrathin, flexible and multimodal tactile sensors based on organic field-effect transistors. *Scientific Reports* 8 (1): 8073.

16 Hunstig, M., Hemsel, T., and Sextro, W. (2013). Stick-slip and slip-slip operation of piezoelectric inertia drives. Part I: Ideal excitation. *Sensors and Actuators A: Physical* 200: 90–100.

17 Judy, J.W., Polla, D.L., and Robbins, W.P. (1990). A linear piezoelectric stepper motor with submicrometer step size and centimeter travel range. *IEEE Transactions on Ultrasonics, Ferroelectrics, and Frequency Control* 37 (5): 428–437.

18 Patel, P.P. and Manohar, P. (2012). Design and simulation of a piezoelectric ultrasonic micro motor. *COMSOL Conference INDIA 2012*, Bangalore (2–3 November 2012).

19 DeAngelis, D.A. and Schulze, G.W. (2015). Performance of PIN-PMN-PT single crystal piezoelectric versus PZT8 piezoceramic materials in ultrasonic transducers. *Physics Procedia* 63: 21–27.

20 Braun, H. (1997). Piezoelectric ink jet print head and method of making. Google Patent, U.P. office, Editor Eastman Kodak Co, US.

21 Hennet, P. (2015). Piezoelectric bone surgery: a review of the literature and potential applications in veterinary oromaxillofacial surgery. *Frontiers in Veterinary Science* 2: 8–8.

22 Vallancien, G., Silbert, L., Borghi, M. et al. (1988). Outpatient piezoelectric lithotripsy. In: *Shock Wave Lithotripsy* (ed. J.E. Lingeman and D.M. Newman). Boston, MA: Springer.

23 Setyawan, D. (2021). Analysis of the use of extracorporeal shock wave lithotripsy (ESWL) based on piezoelectric lithotripter for kidney stone. *Journal of Medical Physics and Biophysics* 8 (1): 1–11.

24 Aly, L.A.A. (2018). Piezoelectric surgery: applications in oral & maxillofacial surgery. *Future Dental Journal* 4 (2): 105–111.

25 Tandon, B., Blaker, J.J., and Cartmell, S.H. (2018). Piezoelectric materials as stimulatory biomedical materials and scaffolds for bone repair. *Acta Biomaterialia* 73: 1–20.

26 Jacob, J., More, N., Kalia, K., and Kapusetti, G. (2018). Piezoelectric smart biomaterials for bone and cartilage tissue engineering. *Inflammation and Regeneration* 38 (1): 2.

27 Zhang, L., Yang, G., Johnson, B.N., and Jia, X. Three-dimensional (3D) printed scaffold and material selection for bone repair. *Acta Biomaterialia* 84 (10): 16–33.

28 Damaraju, S.M., Shen, Y., Elele, E. et al. (2017). Three-dimensional piezoelectric fibrous scaffolds selectively promote mesenchymal stem cell differentiation. *Biomaterials* 149: 51–62.

29 Vasquez-Sancho, F., Abdollahi, A., Damjanovic, D., and Catalan, G. Flexoelectricity in bones. *Advanced Materials* 30 (9): 1705316.

30 Chopra, S. and Gravish, N. (2019). Piezoelectric actuators with on-board sensing for micro-robotic applications. *Smart Materials and Structures* 28 (11): 115036.

31 Ma, K.Y., Chirarattananon, P., Fuller, S.B., and Wood, R.J. (2013). Controlled flight of a biologically inspired. *Insect-Scale Robotic Science* 340 (6132): 603–607.

32 Chen, Y., Wang, H., Helbling, E.F. et al. (2017). A biologically inspired, flapping-wing, hybrid aerial-aquatic microrobot. *Science robotics* 2 (11): 5619.

33 Yu, P., Wang, L., Jin, J. et al. (2019). A novel piezoelectric actuated underwater robotic finger. *Smart Materials and Structures* 28 (10): 105047.

34 Niu, R. and Guo, Y. (2021). Novel piezoelectric crawling robot with multiple degrees of freedom. *AIP Advances* 11 (7): 075306.

35 Barcus, L.M. (1985). String instrument pick up system. Google Patents, Lester M, Barcus.

36 Lee, W.S. and Lee, S.S. (2008). Piezoelectric microphone built on circular diaphragm. *Sensors and Actuators A: Physical* 144 (2): 367–373.

37 Shin, D.H. (2021). Design study of a round window piezoelectric transducer for active middle ear implants. *Sensors (Basel, Switzerland)* 21 (3): 946.

38 Ishida, K., Huang, T.C., Honda, K. et al. (2013). Insole pedometer with piezoelectric energy harvester and 2 V organic circuits. *IEEE Journal of Solid-State Circuits* 48 (1): 255–264.

39 Zhao, J. and You, Z. (2014). A shoe-embedded piezoelectric energy harvester for wearable sensors. *Sensors (Basel, Switzerland)* 14 (7): 12497–12510.

40 Zhang, Y.-H., Lee, C.-H., and Zhang, X.-R. (2019). A novel piezoelectric power generator integrated with a compliant energy storage mechanism. *Journal of Physics D: Applied Physics* 52 (45): 455501.

41 Kanno, I. (2015). Piezoelectric MEMS for energy harvesting. *Journal of Physics: Conference Series* 660: 012001.

42 Ansari, M.H. and Karami, M.A. (2017). Experimental investigation of fan-folded piezoelectric energy harvesters for powering pacemakers. *Smart Materials and Structures* 26 (6): 065001.

43 Dagdeviren, C., Yang, B.D., Su, Y. et al. (2014). Conformal piezoelectric energy harvesting and storage from motions of the heart, lung, and diaphragm. *Proceedings of the National Academy of Sciences of the United States of America* 111 (5): 1927–1932.

44 Hwang, G.-T., Park, H., Lee, J.-H. et al. (2014). Self-powered cardiac pacemaker enabled by flexible single crystalline PMN-PT piezoelectric energy harvester. *Advanced Materials* 26 (28): 4880–4887.

45 Kim, D.H., Shin, H.J., Lee, H. et al. (2017). In vivo self-powered wireless transmission using biocompatible flexible energy harvesters. *Advanced Functional Materials* 27 (25): 1700341.

46 Deterre, M., Lefeuvre, E., Zhu, Y. et al. (2014). Micro blood pressure energy harvester for intracardiac pacemaker. *Journal of Microelectromechanical Systems* 23 (3): 651–660.

47 Zhang, H., Zhang, X.-S., Cheng, X. et al. (2015). A flexible and implantable piezoelectric generator harvesting energy from the pulsation of ascending aorta: in vitro and in vivo studies. *Nano Energy* 12: 296–304.

48 Edlund, C. and Ramakrishnan, S. (2019). An analytic study of vibrational energy harvesting using piezoelectric tiles in stairways subjected to human traffic. *European Journal of Applied Mathematics* 30 (5): 968–985.

49 Liu, H., Hua, R., Lu, Y. et al. (2019). Boosting the efficiency of a footstep piezoelectric-stack energy harvester using the synchronized switch technology. *Journal of Intelligent Material Systems and Structures* 30 (6): 813–s822.

50 Hobeck, J.D. and Inman, D.J. (2012). Artificial piezoelectric grass for energy harvesting from turbulence-induced vibration. *Smart Materials and Structures* 21 (10): 105024.

5

Capacitive Sensors

5.1 Overview

In the previous chapter, we have discussed about the piezo sensors which work on the generation of voltage on the application of external stress. The capacitive sensors on the other hand work on the principle of variation in relative displacement of the electrodes or the displacement of membrane which affects the capacitance of the device. Capacitive sensors have been around for decades [1–6], but have not gained significant interest in comparison to the piezo sensors which are known to be ultrafast devices. However, with the advent of new applications, the principle of capacitive sensors is widely used in measuring systems that have revolutionized robotics [7], wearable devices [8, 9], health care [10], automobiles [11], home applications [12, 13], and security [14, 15]. This is due to the fact that capacitive sensor provides definite advantages like high sensitivity, low power consumption, enhanced temperature performance, and reduced drift in performance over time over its piezoelectric counterpart. Capacitive sensors are mostly used as pressure sensors and gyroscope [16, 17] which are designed by utilizing relative displacement of electrodes with applied pressure [18], touch [19], proximity [20], and liquid-level sensors [21] through variation of electric lines of force on introduction of metallic and nonmetallic target and chemical sensors which utilizes the change in dielectric property of the material between the electrodes. However, only a small share (2.58 billion of 11.38 billion in 2019) of the market for pressure sensors is allocated to capacitive-type sensors [22, 23]. Although having learnt the astonishing properties of the capacitive sensors, it seems strange that the utilization of capacitive sensing technology is so restricted in a few applications and devices. This is due to its low sensitivity, and in some cases high parasitic capacitance, which have resulted in limited adoption of the capacitive technology of sensing. In addition to these shortcomings, design complexities and architectural constraints in capacitive sensors have limited their use in devices. There are significant

Solid-State Sensors, First Edition. Ambarish Paul, Mitradip Bhattacharjee, and Ravinder Dahiya.
© 2024 The Institute of Electrical and Electronics Engineers, Inc.
Published 2024 by John Wiley & Sons, Inc.

research challenges related to fabrication procedures, design of matched sensing circuits, and manufacturing procedures which must be addressed in order to bring capacitive sensing technology in industrial practice.

To understand how capacitive sensors operate, it is important to understand the fundamental properties and principles of capacitors. This chapter focuses on solid-state capacitive sensor, i.e. the devices which are devoid of any movable parts. Thus, this chapter incorporates various touch and proximity sensing and chemical and/or gas sensing technologies that are based on changes in capacitance of the active sensing component. Since accelerometers, gyroscopes, and microelectromechanical systems (MEMS) based sensors contain movable parts, this chapter does not include them here, though being a well-celebrated part of capacitive sensor family. This chapter provides a basic introduction to capacitive sensing, discusses the different types of capacitive architecture in devices, various dielectric materials used and their advantages, device fabrication technologies, and various applications of solid-state capacitive sensors. The chapter concludes with recent research challenges and the possible strategies to overcome them. The below section provides details on the underlying principles of the capacitor.

5.1.1 A Capacitor

The capacitor is an electronic component which has the ability or "capacity" to store energy in the form of an electrical charge producing a potential difference across its plates. Capacitors are generally composed of two conducting plates separated by a nonconducting substance called dielectric (with dielectric constant ε_r as shown in Figure 5.1). The dielectric may be air, mica, ceramic, paper, or other

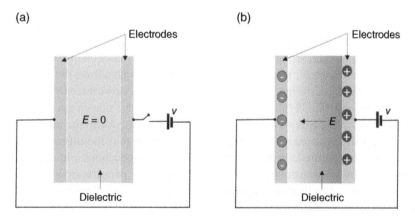

Figure 5.1 Schematic representation of a parallel electrode capacitor under (a) unbiased and (b) biased conditions.

suitable insulating material, while the electrodes are made of metals. Initially under unbiased state, no charge stored at the electrodes and the electric field E in the dielectric layer is zero as shown in Figure 5.1a. When an external voltage in applied across the terminals of the capacitor, the electrodes store the charges on their surface in the form of electrical energy until the potential difference across the electrodes of the capacitor reaches the externally applied voltage as shown in Figure 5.1b. The electrical charges continue to reside on the surface of the electrode even after disconnecting the circuit unless it is consumed by other electronic components of the circuit or it is lost due to leakage only because all practical dielectric is prone to charge leakages. Capacitors with little leakage can hold their charge for a considerable period of time. The plate connected with the positive terminal stores positive charge (or $+Q$) on its surface and the plate connected to the negative terminal stores negative charge (or $-Q$). However, a capacitor is significantly different from a rechargeable battery in terms of working and performance. A rechargeable battery stores energy in the form of electrochemical energy as compared to electrostatic energy in capacitor. The former provides 10^3 times more energy density as compared to the latter and is used for providing constant DC voltage to the circuit.

5.1.2 Capacitance of a Capacitor

Capacitance is the electrical property of capacitors. It is the measure of the amount of charge that a capacitor can hold at a given voltage. Capacitance is measured in farad (F) and named after the British physicist Michael Faraday. Capacitance can be defined in the unit coulomb per volt as:

$$C = Q/V \tag{5.1}$$

where C is the capacitance in farad (F), Q is the magnitude of charge stored on each plate (coulomb), and V is the voltage applied to the plates (volts). A capacitor with the capacitance of one farad can store one coulomb of charge when the voltage across its terminals is 1 V. Note that capacitance C is always positive in value and has no negative units. However, the farad is a very large unit of measurement to use on its own, so sub-multiples of the farad are generally used such as microfarads, nanofarads, and picofarads to express the capacitances of commercially available capacitors of 1 pF to about 1000 µF.

The capacitance of a capacitor depends on the geometry of the electrodes and the architecture of the capacitor and not on an external source of charge or potential difference between the electrodes. For a parallel plate capacitor, the electric field between the plates due to charges $+Q$ and $-Q$ remains uniform, but at the edges, the electric lines of force deviate outward. If the separation between the plates is much smaller than the size of plates, the electric field strength between the plates may be assumed uniform. The capacitance of a parallel plate capacitor is proportional to the area A of the smallest of the two plates and inversely proportional to the distance or separation d (i.e. the dielectric thickness) between these two electrodes, but independent of the electrode metal. The capacitance of a capacitor can be expressed in terms of its geometry and dielectric constant as:

$$C = \frac{\varepsilon_0 \varepsilon_r A}{d} \tag{5.2}$$

where C is the capacitance in farads (F), ε_r is the relative static permittivity (dielectric constant) of the dielectric medium, $\varepsilon_0 (=8.854 \times 10^{-12}$ F/m) is the permittivity of free space, A is the overlapping area between the electrodes (in meters), and d is the separation distance (in meters) of the two plates. It is observed that there are three parameters viz. the overlapping electrode area A, effective dielectric thickness d, and the dielectric constant ε_r that determines the capacitance of the capacitor. Thus, for the construction of capacitive sensor element, we need to fabricate a capacitor whose capacitance can be influenced by an externally applied stimulus called the measurand. This variable capacitor through changes in its capacitance can respond to changes in external stimuli in the form of force, chemical environment, and foreign objects, when there is a variation in its electrode area, effective dielectric thickness, or dielectric constant. The changes in capacitance when measured provide information on the strength of external stimulus. It is to be noted that any change in A, d, and ε_r is associated with the change in electric field E between the electrodes and forms the distinctive feature of capacitive sensing technology.

5.2 Sensor Construction

A capacitive sensor converts a change in position or properties of the dielectric material into an electrical signal. According to Eq. (5.3), capacitive sensors are realized by varying any of the three parameters of a capacitor: distance (d), area of capacitive plates (A), and dielectric constant (ε_r); therefore:

$$C = f(d, A, \varepsilon_r) \tag{5.3}$$

A wide variety of different kinds of sensors has been developed that is primarily based on the capacitive principle described in Eq. (5.3). These sensors' functionalities range from humidity sensing [24], through level sensing, to displacement sensing.

5.2.1 Overlapping Electrode Area A

As discussed earlier, the capacitive phenomenon is related to the electric field between the two overlapping electrodes of the capacitor. The electrodes with larger surface area are able to store more electrical charge; therefore, a larger capacitance value is obtained with greater surface area. For a capacitive architecture with dissimilar electrode dimension, a higher electric field $E(=V/d)$ is developed between the overlapping electrodes, as compared to that at the edges where the electrodes are nonoverlapping, only because the effective distance d between the electrodes increases in the later case. Since in a parallel plate capacitor the quantity of charge on one plate is balanced by an equal quantity of opposite charge in the other plate, the electrical field E between the two plates is normal to the electrode surfaces only at the overlapping area. However, at the nonoverlapping area between the two electrodes, the electric field is sparse with fringing electric lines of force emanating from the positive electrode. The opposite charges in the larger electrode become concentrated in the region that directly faces and overlaps the smaller plate as shown in Figure 5.2a. Calculating the capacitance of the capacitor with dissimilar electrode dimensions, having the fringing fields lines that loop around from the sides and top surface of the smaller plate and then terminate either on the larger plate or other surrounding conductive items, is quite complex.

Again the capacitance can also change when one electrode of the parallel plate capacitor with same dimension electrodes is displaced parallelly relative to the other electrode as shown in Figure 5.2b, c. The dielectric between the overlapping electrodes experience uniform electric field E, whereas the field in the nonoverlapping region is reduced, thereby decreasing the effective capacitance of the capacitor (Figure 5.2c). This mechanism of capacitive sensing technology is utilized for the detection of shear forces in the direction parallel to the plane of the electrodes [25]. As the overlapping area between the electrodes decreases in response to the shearing force [26], the effective capacitance decreases.

5.2.2 Dielectric Thickness d

The capacitance of a parallel plate capacitor is indirectly proportional to the distance between the electrodes or the dielectric thickness as it is sandwiched between the electrodes (Figure 5.3a) [27]. The electric field between the two electrodes increase as the dielectric thickness between the electrodes decreases [28], thereby increasing the capacitance as shown in Figure 5.3b. Any external stimulus

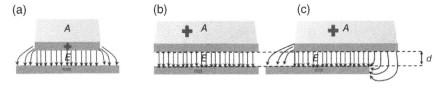

Figure 5.2 Schematic representation of parallel electrode capacitor with (a) *top* electrode area smaller than *bottom* electrode. (b) *Top* and *bottom* electrodes with equal area and aligned to obtain maximum overlap and (c) equal area but misaligned top and bottom electrodes showing the intensities of electric field at the edges.

(a)   (b)

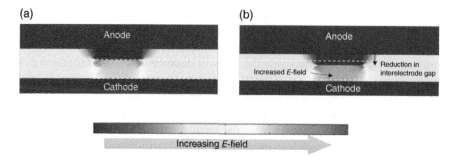

Figure 5.3 COMSOL simulation showing the variation of electric field with distance d between the electrodes: (a) increased d and (b) reduced d.

that is capable of changing the dielectric thickness of the capacitor can be measured by recording the variation in dielectric thickness of the sensor. However, it has to be ensured that the operation of the sensor does not produce change in the area of electrodes during operation. Most common sensors which work on this principle are pressure sensors, force sensors, vacuum gauges, and accelerometers. A large number of devices which use air as the dielectric material involve ultrathin membranes as one of the electrodes as it can be easily displaced under mechanical stimuli like pressure and vacuum.

The variation of overlapping area of the electrodes and dielectric thickness can be utilized in the construction of three-axis force measurement device [29] as depicted in Figure 5.4. The construction of the device consists of four separate bottom electrodes (BEs) $B1$, $B2$, $B3$, and $B4$ which are held at negative potential relative to the top electrode (TE) T. The TE T is separated from the BEs B by an elastomer dielectric of thickness d, which can deform itself in response to shear forces parallel to the x–y plane and also in the perpendicular z direction resulting in the decrease in the dielectric thickness. The TE T is so placed over the BEs such

(a)   (b)

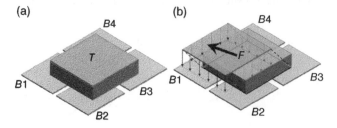

Figure 5.4 Schematic representation of three-axis force sensor under (a) initial and (b) displaced conditions.

that the overlapping areas with separate BEs $B1$, $B2$, $B3$, and $B4$ are equal. The capacitive construction without the response of the external stimuli is shown in Figure 5.4a. In response to a shear force in the x–y plane, the elastomer deforms itself such that the overlapping area between the TE and the BEs segments ($B1$ and $B4$) in the direction of the shear are separately increased, whereas the overlap on the bottom area segments ($B2$ and $B3$) opposite to the shear direction are separately reduced as shown in Figure 5.4b. Thus, the increase in capacitance due to increased overlapping area between the top and separate segments of BEs and the decrease in capacitance due to increased nonoverlapping area for the other BE segments determine the direction of the shear force in the x–y plane. Again, when a force exerted perpendicular to the TE reduces the dielectric thickness of the capacitor, a change in the capacitance is observed. Thus, the perpendicular force on the TE is manifested as an increase in capacitance, and thus can be detected.

5.2.3 Dielectric Material

The conductive electrodes of the capacitor are generally made of a metal allowing for the flow of electrons and charge, whereas the dielectric material is always an insulator. The various insulating materials are utilized as capacitor dielectric and vary significantly in their dielectric behavior, i.e. their ability to block or pass an electrical charge. When an external voltage V is applied between the terminals, the electric field generated between the electrodes distorts the atoms in the dielectric material causing separation of negative (electrons) and positive charges. This relative separation of opposite charges to form a dipole under the influence of electric field E and subsequently aligning themselves along the direction of E is known as the polarization P. The amount of polarization of the dielectric depends on the nature of the dielectric material and this is measured by the permittivity ε ($=\varepsilon_0\varepsilon_r$) of the material.

Dielectric constant ε_r: Let us consider a parallel plate capacitor in vacuum with interelectrode separation of d and biased with an applied voltage V as shown in Figure 5.5a. The electrodes of the capacitor are separated by free space having permittivity ε_0. The electric field E developed between the electrodes is given by

$$E = \frac{\sigma}{\varepsilon_0} = \frac{V}{d}, \tag{5.4}$$

where σ denotes the charge density in the dielectric material.

Now, when a dielectric material with permittivity ε is introduced between the two parallel electrodes of the capacitor, the effective electric field E_{eff} is changed to:

$$E_{\text{eff}} = \frac{\sigma}{\varepsilon} = \frac{\sigma}{\varepsilon_0\varepsilon_r}. \tag{5.5}$$

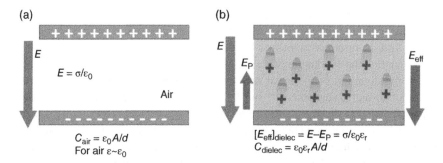

Figure 5.5 Parallel electrode capacitor with (a) air ε_0 and (b) dielectric material with relative permittivity ε_r depicting the decrease in effective electric field $E_{eff} = E - E_P$ due to the introduction of the material.

The reduction in electric field E by a factor ε_r upon introduction of the dielectric is attributed to the opposing electric field E_P generated due to the polarized charges in the dielectric as shown in Figure 5.5b. The dielectric constant ε_r is defined to justify the reduction of E_{eff} in the dielectric. The capacitance is increased from $C_{free} = (\varepsilon_0 A)/d$ in vacuum to $C_{dielec} = (\varepsilon_r \varepsilon_0 A)/d$ after the introduction of the dielectric. The capacitance is inversely proportional to the electric field between the electrodes, and the presence of the dielectric reduces the effective electric field. Permittivity ε relates to a material's ability to transmit an electric field. In the capacitors, an increased permittivity allows the same charge to be stored with a smaller electric field, leading to an increased capacitance. The dielectric material is recognized in terms of the dielectric constant and the effective capacitance with the dielectric material is the capacitance in free space multiplied with the dielectric constant. Different materials have different magnitudes of dielectric constant [18, 30, 31]. For example, air has a dielectric constant equal to 1.0006 at 1 atm. pressure, and solid dielectric such as paraffin has a dielectric constant of 2.5. If paraffin is used as dielectric instead of air, the capacitance value using the paraffin as dielectric will increase by a factor of 2.5. This factor is called relative dielectric constant or relative electric permittivity ε_r. Some commonly used dielectric materials and their corresponding dielectric values are listed in Table 5.1.

Dielectric strength: The electrical insulating properties of any material are dependent on dielectric strength. The dielectric strength of an insulating material describes the maximum electric field of that material. If the magnitude of the electric field across the dielectric material exceeds the value of the dielectric strength, the insulating properties of the dielectric material will breakdown and the dielectric material will begin to conduct. The breakdown voltage or rated voltage of a capacitor represents the largest voltage that can be applied to the capacitor without exceeding the dielectric strength of the dielectric material. The applied voltage

Table 5.1 Dielectric constants of different materials.

Materials	Dielectric constant ε_r	Materials	Dielectric constant ε_r
Mica	5.7–6.7	Plexiglas	3.4
Paper	1.6–2.6	Polyethylene	2.25
Bakelite	4.9	Polyvinyl chloride	3.18
Porcelain	6	Teflon	2.1
Glass	5–10	Strontium titanate	310
Mylar	3.1	Paraffin	2.5
Neoprene	6.7	Rubber	3

More info at http://www.clippercontrols.com/pages/Dielectric-Constant-Values.html.

Table 5.2 Dielectric strengths of different materials.

Materials	Dielectric strength ($\times 10^6$ V/m)	Materials	Dielectric strength ($\times 10^6$ V/m)
Mica	118	Plexiglas	14
Paper	16	Polyethylene	21.7
Bakelite	24	Polyvinyl chloride	40
Porcelain	12	Teflon	60
Glass	14	Strontium titanate	8
Mylar	7	Paraffin	4060
Neoprene	12	Rubber	12

across a capacitor must be less than its rated voltage. The operating voltage across a capacitor can be increased depending on the insulating material or the dielectric constant. Teflon and polyvinyl chloride have greater dielectric strength. The dielectric constant can be increased by adding high dielectric constant filler material. Table 5.2 lists the dielectric strength values for different types of materials at room temperature.

5.2.4 Parallel Fingers and Fringing Fields

The parallel fingers architecture, consisting of a positively biased and a ground electrode, works under the principle of fringing capacitance [32]. High sensitivity along the direction perpendicular to the parallel electrodes of the sensor allows the

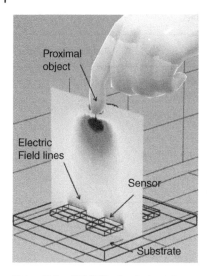

Proximal
object

Electric
Field lines

Sensor

Substrate

Figure 5.6 COMSOL simulation of fringing electric field developed near the active electrodes of proximity sensor in response to an approaching finger.

topology for possible application in motion detection, liquid-level sensing, proximity detection, and touch sensors. Due to the edge effects of the parallel electrode capacitive arrangement in sensor, the electric field lines are more spanned but less intense near the edges between the sensor and ground plates as shown in Figure 5.6. The calculations of fringing electric field emanating from the positive electrode and the capacitance between them are complex. However, it is seen that the sensitivity of this capacitive arrangement depends on the electrode size and the distance between the electrodes. A shield electrode on the under surface of the main sensor and ground electrode provides enhance E-field intensification toward the target.

The capacitance between the positively bias active electrode and the ground electrode is monitored for significant change which reflects the approach of an external object toward the device. The change in capacitance in response to an approaching object is attributed to the absorption of the E lines of force by the advancing object. The extent of capacitance change depends on the dielectric permittivity of the approaching object and the distance of the object from the sensor. The capacitance change due to an object in proximity will be significantly smaller than that due to a touch signal only because the E-field at longer distances is sparser and hence the change in capacitance is low. To maintain a reliable detection set up the sensor has to be designed so that the signal-to-noise ratio is low to achieve low resolution in approach distance detection. Several variant configurations can be designed with the parallel fingers configuration. Multiple positively biased electrode and ground electrodes can be alternated to have a central ground electrode. A central ground electrode provides wide window for sensing along the width of the electrodes and gives the sharpest response. The comb configuration comprised both of these variants, and very effective for wide and high direction based detection. The comb configuration is typically used in rain sensor applications and other applications that require a large sensing area and high sensitivity/ resolution.

5.3 Sensor Architecture

The architecture of the fabricated capacitor influences the performance of the sensor under response to an external stimulus. The sensor is designed to provide high sensitivity, low limit of detection, and low resolution of electrical signal output against applied external stimuli. In order to design a portable and small size capacitive sensor with high performance, the choice of dielectric, shape of electrodes, and device architecture play an important role in the outcome. The effect of dielectric and the shape and size of the electrodes were discussed in details in the previous sections. Here we discuss about the sensor design to achieve enhanced overall capacitance that helps in the increase in the signal-to-noise ratio and reducing the limit of detection.

5.3.1 Mixed Dielectrics

The advent of mixed dielectric capacitors has highly revolutionized electronics. The mixed dielectric capacitors incorporate two or more dielectrics between the electrodes and found to the more stable and possess longer life as compared to the single dielectric type capacitors [33–35]. The dual layered construction provides them with lower voltage limit, as well as acting as a reliable middle option solution between electrolytic capacitors and the general rechargeable battery. In addition, energy storage and tolerance capabilities are also quite high. They can even operate at higher temperatures where ordinary capacitors suffer. Their ability to discharge a large amount of electric load in such a short space of time has allowed them to be considered one of the absolute products to be found out there. A few of the sensor elements using multiple dielectric components have benefits in terms of detection range and resolution. The net capacitance of multiple capacitors, connected next to each other, depends on their connection configurations.

Let us consider a case where the two electrodes are separated by two dielectrics, each of thickness d, but one has area A_1 and permittivity ϵ_1, while the other has area A_2 and permittivity ϵ_2 as shown in Figure 5.7a. Since the two capacitors are connected in parallel, they both will have the same voltage across them; therefore, their net capacitance will be the sum of the two capacitances. The net capacitance of a parallel combination of capacitors is given as:

$$C_{\mathrm{T}} = \frac{Q_1}{V} + \frac{Q_2}{V}, \tag{5.6}$$

$$C_{\mathrm{T}} = C_1 + C_2, \tag{5.7}$$

where C_{T} is the total capacitance of the capacitors connected in parallel and Q_1 and Q_2 depict the charges across the first and second dielectrics with dielectric constant

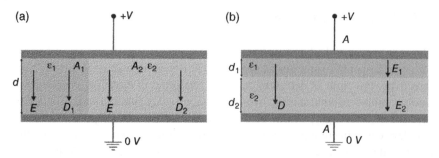

Figure 5.7 Parallel electrode capacitor with combination of two capacitances in (a) parallel and (b) series arrangement.

ε_1 and ε_2, respectively. $C_1(=(\varepsilon_1 A_1)/d)$C1 and $C_2(=(\varepsilon_2 A_2)/d)$ are their individual capacitances.

Since the E-field is independent of any medium between the plates, $E = V/d$ is constant in each of the two dielectrics. Thus, the electric displacement field D in the first, and second dielectric is given by $D_1 = \varepsilon_1 E$ and $D_2 = \varepsilon_2 E$. The charge density on the plates is given by Gauss's law as $\sigma = D$, so that, if $\varepsilon_1 < \varepsilon_2$, the charge density across the first dielectric is less than that across the second even if the potential is the same throughout the electrode. The two different charge densities on each plate are attributed to the different polarizations of the two dielectrics.

Figure 5.8a The circuit configuration of multiple capacitors having capacitances $(C_1, C_2, ..., C_n)$ with equivalent capacitance C_T, which is the sum of all capacitances.

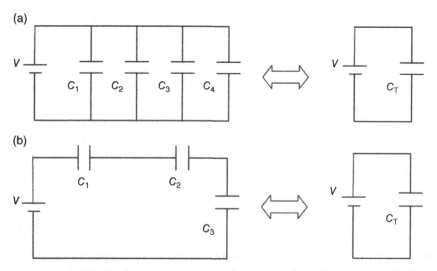

Figure 5.8 Equivalent circuit diagram for combination of capacitors connected in (a) parallel and (b) series.

Thus, if a sensor element contains n parallelly connected capacitances, then the total capacitance C_T is the sum of all the individual capacitances as:

$$C_T = C_1 + C_2 + \cdots + C_n \tag{5.8}$$

However, let us now consider that the dielectrics are stacked over each other in series between the parallel electrodes where the first dielectric of dielectric constant ε_1 has a thickness of d_1 and the second dielectric with dielectric constant ε_2 has a thickness of d_2 as shown in Figure 5.7b. It is assumed that the thicknesses of the dielectrics are supposed to be small so that the fields within them are uniform. It is evident that the individual capacitances $C_1(=(\varepsilon_1 A_1)/d)$ and $C_2(=(\varepsilon_2 A_2)/d)$ between the electrodes are in series and the total capacitance C_T is given by:

$$\frac{1}{C_T} = \frac{V_1}{Q} + \frac{V_2}{Q} \text{ or} \tag{5.9}$$

$$\frac{1}{C_T} = \frac{1}{C_1} + \frac{1}{C_2} = \frac{\varepsilon_1 \varepsilon_2 A}{\varepsilon_2 d_1 + \varepsilon_1 d_2} \tag{5.10}$$

If one electrode is held at a potential V and the other at 0 V, then the charge Q held by each of the positive and negative electrode of the capacitor is given by $Q = CV$ and hence the surface charge density is $\sigma = CV/A$. Following the Gauss's law, the total D-flux arising from a charge is equal to the charge density σ, such that for this architecture $D = \sigma$ and this is not changed by the nature of the dielectric materials between the plates. Thus, in both media, $D = CV/A = Q/A$ holds good. Thus, D is continuous across the boundary. Using $D = \epsilon E$ to each dielectric, the electric field E in the first and second dielectric is given as $E_1 = Q/\epsilon_1 A$ and $E_2 = Q/\epsilon_2 A$, respectively. Since $\epsilon_1 < \epsilon_2$, E-field is found to be high in the dielectric with small dielectric constant.

Thus, when n capacitances are connected in series, the potential at their interfaces are different and the total capacitance C_T is given by:

$$\frac{1}{C_T} = \frac{1}{C_1} + \frac{1}{C_2} + \cdots + \frac{1}{C_n} \tag{5.11}$$

The equivalent capacitance of capacitors connected in series can be stated as (Figure 5.8b).

Now let us consider a separate case where the mixed dielectric is in the form of a combination of two wedge-shaped dielectric materials of dielectric constants ε_1 and ε_2 as shown in the Figure 5.9. If the length and the width are denoted as ℓ and w, respectively, then the area of the electrodes is given by $A = \ell \times w$. The wedge angle is represented as θ such that $\tan \theta = d/\ell$. To determine the overall capacitance of the capacitive architecture, we choose an infinitesimal section length dx and width w. at a distance x from the edge of the capacitor. Let the capacitances of the first and second infinitesimal sections are dC_1 and

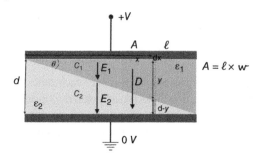

Figure 5.9 Capacitance with mixed dielectric material.

dC_2, respectively, and the total capacitance of the section is obtained to be dC_T. The infinitesimal capacitance dC_1 and dC_2 are given by $dC_1 = ((\varepsilon_1\varepsilon_0 w)/(y))dx$ and $dC_2 = ((\varepsilon_2\varepsilon_0 w)/(d-y))dx$. Rearranging and substituting $y = x\tan\theta = xd/\ell$, we get,

$$dC_1 = \frac{\varepsilon_1\varepsilon_0(w\ell)}{xd}dx \tag{5.12a}$$

$$dC_2 = \frac{\varepsilon_2\varepsilon_0(w\ell)}{d(\ell-x)}dx \tag{5.12b}$$

Since the capacitances are connected in series, the total infinitesimal capacitance dC_T can be expressed as:

$$\frac{1}{dC_T} = \frac{1}{dC_1} + \frac{1}{dC_2} \tag{5.13}$$

$$\frac{1}{dC_T} = \frac{xd}{\varepsilon_1\varepsilon_0(w\ell)} \cdot \frac{1}{dx} + \frac{d(\ell-x)}{\varepsilon_2\varepsilon_0(w\ell)} \cdot \frac{1}{dx}$$

$$\frac{1}{dC_T} = \frac{xd}{\varepsilon_1\varepsilon_0(w\ell)} \cdot \frac{1}{dx} + \frac{d(\ell-x)}{\varepsilon_2\varepsilon_0(w\ell)} \cdot \frac{1}{dx}$$

$$dC_T = \frac{(w\ell)\varepsilon_1\varepsilon_2\varepsilon_0}{(\varepsilon_2-\varepsilon_1)xd + \varepsilon_1\ell d}dx$$

$$C_T = (w\ell)\varepsilon_1\varepsilon_2\varepsilon_0 \int_0^l dx/(\{(\varepsilon_2-\varepsilon_1)d\}x + \varepsilon_1\ell d)$$

$$C_T = \frac{\varepsilon_1\varepsilon_2\varepsilon_0 A}{(\varepsilon_2-\varepsilon_1)d} \ln\left[(\varepsilon_2-\varepsilon_1)xd + \varepsilon_1\ell d\right]|_0^\ell$$

$$C_T = \frac{\varepsilon_1\varepsilon_2\varepsilon_0 A}{(\varepsilon_2-\varepsilon_1)d} \ln\left(\frac{\varepsilon_2}{\varepsilon_1}\right)$$

$$C_T = \left(\frac{\ln(\varepsilon_2/\varepsilon_1)}{((1/\varepsilon_1)-(1/\varepsilon_2))}\right)\frac{\varepsilon_0 A}{d} \tag{5.14}$$

When the dielectrics are replaced by a single dielectric of dielectric constant such that $\varepsilon_1 = \varepsilon_2$, we have below limiting condition,

$$C_T = C_T\big|_{\varepsilon_2 \to \varepsilon_1} = \lim_{\varepsilon_1 \to \varepsilon_2} \frac{\varepsilon_1 \varepsilon_2 \varepsilon_0 A}{(\varepsilon_2 - \varepsilon_1)d} \ln\left(\frac{\varepsilon_2}{\varepsilon_1}\right)$$

$$C_T = C_T\big|_{\varepsilon_2 \to \varepsilon_1} = \lim_{\varepsilon_1 \to \varepsilon_2} \frac{\varepsilon_2 \varepsilon_0 A}{(((\varepsilon_2)/(\varepsilon_1)) - 1)d} \ln\left(\frac{\varepsilon_2}{\varepsilon_1} + 1 - 1\right)$$

$$(5.15)$$

$$\text{If } p = \frac{\varepsilon_2}{\varepsilon_1} - 1, \text{we get, } C_T = C_T\big|_{p \to 0} = \lim_{p \to 0} \frac{\varepsilon_2 \varepsilon_0 A}{d} \frac{\ln(p - 1)}{p} = \frac{\varepsilon_2 \varepsilon_0 A}{d}$$

$$\left[\lim_{p \to 0} \frac{\ln(p - 1)}{p}\right]$$

$$(5.16)$$

Thus, when $\varepsilon_1 = \varepsilon_2 = \varepsilon_r$, we have $C_T = (\varepsilon_r \varepsilon_0 A)/d$, which is identical to Eq. (5.2).

5.3.2 Multielectrode Capacitor

The multielectrode capacitor architecture is also termed as the comb-type device which uses a series of parallel electrodes of alternate negative and positively biased terminals [36]. Due to the introduction of a series of parallel electrodes the effective overlapping area of the electrodes increases, thereby increasing the overall capacitance, keeping the net interelectrode distance d unchanged. For a standard parallel plate capacitor as shown in Figure 5.1, the capacitor has two plates. Since the number of electrodes are two, we can say that $n = 2$, where "n" represents the number of parallel electrodes. The capacitance can be expressed in terms of parallel electrodes n as $C_T = ((n-1)\varepsilon_0 \varepsilon_r A)/d$ (in farads). For a parallel plate capacitor with two electrodes $n = 2$, generating $C_T = (\varepsilon_0 \varepsilon_r A)/d$, which is the standard equation in (5.2).

Let us consider a multielectrode capacitor with 10 interdigitated electrodes (IDEs) in which there is an alternate arrangement of five electrodes each biased at positive and negative potential as shown in Figure 5.10a. Both sides of four positively and negatively biased electrodes and one side of each of the peripheral electrodes are in contact with the dielectric. The electrodes are equidistant from each other by a gap $d/(n-1)(=d/9)$ and separated by a material of dielectric constant ε_1. The individual capacitances developed between respective electrodes are depicted in Figure 5.11a. Since the capacitances are connected in parallel the total capacitance is obtained from the equivalent circuit diagram as $C_T = (9\varepsilon_0 \varepsilon_r A)/d$. However, if the capacitive architecture is planar found in microdevices and bendable sensors where the thickness of the electrodes are in nanometers (i.e. the width of the capacitive arrangement is a few orders lower than the length ℓ), the C_T is modified as $C_{T1} = (9\varepsilon_r \varepsilon_0 \ell)/d$. Now a situation may arise when one or more of the electrodes are disconnected/damaged due to faulty fabrication procedure. In such the case, the C_T of the capacitor decreases significantly as compared to the well-fabricated device to desired architecture as shown in

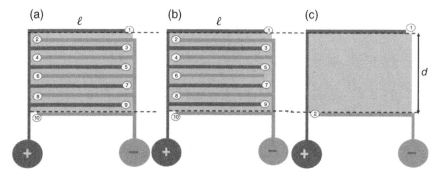

Figure 5.10 Multielectrode capacitor design with dielectric material of relative permittivity ε_r having (a) well fabricated electrodes and (b) Electrode No. 6 disconnected due to faulty fabrication process. (c) Two electrode capacitor design with equal d and ℓ.

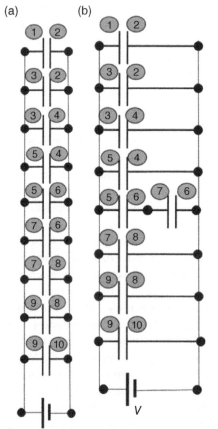

Figure 5.11 (a) and (b) representing equivalent circuit diagram for the capacitor combination shown in Figure 5.10a, b, respectively.

Figure 5.10a. Figure 5.10b shows an identical capacitive electrode architecture, but with disconnected electrode 6. When electrode 6 is not connected to any of the terminals, then positive charge on electrode 5 will induce negative charge on one side of electrode 6, and negative charge on electrode 7 will induce positive charge on the other side of electrode 6 as depicted in the equivalent circuit shown in Figure 5.11b. Thus, using the concepts of parallel and series combination, the equivalent C_T is given by: $C_{T2} = (7.5\varepsilon_r\varepsilon_0\ell)/d$.

When these results are compared with capacitance C_{T3} with two electrodes separated by identical distance d and dielectric ε_1, it is seen that the capacitance in the multielectrode architecture is increased by the factor equal to the number of gap segments $(n - 1)$ sandwiched between the electrodes. However, there is a reduction in capacitance C_{T2} $(<C_{T1})$ due to the improper IDE architecture.

Since the multielectrode capacitances are a combination of numerous capacitive elements connected in series or in parallel, their use as sensors may have limiting results if not used in dedicated applications. For example, when the IDE capacitive architecture is used as cell counter based on the principle of variation of dielectric when a biological cell is exposed to the fringing field emanating from the electrodes, the microsized cell passes over numerous capacitive arrangements connected in parallel. The individual capacitive element produces their respective field variations which interfere between themselves to provide a dull output signal with significant noise. In such a case, two-electrode capacitance with microsized inter-electrode gap comparable to the size of the cell generate sharp signal with significant resolution. However, this IDE capacitors architecture is suitable for touch sensor application where the area of IDE, being similar to the size of the fingertip, will generate enhanced variation in capacitive output. Thus, the design of the electrodes in a capacitive sensor is dominantly dependent on the desired application.

5.3.3 Geometry

The performance of the sensor against externally applied stimulus is dependent on the device architecture including the sensor design, shape, and dimension of the electrodes [37]. To explain the role of sensor geometry on the sensor performance we choose the example of a capacitive pressure sensor (CPS) which has a fixed BE but possess a TE patterned on a thin membrane (commonly called the diaphragm) and capable of vertical displacement under applied pressure P as shown in Figure 5.12a. The BE and the TE are separated by a dielectric material of dielectric constant ε_r of thickness d under initial condition $P = 0$. If the A $(=\pi r^2$, where r is the radius of the circular electrodes) denotes the area of the TE and BE, the capacitance developed between TE and BE is given Eq. (5.2) (Figure 5.12b). Under the application of external pressure P, the diaphragm is displaced from its original position leading to the reduction in d which in turn increases the capacitance of

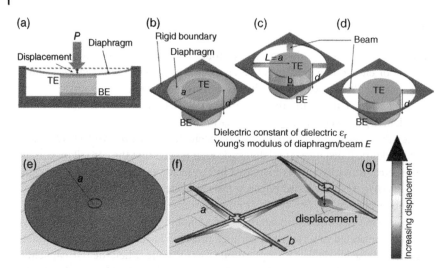

Figure 5.12 (a) Side view of capacitive sensor with displaceable diaphragms. Capacitive displacement sensors with (b) circular diaphragms, (c) quad beam, and (d) dual beam architecture. COMSOL representation of the devices showing the comparative displacements in (e) circular, (f) quad beam and (g) dual beam devices.

the sensor. Higher the displacement of the diaphragm from its original position, higher will be the capacitance of the sensor due to the increased reduction in d. Now if the radius, thickness, and Young' modulus of the diaphragm is a, h, and Y, respectively, then the spring constant of the circular diaphragm is given by:

$$k_{\text{Circular}} = \frac{16\pi E h^3}{a^2(1-v^2)},$$ (5.17)

where v is the Poisson ratio of the diaphragm. Now if the circular diaphragm is replaced by a quad beam structure and a dual beam structure supporting the TE as shown in Figure 5.12c, d, respectively, their spring constants are given by:

$$k_{\text{quad-beam}} = \frac{4bEh^3}{a^3},$$ (5.18)

and

$$k_{\text{dual-beam}} = \frac{2bEh^3}{a^3},$$ (5.19)

respectively,

where b denotes the width of the beams and the length of the beam is assumed to be equal to the radius a of the circular diaphragm. It is evident from the Eqs. (5.17), (5.18), and (5.19) that since the factor $(1 - v^2) < 1$, $k_{\text{Circular}} > k_{\text{quad-beam}} > k_{\text{dual-beam}}$.

As higher spring constant is manifested by stiffer architecture, the displacement produced in a highly stiff structure is low, producing small capacitance variation under applied pressure P as shown in Figure 5.12e. On the other hand, since the dual beam architecture has a low spring constant $k_{\text{dual-beam}}$, it produces enhanced displacement as compared to quad beam structure (Figure 5.12f, g) under identical applied pressure P, thereby generating increased variation in capacitance. Thus, the variation in capacitance due to an applied pressure P in the order $\Delta C_{\text{dual-beam}} > \Delta C_{\text{quad-beam}} > \Delta C_{\text{circular}}$, where $\Delta C_{\text{dual-beam}}$, $\Delta C_{\text{dual-beam}}$, and $\Delta C_{\text{circular}}$ represent the variation in capacitance in dual beam, quad beam, and circular diaphragm structure, respectively, due to an applied pressure P. It can be further noticed that the sensitivity of the sensor in response to applied pressure can be enhanced by increasing the length a, reducing the width b and height h of the beam, and using a low Y material such as elastomers as the beam/diaphragm.

5.4 Classifications of Capacitive Sensors

A number of different kinds of capacitance-based sensors used in a variety of industrial and automotive applications are discussed in this section.

5.4.1 Displacement Capacitive Sensor

Capacitive sensors for detection of mechanical quantities all function based on measuring displacement values [38, 39]. The movement of a suspended electrode with respect to a fixed electrode creates a varying capacitor value between the electrodes [40]. It is possible to measure this effect, and in the case where the mechanical quantity controls the movable electrode, a sensor can be realized [41]. Owing to the fact that the value of the capacitor is directly related to its dimensions, and a tiny capacitor involves high noise susceptibility, capacitive sensors are required to be as large as possible. The most common displacement-based sensors are pressure sensors where the displacement of the electrodes leads to the decrease in the dielectric thickness and hence increase in the capacitance of the sensor. Relatively low sensitivity, and in some cases high parasitic capacitances, have slowed the adoption of the CPS. These sensors are essentially two parallel plates, separated by a gap that varies with pressure. The sensors contain an intermediate elastomeric layer sandwiched between two electrodes as shown in Figure 5.1. The soft elastomer deforms in response to target pressure leading to change in thickness which in turn changes the capacitance of the gauge. The electrodes of the capacitive sensors can be displaced vertically perpendicular to the plane of the electrodes [42], or the electrodes fabricated on a planar elastomeric substrate/dielectric can be displaced horizontally under strain [43]. The output signal of the sensor may suffer from temperature drift.

To combat the drift in output signal due to variation in temperature, the capacitive sensors with diaphragms are introduced. The area of the diaphragm, dielectric gap, and diaphragm thickness are the first-order design parameters controlling the performance of a CPS. The shape, size, and architecture of the diaphragm control the sensitivity and reliability of displacement type, capacitive pressure. An electrode with stable deflection properties can measure pressure with a spacing-sensitive detector. The performance of the CPS can be enhanced by using appropriate geometry, elastomeric dielectric material, and architecture of the deflecting diaphragm. The CPS architecture can be classified into: (i) absolute, (ii) differential, and (iii) comb-type pressure sensors, as discussed below.

Miniaturized absolute CPS with cavity for pressure sensing was first developed in 1960 using conventional Si bulk micromachined technique and is shown in Figure 5.13a. The output capacitance varies with gas pressure due to mechanical

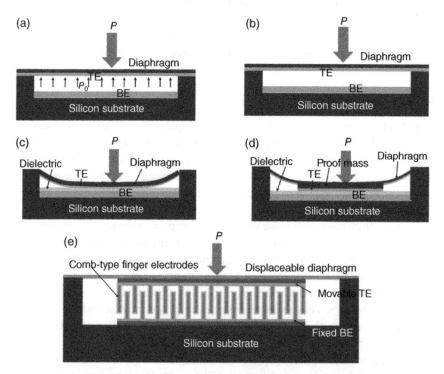

Figure 5.13 Schematic representation of (a) absolute, (b) differential, (c) contact mode without proof mass, (d) contact mode with proof mass, and (e) comb-type capacitive pressure sensor.

downward deflection of diaphragm. Although the absolute CPS provides a dynamic range of 10^2–10^3 mbar, it suffers from low sensitivity of 1.58 fF/mbar, nonlinearity in output signal due to parasitic capacitances, and lacks scope of commercialization because of its complex structure [44].

The differential CPS consists of a sealed vacuum cavity between the diaphragms in the form of TE and fixed BE as shown in Figure 5.13b. The pressure of the sealed cavity was considered as reference pressure [45, 46]. The relative variation of gas pressure between top and bottom diaphragm changes the capacitance of the gauge. However, the calibration curves for such differential and absolute capacitive-type sensors, which work on the principle of variation in dielectric thickness under applied P, are not linear.

5.4.2 Overlapping Area Variation Based Capacitive Sensor

Since the capacitance resulting from the deflection of a circular diaphragm is not linear to the pressure, a "contact mode" was introduced for the capacitive sensor to increase linearity. The contact mode CPS contains a fixed BE, a thin insulating layer on the top of BE, and a diaphragm as TE, which was designed to work in a region where it touches the insulating layer mechanically and is shown in Figure 5.13c. The capacitance becomes nearly proportional to the area of contact, and therefore exhibits good linearity to applied pressures [47]. The output capacitance varies as the touched area of diaphragm with insulating layer varies with the applied pressure. The nonlinearity in the output signal characteristics was reduced from 6.47% in absolute CPS to 0.68% in touch mode CPS by capturing the output signal from the touched area of diaphragm that was separated from the BE only by the thickness of thin insulating layer. The touch mode CPS exhibited a dynamic range of 6×10^2–10^3 mbar and sensitivity of 0.12 fF/mbar.

The performance of the sensor can be improved by an alternative strategy through the use of proof mass diaphragms where the thickness at the edges of the diaphragm are reduced providing increased sensitivity in addition to linear calibration curve of the sensor as evident from the Eqs. (5.17), (5.18), and (5.19). When the thickness h of the diaphragm is reduced, the spring constant of the diaphragm is reduced significantly due to its h^3 dependence on thickness. The thicker center portion (proof mass) is much stiffer than the thinner portion of the diaphragm on the outside. The center boss contributes most of the capacitance of the structure and its shape does not distort appreciably under an applied load. Hence, the capacitance–pressure characteristics are more linear as shown in Figure 5.13d.

The sensor performance can even be improved by increasing the overlapping area between the electrodes using comb-type capacitive sensor. The comb drive CPS consists of movable membrane-shaped TE, a fixed BE, and a number of comb drives between the two electrodes, and is shown in Figure 5.13e. The capacitance varies with area of overlap between the TE and BE, and also on the separation between the two electrodes when the TE is deflected by pressure. The 40 comb drives of the gauge enhance the sensitivity by 2 times and dynamic range by one order than that of the absolute CPSs. The comb drive CPSs offer an average sensitivity of 14 µF/mbar and dynamic range of 10^1–10^3 mbar.

5.4.3 Effective Dielectric Permittivity Variation Based Capacitive Sensor

In this section, we discuss on the type of capacitive sensors which detects the measurand by the virtue of the variation of effective dielectric permittivity between the two electrodes. The most common among this class are the fluid-level sensors [48, 49], contaminant detectors [50–52], chemical detectors [53, 54], and measurement of dielectric thickness of unknown fluid [55]. Capacitive liquid-level detectors sense the liquid level in a reservoir by measuring changes in capacitance between conducting electrodes which are immersed in the liquid. These sensors have been frequently used for wide range of solids, aqueous, organic liquids, and slurries. This technique is frequently stated as the radiofrequency signals applied to a capacitance circuit. The capacitive sensors are designed to sense material with dielectric constants as low as 1.1 for coke and fly ash, and as high as 88 for water or other liquids.

Capacitive liquid-level detectors sense the liquid level in a reservoir by measuring changes in capacitance between conducting electrodes which are immersed in the liquid [56]. A simple capacitive liquid-level sensor consists of two conducting electrodes establishing a variable capacitor which increases as the liquid level rises. If the gap between the two electrodes is fixed, the liquid level can be determined by measuring the capacitance between the electrodes immersed in the liquid. Since the capacitance is proportional to the dielectric constant, fluids rising between the two parallel electrodes will increase the net capacitance of the measuring cell as a function of liquid height. To measure the liquid level, an excitation voltage is applied with a drive electrode and detected with a sense electrode. Figure 5.14a illustrates a basic setup of a liquid-level measurement system. Since the capacitance between the pair of electrodes is dependent on the architecture, shape, and alignment of the electrodes, the sensor performance will be modified in accordance with the electric field developed between the electrodes. However, the working principle for capacitive liquid-level sensing remains the same.

To understand the working of a capacitive liquid-level sensor let us assume two parallel long rods of equal length L and radius r as the electrodes which are mutually separated from each other by a fixed distance d. The two electrodes are immersed in the tank which is partially filled with fluid. The part of the rods

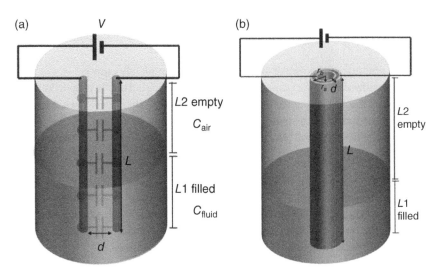

Figure 5.14 Basic liquid-level sensing system with (a) parallel rod and (b) coaxial tube capacitance.

immersed in the fluid and that above it can be thought of as separate combinations of parallel capacitances with effective capacitances C_1 and C_2. If L_1 and L_2 are the lengths of the rod immersed and above the fluid level, then the dielectric constants experienced by the former and later part of the rods are $\varepsilon_{\text{fluid}}$ and ε_{air}, respectively. Thus, the total capacitance of the system is given by:

$$C_T = C_{\text{fluid}} + C_{\text{air}}, \tag{5.20}$$

$$\text{where } C_{\text{fluid}} = \frac{\pi \varepsilon_0 \varepsilon_{\text{fluid}}}{\ln(d/r)} L_1, \quad d \gg r, \text{and } C_{\text{air}} = \frac{\pi \varepsilon_0 \varepsilon_{\text{air}}}{\ln(d/r)} L_2, \quad d \gg r \tag{5.21}$$

$$C_T = C_{\text{fluid}} + C_{\text{air}} = \frac{\pi \varepsilon_0}{\ln(d/r)} [\varepsilon_{\text{fluid}} L_1 + \varepsilon_{\text{air}} L_2] \tag{5.22}$$

$$C_T = C_{\text{fluid}} + C_{\text{air}} = \frac{\pi \varepsilon_0}{\ln(d/r)} [\varepsilon_{\text{fluid}} L_1 + L_2], \quad \varepsilon_{\text{air}} = 1 \tag{5.23}$$

Thus, the normalized relative change in capacitance with increasing fluid level is given by:

$$\Delta C_T = \frac{(C_{\text{air}} + C_{\text{fluid}}) - C_0}{C_0} = \frac{L_2 + \varepsilon_{\text{fluid}} L_1 - L}{L} = \frac{L_1}{L} (\varepsilon_{\text{fluid}} - 1), L_2 = L - L_1 \tag{5.24}$$

$$\text{where } C_0 \left\{ = \frac{\pi\varepsilon_0\varepsilon_{\text{air}}}{\ln(d/r)}(L_1 + L_2) = \frac{\pi\varepsilon_0}{\ln(d/r)}L \right\} \tag{5.25}$$

is the capacitance of the sensor when the tank is completely empty.

The normalized change in capacitance C_T of the device is directly proportional to the length of the rod submersed under the fluid. As the fluid level rises, L_1 rises, and hence increases the response due to the increasing fluid level.

However, when the rod electrodes are kept in proximity such that $d \ll r$, Eq. (5.25) is modified to $C_T = \left((\pi\varepsilon_0\varepsilon_r) / \left(\ln\left(\left(d + \sqrt{d^2 - 4r^2} \right) / 2r \right) \right) \right) L$. The signals induced on the cable or wire electrodes connecting a probe could disturb the analog measurement signal. The signal disturbances can be caused by an external electromagnetic field, such as generated by a vehicle radio set. Researchers have utilized various electrode geometry including coaxial cables, IDEs, and helical capacitive arrangements to combat these shortcomings.

With the growing demand for robustness, portability, and enhancement in sensor performance, the coaxial-type capacitive sensors are utilized over rod-type capacitive sensors as shown in Figure 5.14b. Such sensors are often used to detect gas leakages and measure contaminants in fluid using the following procedures. A cylindrical capacitor can be thought of as having two cylindrical tubes, inner and outer. The inner cylinder can be connected to the positive terminal, whereas the outer cylinder can be connected to the negative terminal. An electric field will exist if a voltage is applied across the two terminals. If r_a is the radius of the inner cylinder and r_b is the radius of the outer cylinder, then the capacitance can be calculated by using:

$$C = \frac{\pi\varepsilon_0\varepsilon_r}{\ln(r_b/r_a)}L. \tag{5.26}$$

When the coaxial electrodes are in air and exposed to a fluid medium, the capacitance developed between the electrodes are given by $C_{\text{air}} = ((\pi\varepsilon_0\varepsilon_{\text{air}})/(\ln(r_b/r_a)))L$ and $C_{\text{gas}} = ((\pi\varepsilon_0\varepsilon_{\text{gas}})/(\ln(r_b/r_a)))L$, respectively. The normalized change in capacitance ΔC_T from the initial value (i.e. in air) reflects the presence of a leakage gas/fluid. Thus,

$$\Delta C_T = \frac{C_{\text{gas}} - C_{\text{air}}}{C_{\text{air}}} = \frac{\varepsilon_{\text{gas}} - \varepsilon_{\text{air}}}{\varepsilon_{\text{air}}} = \frac{\varepsilon_{\text{gas}}}{\varepsilon_{\text{air}}} - 1. \tag{5.27}$$

where ε_{gas} is the dielectric constant of leakage gas.

A leakage gas is detected when $\Delta C_T > 1$. In order to use the same sensor architecture for measuring the concentration of a contaminant gas, the equation (5.26) is modified as follows.

The capacitance of the coaxial tube architecture in the presence of only pure gas is given by:

$$C_{gas} = \frac{\pi \varepsilon_0 \cdot (\varepsilon_{gas})}{\ln(r_b/r_a)} L \tag{5.28}$$

while under the exposure to x parts concentration of contaminant gas (c-gas), the capacitance changes to:

$$C'_{gas} = \frac{\pi \varepsilon_0 \cdot (1-x)\varepsilon_{gas}}{\ln(r_b/r_a)} L + \frac{\pi \varepsilon_0 \cdot (x)\varepsilon_{c\text{-}gas}}{\ln(r_b/r_a)} L \tag{5.29}$$

Thus, the normalized change in capacitance $\Delta C_T = \left(C'_{gas} - C_{gas} \right)/C_{gas} = \left(x(\varepsilon_{c\text{-}gas} - \varepsilon_{gas}) \right)/\varepsilon_{gas}$. As $\varepsilon_{c\text{-}gas}$ and ε_{gas} are constants, the change in capacitance ΔC_T, which is experimentally measured, is directly proportional to the concentration x of c-gas. The concentration of the c-gas is obtained from equation:

$$x = \frac{\Delta C_T \varepsilon_{gas}}{\left(\varepsilon_{c\text{-}gas} - \varepsilon_{gas} \right)} \tag{5.30}$$

Relative permittivity of a dielectric changes when a target gas or water molecules are absorbed or adsorbed on the dielectric material [57]. This change in the relative permittivity of the dielectric material after exposure to target gas changes the effective capacitance of the device and thus can be used as a sensor. Humidity gas sensor is a very good capacitive-type sensor because of high dielectric constant of water. It has a large dielectric constant of 78.5 at 298 K. By adsorption of water with humidity sensor there is a change in relative permittivity, which provides a simplest detection mechanism [58]. Capacitive humidity sensors commonly contain layers of hydrophilic inorganic oxides which act as a dielectric [59]. Absorption of polar water molecules has a strong effect on the dielectric constant of the material [60]. The magnitude of this effect increases with a large inner surface which can accept large amounts of water. Polymer [61] and ceramics [62] are commonly used materials used in capacitive-type humidity sensor. Dielectric constant of polymer such as polyimides is of 3–6, whereas water has a dielectric constant of 80. On the adsorption of water on polymer and ceramic surface, the capacitance of the device increases and hence humidity can be sensed. The most commonly used ceramic dielectric material for humidity sensing is alumina film [63]. The microporous morphology of alumina traps in water molecules, which increases the effective relative permittivity/dielectric constant of the dielectric, which in turn increases in capacitance of the film [64].

5.4.4 Fringing Field Capacitive Sensor

As seen in Section 5.1, a simple capacitive sensor consists of two electrodes separated by a dielectric. Under an applied bias between the electrodes the electric field developed between them exists not just directly between the plates, but extends some distance away. This is called the *edge effect* and the field which extends beyond the overlapping area is called the *fringing* field as shown in Figure 5.15a. To accurately measure the capacitance of a capacitor, the domain used to assume the fringing field must be sufficiently larger than the actual visible region, and the appropriate boundary conditions must be used.

In contrast to an infrared and ultrasonic sensor which emits electromagnetic radiation and acoustic waves, a capacitive proximity sensor emits an electrostatic field, and detects any change in the field. A capacitive proximity sensor utilizes the fringing field of the capacitor to detect the presence of an approaching or nearby stationery objects without any physical contact and also measure its distance from the sensor [65]. The capacitor model equation (5.1) demonstrates that the capacitance is inversely proportional to the distance between the two capacitor plates ($1/d$). In a typical application, a conductive electrode pad will be one plate of the capacitor, while the desired object will act as the other capacitor plate. Capacitive-type proximity sensors consist of an oscillator whose frequency is determined by an inductance–capacitance (LC) circuit to which a metal plate is connected. As the conducting or partially conducting target approaches the sensor, the target enters the electrostatic field of the electrodes and thus changes the mutual capacitance in an oscillator circuit as a result of which the oscillator begins oscillating. The oscillations are detected and sent to the controller unit. The amplitude of oscillation increases as the target approaches the sensor. The trigger circuit reads the oscillator's amplitude and when it reaches a specific level, the output state of the

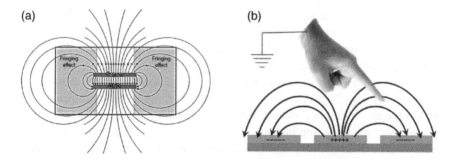

Figure 5.15 Representation of the (a) fringing electric field emanating from the edges of the parallel electrode capacitive arrangement and (b) proximity sensor using the principle of fringing electric field.

sensor changes. Figure 5.15b shows an example of the capacitive proximity sensor. Since the relationship between the distance and the capacitance is asymptotic, this type of sensor is better suited in applications where high resolution in close proximity is desired.

The maximum distance that a proximity sensor can detect is defined as "nominal range." Some sensors have adjustments of the nominal range or ways to report a graduated detection distance. A proximity sensor adjusted to a very short range is often used as a touch switch. Capacitive proximity detectors have a range twice that of inductive sensors, while they detect not only metal objects but also dielectrics such as paper, glass, wood, and plastics [66]. They can even detect through a wall or cardboard box. Because the human body behaves as an electric conductor at low frequencies, capacitive sensors have been used for human activity monitoring [67] and in intrusion alarms [68]. Capacitive-type proximity sensors have a high reliability and long functional life because of the absence of mechanical parts and lack of physical contact between sensor and the sensed object [69].

Proximity sensing technology can play an important part in security and safety applications in a number of markets. For instance, they can be used to detect the presence of seated occupants in a car [70] and even determine the size of the occupants to trigger seat belt alarms and provide valuable data to the airbag deployment system. Again, a proximity sensor can be used as a limit switch, which is a mechanical pushbutton switch that is mounted in such a way that it is activated when a mechanical part or lever arm gets to the end of its intended travel. It can be implemented in an automatic garage door opener, where the controller needs to know if the door is all the way open or all the way closed.

5.5 Flexible Capacitive Sensors

Flexible sensors are widely studied now-a-days due to their promise and potential in wearable and skin electronics applications. Flexible devices supported with artificial intelligence and enabled for Internet of things application have transformed the world of smart sensing systems. The emergence of flexible electronics has led to the fabrication of slender, lightweight, stretchable, and foldable sensors [71]. The advantages of capacitive sensing technologies in convergence with flexible electronics have led to burst in research activities in flexible capacitive sensors and their rise in global market revenue since 2010. A capacitive flexible sensor finds applications as tactile, pressure, proximity sensors and offers quick response, a broad dynamic range and high sensitivity.

Ma et al. [18] designed a highly flexible CPS, with barium titanate-reinforced polyvinylidene fluoride (PVDF) dielectric layer between top and bottom microarray structured Au electrodes on polydimethylsiloxane (PDMS) substrate, which operates in the pressure range of 0–2500 Pa. The device offers high sensitivity of 4.9 kPa^{-1}, low detection limit of 1.7 Pa, short response time of <50 ms, a stable response over 5000 loading–unloading cycles, and bending stability. The sensor performance of the device can be tuned by varying the BaTiO$_2$ filler content in the PVDF matrix dielectric. However, the device suffers from nonlinear calibration curve as the sensitivity drops dramatically toward high pressure range. To overcome this shortcoming in CPSs and to attain high linearity over a broad sensing range, Wu et al. [31] used percolative composites as the dielectric layer where the linear response was attributed to the fast increase in dielectric constant that was able to compensate for the sensitivity drop caused by the decreased compressibility during compression. The CPS fabricated with spiky nickel/polydimethyl siloxane composite as the dielectric layer exhibits excellent linearity ($R^2 = 0.999$) up to 1.7 MPa. Flexible tactile sensors can perform the synchronized interactions with surrounding environment using the sense of touch. Li et al. [34] designed and fabricated a high-performance flexible capacitive tactile sensor utilizing the bionic microstructures on natural lotus leaves. They biomimicked the unique surface micropattern of lotus leave to develop a template for electrodes and using polystyrene microspheres as the dielectric layer. The device offered high sensitivity (0.815 kPa^{-1}), wide dynamic response range (from 0 to 50 N), and fast response time ($\approx$38 ms). The device owing to the responsivity toward pressure, bending, and stretching forces, finds application in electronic skins, wearable robotics, and biomedical devices. Researchers have utilized various capacitive architecture to enhance the sensitivity as well as reduce the detection limit of the device. Luo et al. [35] fabricated a flexible capacitive sensor with tilted micropillar array-structured dielectric layer sandwiched between two electrodes and offered high pressure sensitivity of 0.42 kPa^{-1} and very small detection limit of 1 Pa. The high sensitivity of the device was attributed to the bending deformation of the tilted micropillars rather than compression deformation as a result of which the distance between the electrodes changed easily, even discarding the contribution of the air gap at the interface of the structured dielectric layer and the electrode. The device was claimed to be highly robust with high stability and reliable capacitance response which was attributed to eliminating the presence of uncertain air gap through the fabrication of tilted micropillars that connects the two electrodes, which allows the dielectric layer is strongly bonded with the electrode. In a nutshell, flexible capacitive sensors utilizes either the parallel plate [47] or the IDE [72]-type architecture, to fabricate devices with diverse dielectric layers with variety of materials [19, 27, 30, 73] and microstructures [10, 34, 74] and also with different electrode designs [33, 75–78] to enhance the sensor performance in terms of sensitivity [79], lower detection limit, and response time.

5.6 Applications

With the knowledge of different architectures of capacitive sensors fabricated for different applications using diverse elastomeric dielectric materials and microstructures electrodes, such devices have been explored for various uses in its suitable form. Some of such applications are discussed below.

5.6.1 Motion Detection

5.6.1.1 Displacement Motion (z-Direction)

Parallel plate capacitive arrangement can be used as a motion sensor utilizing the spacing variation in the dielectric material when the spacing change is less than the electrode size. The motion detector is designed in such a way that one plate of the capacitor is fixed, while the other is movable perpendicular to the plane of the electrodes. The parallel plate capacitance formula in Eq. 5.2 shows that capacitance is inversely related to spacing d. Since the sensitivity of the sensor dC/dd is large for small d following the parabolic capacitance–motion displacement relationship, the sensor conveniently yields large variation in capacitance at small displacement due to motion. However, the nonlinearity of the sensor poses problems during calibration. The linearity in sensor during motion detection through displacement measurement can be solved by plotting the impedance Z_C–displacement d curve which is linear following the relation $Z_C = d/(j\omega\varepsilon_0\varepsilon_r A)$.

5.6.1.2 Shear Motion (x Direction)

The shear motion detector utilizes the overlapping electrode of the parallel plate capacitor architectural design as described in Section 5.4.2. The overlapping area principle is suitable when the displacement incurred due to motion is larger than the dimensions of the electrodes. Here, the transverse motion of the electrodes relative to the other changes the capacitance linearly with shear displacement. Quite long excursions are possible with good linearity, but the gap needs to be small and well controlled. As with spacing variation, overlap is needed so that unwanted sensitivities are minimized.

5.6.1.3 Tilt Sensor

Tilt sensors are devices that produce an electrical signal that varies with an angular movement. These sensors are used to measure slope and tilt within a limited range of motion. Sometimes, the tilt sensors are referred to as inclinometers. These devices consist of a multiple pair of comb electrodes based capacitors and central proof mass or other dielectric liquid. When a tilt occurs, the central mass/liquid moves toward one of the combs so the capacitance increases at one side and decreases at the other side.

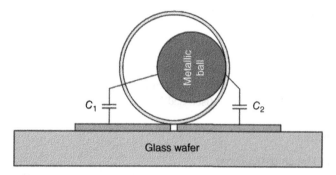

Figure 5.16 Schematic representation of metallic ball-based rotary motion sensor.

5.6.1.4 Rotary Motion Sensor

The rotary motion detector was designed by Lee et al. and consists of movable TE in the form of a metallic ball and separate and fixed BEs as shown in Figure 5.16. The top metallic ball and the BEs are separated by a plastic pipe along whose inner walls the metallic ball rotates through an angle of 360°. The sensor uses the spatial distance between the metallic ball and each of the BEs to determine the respective capacitance developed between them. The capacitances developed between the metallic ball and each of the BEs due to the spatial position of the metallic ball under tilt gives the tilt in the sensor.

5.6.1.5 Finger Position (2D, *x–y* Direction)

Capacitive graphic input tablets of different sizes can replace the computer mouse as an *x–y* coordinate input device. Finger-touch-sensitive, *z*-axis-sensitive, and stylus-activated devices are available. This device, often found just in front of the keyboard on a laptop, drives a pulse in succession on each column and measures coupling to each row. By locating the coordinates of peak coupling and interpolating between adjacent rows, the location of a shielding human finger is measured to a fraction of a millimeter.

5.6.2 Pressure

A stretchable diaphragm with stable deflection properties can measure pressure as a spacing-sensitive pressure detector. Moreover, compressible elastomeric dielectrics were also used in such applications.

5.6.3   Liquid Level

Capacitive liquid-level detectors sense the liquid level in a reservoir by measuring changes in capacitance between conducting plates which are immersed in the liquid, or applied to the outside of a nonconducting tank.

5.6.4   Spacing

If a metal object is near a capacitor electrode, the mutual capacitance is a very sensitive measure of spacing.

5.6.5   Scanned Multiplate Sensor

The single-plate spacing measurement can be extended to contour measurement by using many plates, each separately addressed. Both conductive and dielectric surfaces can be measured.

5.6.6   Thickness Measurement

Two plates in contact with an insulator will measure the insulator thickness if its dielectric constant is known, or the dielectric constant if the thickness is known.

5.6.7   Ice Detector

Airplane wing icing can be detected using insulated metal strips in wing leading edges.

5.6.8   Shaft Angle or Linear Position

Capacitive sensors can measure angle or position with a multiplate scheme giving high accuracy and digital output, or with an analog output with less absolute accuracy but faster response and simpler circuitry.

5.6.9   Lamp Dimmer Switch

The common metal-plate soft-touch lamp dimmer uses 60 Hz excitation and senses the capacitance to a human body.

5.6.10   Key Switch

Capacitive key switches use the shielding effect of a nearby finger or a moving conductive plunger to interrupt the coupling between two small plates.

5.6.11 Limit Switch

Limit switches can detect the proximity of a metal machine component as an increase in capacitance, or the proximity of a plastic component by virtue of its increased dielectric constant over air.

5.6.12 Accelerometers

Analog devices have introduced integrated accelerometer circuits with a sensitivity of 1.5 g. With this sensitivity, the device can be used as a tiltmeter.

5.6.13 Soil Moisture Measurement

A capacitive moisture sensor [80] works by measuring the capacitance changes due to the changes in the dielectric property of the soil. Instead of measuring the soil directly (pure water does not conduct electricity well), the sensor measures the ions that are dissolved in the moisture. The concentration of the ions can change with the amount of fertilizer in the soil and its moisture retention. Capacitive measuring basically measures the dielectric that is formed by the soil, and the water is the most important factor that affects the dielectric.

5.7 Prospects and Limitations

5.7.1 Prospects

The theoretical relation between displacement and capacitance is governed by a simple expression, which in practice can be approximated with high accuracy, resulting in a very high linearity. Using special constructions, the measurement range of capacitive sensors can be extended almost without limit while maintaining the intrinsic accuracy. Moreover, because of the analog nature of the capacitive principle, the sensors have excellent resolution. The miniaturized capacitive sensor is robust in architecture and offers high sensitivity. High sensitivity and robust architecture make it suitable for system on-chip application and use as wearable medical devices. Again, the performance of the capacitive sensor is less suffered from the thermal drift. Thus, the capacitive sensor can be used at high-temperature operation.

5.7.2 Limitations

The performance of the capacitive sensor may be affected by the noise generated at the device connections, stray capacitance, and environmental changes. The effect of noise and stray capacitance on the sensor's sensitivity and environmental changes on its interelectrode spacing are discussed as follows.

Sensor plates may have signal capacitances in the fractional picofarad (pF) range, and connecting to these plates with a 60 pF/m coaxial cable could totally obscure the signal. However, with correct shielding of the coaxial cable as well as any other stray capacitance one can almost completely eliminate the effects of noise.

One hazard of the oscillator circuits is that the frequency is changed if the capacitor picks up capacitively coupled crosstalk from nearby circuits. The sensitivity of an RC oscillator to a coupled narrow noise spike is low at the beginning of a timing cycle, but high at the end of the cycle. This time variation of sensitivity leads to beats and aliasing where noise at frequencies which are integral multiples of the oscillator frequency is aliased down to a low frequency. This problem can usually be handled with shields and careful power supply decoupling.

The capacitance is dependent on the gap or distance between the conducting electrodes. This distance can, however, increase or decrease depending on the environmental conditions and the material, which could incorporate inaccuracies in the level readings. In some cases, movement of the fluid container can skew or bend the sensor, which will alter the distance between the electrodes, thereby errors will be produced in the capacitance value, and hence the fluid level.

Static charge can build up on the insulators near capacitive sense plates due to triboelectric charging, causing a tiny spark in serious cases. In less serious cases, a charge buildup of 50–100 V will not arc over but can cause an unwanted sensitivity to mechanical vibration. This sensitivity results from the voltage $V = Q/C$ caused by constant capacitor charge and a capacitance that varies with spacing change due to vibration and can drive the amplifier input over its rails. To reduce static charge, capacitors must operate under high humidity, circuit must be designed with low-input impedance AC amplifier, bare metal electrodes must be used to minimize triboelectric effect static discharge problems, and high pass filter must be used to minimize mechanical resonance and static charge effects.

List of Abbreviations

BE	Bottom Electrode
CPS	Capacitive Pressure Sensor
IDE	Interdigitated Electrode
LC circuit	Inductance–Capacitance Circuit
MEMS	Microelectromechanical Systems
PDMS	Polydimethylsiloxane
PVDF	Polyvinylidene Fluoride
TE	Top Electrode

References

1 Goustouridis, D., Normand, P., and Tsoukalas, D. (1998). Ultraminiature silicon capacitive pressure-sensing elements obtained by silicon fusion bonding. *Sensors and Actuators A: Physical* 68: 269–274.

2 Zhang, Y., Howver, R., Gogoi, B., and Yazdi, N. (2011). A high-sensitive ultra-thin MEMS capacitive pressure sensor. *2011 16th International Solid-State Sensors, Actuators and Microsystems Conference, TRANSDUCERS'11*, Beijing, China (5–9 June 2011), pp. 112–115. IEEE.

3 Sparks, D.R., Baney, W.J., Staller, S.E., Wesley, D., James, C., and Siekkinen, W. (1997). All-silicon capacitive pressure sensor. Google Patents, US Patent office, US5936164A, Delphi Technologies Inc.

4 Philipp, H. (2005). Capacitive sensor. Google Patents, US Patent office, US7515140B2, Atmel Corp.

5 Badaye, M. and Landry, G. (2011). Flexible capacitive sensor array. Google Patents, US Patent office, US9459736B2, Parade Technologies Ltd.

6 Philipp, H. (1997). Capacitive position sensor. Google Patents, US Patent office, US6288707B1, Neodron Ltd.

7 Ramalingame, R., Lakshmanan, A., Māller, F. et al. (2019). Highly sensitive capacitive pressure sensors for robotic applications based on carbon nanotubes and PDMS polymer nanocomposite. *Journal of Sensors and Sensor Systems* 8: 87–94.

8 Kwon, D., Lee, T.-I., Shim, J. et al. (2016). Highly sensitive, flexible, and wearable pressure sensor based on a giant piezocapacitive effect of three-dimensional microporous elastomeric dielectric layer. *ACS Applied Materials & Interfaces* 8: 16922–16931.

9 Zhao, S., Ran, W., Wang, D. et al. (2020). 3D dielectric layer enabled highly sensitive capacitive pressure sensors for wearable electronics. *ACS Applied Materials & Interfaces* 12: 32023–32030.

10 Xiong, Y., Shen, Y., Tian, L. et al. (2020). A flexible, ultra-highly sensitive and stable capacitive pressure sensor with convex microarrays for motion and health monitoring. *Nano Energy* 70: 104436.

11 Matsuzaki, R. and Todoroki, A. (2007). Wireless flexible capacitive sensor based on ultra-flexible epoxy resin for strain measurement of automobile tires. *Sensors and Actuators A: Physical* 140: 32–42.

12 Millet, T. (1987). Door alarm with infrared and capacitive sensors. Google Patents, US Patent office, US4804945A.

13 Iqbal, J., Lazarescu, M.T., Tariq, O.B. et al. (2018). Capacitive sensor for tagless remote human identification using body frequency absorption signatures. *IEEE Transactions on Instrumentation and Measurement* 67: 789–797.

14 Fergusson, R.T. (2005). Digital capacitive sensing device for security and safety applications. Google Patents, US Patent office, US7187282B2, Uniroyal Global Engineered Products Inc.

15 Christian, M., Elgar, F., Jan, M. et al. (2008). Flexible-foam-based capacitive sensor arrays for object detection at low cost. *Applied Physics Letters* 92: 013506.

16 Selvakumar, A. and Najafi, K. (1998). A high-sensitivity z-axis capacitive silicon microaccelerometer with a torsional suspension. *Journal of Microelectromechanical Systems* 7: 192–200.

17 Zega, V., Invernizzi, M., Bernasconi, R. et al. (2019). The first 3D-printed and wet-metallized three-axis accelerometer with differential capacitive sensing. *IEEE Sensors Journal* 19: 9131–9138.

18 Ma, L., Yu, X., Yang, Y. et al. (2020). Highly sensitive flexible capacitive pressure sensor with a broad linear response range and finite element analysis of micro-array electrode. *Journal of Materiomics* 6: 321–329.

19 Sun, X., Sun, J., Li, T. et al. (2019). Flexible tactile electronic skin sensor with 3D force detection based on porous CNTs/PDMS nanocomposites. *Nano-Micro Letters* 11: 57.

20 Hoch, D. (2011). Proximity sensing for capacitive touch sensors. Google Patents, US Patent office, US8902191B2, Synaptics Inc.

21 Gong, C.A., Chiu, H.K., Huang, L.R. et al. (2016). Low-cost comb-electrode capacitive sensing device for liquid-level measurement. *IEEE Sensors Journal* 16: 2896–2897.

22 Tewari, D. (2020). Pressure Sensor Market by Type (Absolute Pressure Sensor, Gauge Pressure Sensor, and Differential Pressure Sensor), Technology (Piezoresistive, Electromagnetic, Capacitive, Resonant Solid State, Optical, and Others), and Application (Automotive, Oil & Gas, Consumer Electronics, Medical, Industrial and Others): Global Opportunity Analysis and Industry Forecast, 2020–2027. Allied Market Research.

23 Company (2020). Industry Reports, Electronics, Capacitive Pressure Sensor Market – Growth, Trends, and Forecast (2020–25). *Mordor Intelligence*.

24 McIntosh, R. and Casada, M. (2008). Fringing field capacitance sensor for measuring the moisture content of agricultural commodities. *IEEE Sensors Journal* 8: 240–247.

25 Kolb, P.W., Decca, R.S., and Drew, H.D. (1998). Capacitive sensor for micropositioning in two dimensions. *Review of Scientific Instruments* 69: 310–312.

26 Roberts, P., Damian, D.D., Shan, W. et al. (2013). Soft-matter capacitive sensor for measuring shear and pressure deformation. *Presented at 2013 IEEE International Conference on Robotics and Automation,* Karlsruhe, Germany (6–10 May 2013). IEEE.

27 Zou, Q., Lei, Z., Xue, T. et al. (2020). Highly sensitive flexible pressure sensor based on ionic dielectric layer with hierarchical ridge microstructure. *Sensors and Actuators A: Physical* 313: 112218.

28 Mukherjee, T. and Paul, A. (2016). Dielectric breakdown-assisted corona discharge-based pressure sensor using poly-Si microtips. *IEEE Transactions on Electron Devices* 63: 2080–2088.

29 Fernandes, J. and Jiang, H. (2017). Three-axis capacitive touch-force sensor for clinical breast examination simulators. *IEEE Sensors Journal* 17: 7231–7238.

30 Zhu, Y., Wu, Y., Wang, G. et al. (2020). A flexible capacitive pressure sensor based on an electrospun polyimide nanofiber membrane. *Organic Electronics* 84: 105759.

31 Wu, J., Yao, Y., Zhang, Y. et al. (2020). Rational design of flexible capacitive sensors with highly linear response over a broad pressure sensing range. *Nanoscale* 12: 21198–21206.

32 Morais, F.V., Carvalhaes-Dias, P., Duarte, L. et al. (2020). Fringing field capacitive smart sensor based on PCB technology for measuring water content in paper pulp. *Journal of Sensors* 2020: 3905804.

33 Dobrzynska, J.A. and Gijs, M.A.M. (2012). Polymer-based flexible capacitive sensor for three-axial force measurements. *Journal of Micromechanics and Microengineering* 23: 015009.

34 Li, T., Luo, H., Qin, L. et al. (2016). Flexible capacitive tactile sensor based on micropatterned dielectric layer. *Small* 12: 5042–5048.

35 Luo, Y., Shao, J., Chen, S. et al. (2019). Flexible capacitive pressure sensor enhanced by tilted micropillar arrays. *ACS Applied Materials & Interfaces* 11: 17796–17803.

36 Filippidou, M.K., Chatzichristidi, M., and Chatzandroulis, S. (2019). A fabrication process of flexible IDE capacitive chemical sensors using a two step lift-off method based on PVA patterning. *Sensors and Actuators B: Chemical* 284: 7–12.

37 Zeinali, S., Homayoonnia, S., and Homayoonnia, G. (2019). Comparative investigation of interdigitated and parallel-plate capacitive gas sensors based on Cu-BTC nanoparticles for selective detection of polar and apolar VOCs indoors. *Sensors and Actuators B: Chemical* 278: 153–164.

38 Kim, M. and Moon, W. (2006). A new linear encoder-like capacitive displacement sensor. *Measurement* 39: 481–489.

39 Restagno, F., Crassous, J., Charlaix, E., and Monchanin, M. (2000). A new capacitive sensor for displacement measurement in a surface-force apparatus. *Measurement Science and Technology* 12: 16–22.

40 Zhu, F., Spronck, J.W., and Heerens, W. (1991). A simple capacitive displacement sensor. *Sensors and Actuators A: Physical* 26: 265–269.

41 Xia, S., Nihtianov, S., Nihtianov, S., and Luque, A. (2018). 4 – Capacitive sensors for displacement measurement in the subnanometer range. In: *Smart Sensors and MEMs*, 2e, 87–99. Woodhead Publishing.

42 Wang, X., Xia, Z., Zhao, C. et al. (2020). Microstructured flexible capacitive sensor with high sensitivity based on carbon fiber-filled conductive silicon rubber. *Sensors and Actuators A: Physical* 312: 112147.

43 Xu, H., Lv, Y., Qiu, D. et al. (2018). An ultra-stretchable, highly sensitive and biocompatible capacitive strain sensor from an ionic nanocomposite for on-skin monitoring. *Nanoscale* 11: 1570–1578.

44 Liu, Z., Pan, Y., Wu, P. et al. (2019). A novel capacitive pressure sensor based on non-coplanar comb electrodes. *Sensors and Actuators A: Physical* 297: 111525.

45 Nguyen, D.-S., Pillatsch, P., Zhu, Y. et al. (2015). MEMS-based capacitive pressure sensors with pre-stressed sensing diaphragms. *2015 IEEE Sensors*, Busan, Korea (South) (1–4 November 2015), pp. 1–4. IEEE.

46 Khan, S.M., Mishra, R.B., Qaiser, N. et al. (2020). Diaphragm shape effect on the performance of foil-based capacitive pressure sensors. *AIP Advances* 10: 015009.

47 Berger, C., Phillips, R., Pasternak, I. et al. (2018). Touch-mode capacitive pressure sensor with graphene-polymer heterostructure membrane. *2D Materials* 5: 015025.

48 Kumar, B., Rajita, G., and Mandal, N. (2014). A review on capacitive-type sensor for measurement of height of liquid level. *Measurement and Control* 47: 219–224.

49 Chetpattananondh, K., Tapoanoi, T., Phukpattaranont, P., and Jindapetch, N. (2014). A self-calibration water level measurement using an interdigital capacitive sensor. *Sensors and Actuators A: Physical* 209: 175–182.

50 Beloglazova, N.V., Lenain, P., De Rycke, E. et al. (2018). Capacitive sensor for detection of benzo(a)pyrene in water. *Talanta* 190: 219–225.

51 Shi, H., Zhang, H., Ma, L., and Zeng, L. (2019). A multi-function sensor for online detection of contaminants in hydraulic oil. *Tribology International* 138: 196–203.

52 Fang, X., Zong, B., and Mao, S. (2018). Metal–organic framework-based sensors for environmental contaminant sensing. *Nano-Micro Letters* 10: 64.

53 Dhanjai, N.Y. and Mugo, S.M. (2019). A flexible-imprinted capacitive sensor for rapid detection of adrenaline. *Talanta* 204: 602–606.

54 Altintas, Z., Kallempudi, S.S., and Gurbuz, Y. (2014). Gold nanoparticle modified capacitive sensor platform for multiple marker detection. *Talanta* 118: 270–276.

55 Howard, J.E., Lieder, O.H., Bowlds, B.B., and Lindsay, P.A. (2004). Capacitive sensor and method for non-contacting gap and dielectric medium measurement. Google Patents, US Patent office, US7256588B2, Baker Hughes Inc.

56 Rogers, C.R. (1998). Reservoir volume sensors. Google Patents, US Patent office, US6542350B1, Medtronic Inc.

57 Tripathy, A., Pramanik, S., Manna, A. et al. (2016). Design and development for capacitive humidity sensor applications of lead-free Ca, Mg, Fe, Ti-oxides-based electro-ceramics with improved sensing properties via physisorption. *Sensors* 16: 1135.

58 Yang, H., Ye, Q., Zeng, R. et al. (2017). Stable and fast-response capacitive humidity sensors based on a ZnO nanopowder/PVP-RGO multilayer. *Sensors* 17: 2415.

59 Zhao, Y., Yang, B., and Liu, J. (2018). Effect of interdigital electrode gap on the performance of SnO_2-modified MoS_2 capacitive humidity sensor. *Sensors and Actuators B: Chemical* 271: 256–263.

60 Sapsanis, C., Omran, H., Chernikova, V. et al. (2015). Insights on capacitive interdigitated electrodes coated with MOF thin films: humidity and VOCs sensing as a case study. *Sensors* 15: 18153–18166.

61 Najeeb, M.A., Ahmad, Z., and Shakoor, R.A. (2018). Organic thin-film capacitive and resistive humidity sensors: a focus review. *Advanced Materials Interfaces* 5: 1800969.

62 Wagner, T., Krotzky, S.R., Weib, A. et al. (2011). A high temperature capacitive humidity sensor based on mesoporous silica. *Sensors* 11: 3135–3144.

63 Zargar, Z.H. and Islam, T. (2020). A thin film porous alumina-based cross-capacitive humidity sensor. *IEEE Transactions on Instrumentation and Measurement* 69: 2269–2276.

64 Almasi Kashi, M., Ramazani, A., Abbasian, H., and Khayyatian, A. (2012). Capacitive humidity sensors based on large diameter porous alumina prepared by high current anodization. *Sensors and Actuators A: Physical* 174: 69–74.

65 Erickson, Z., Collier, M., Kapusta, A., and Kemp, C.C. (2018). Tracking human pose during robot-assisted dressing using single-axis capacitive proximity sensing. *IEEE Robotics and Automation Letters* 3: 2245–2252.

66 Ding, Y., Zhang, H., and Thomas, U. (2018). Capacitive proximity sensor skin for contactless material detection. *Presented at 2018 IEEE/RSJ International Conference on Intelligent Robots and Systems (IROS)*, Madrid Spain (1–5 October 2018). IEEE.

67 Eren, H. and Sandor, L.D. (2005). Fringe-effect capacitive proximity sensors for tamper proof enclosures. *Presented at 2005 Sensors for Industry Conference*, Houston, TX, USA (8–10 February. 2005). IEEE.

68 Rogers, G.W. (1979). Capacitive article removal alarm. Google Patents, US Patent office, US4293852A, Lawrence Security Services Ltd.

69 Stetco, C., Mühlbacher-Karrer, S., Lucchi, M. et al. (2020). Gesture-based contactless control of mobile manipulators using capacitive sensing. *Presented at 2020 IEEE International Instrumentation and Measurement Technology Conference (I2MTC)*, Dubrovnik, Croatia (25–28 May 2020). IEEE

70 Laput, G., Chen, X.A., and Harrison, C. (2016). SweepSense: ad hoc configuration sensing using reflected swept-frequency ultrasonics. *Presented at Proceedings of the 21st International Conference on Intelligent User Interfaces*, Sonoma, California, USA (7–10 March 2016). New York, NY: Association for Computing Machinery.

71 Neely, J.S. and Restle, P.J. (1995). Capacitive bend sensor. Google Patents, US Patent office, US5610528A, International Business Machines Corp.

72 Yagati, A.K., Behrent, A., Beck, S. et al. (2020). Laser-induced graphene interdigitated electrodes for label-free or nanolabel-enhanced highly sensitive capacitive aptamer-based biosensors. *Biosensors and Bioelectronics* 164: 112272.

73 Stojanovska, E., Calisir, M.D., Ozturk, N.D. et al. (2019). 3 – Carbon-based foams: preparation and applications. In: *Nanocarbon and Its Composites*, 43–90. Woodhead Publishing.

74 Yang, X., Wang, Y., Sun, H., and Qing, X. (2019). A flexible ionic liquid-polyurethane sponge capacitive pressure sensor. *Sensors and Actuators A: Physical* 285: 67–72.

75 Kou, H., Zhang, L., Tan, Q. et al. (2018). Wireless flexible pressure sensor based on micro-patterned graphene/PDMS composite. *Sensors and Actuators A: Physical* 277: 150–156.

76 C. Zhan, J. Neal, J. Wu, and D.-e. Jiang, "Quantum effects on the capacitance of graphene-based electrodes," *The Journal of Physical Chemistry C*, vol. 119, pp. 22297–22303, 2015.

77 Kang, M., Kim, J., Jang, B. et al. (2017). Graphene-based three-dimensional capacitive touch sensor for wearable electronics. *ACS Nano* 11: 7950–7957.

78 Yang, J., Luo, S., Zhou, X. et al. (2019). Flexible, tunable, and ultrasensitive capacitive pressure sensor with microconformal graphene electrodes. *ACS Applied Materials & Interfaces* 11: 14997–15006.

79 Pyo, S., Choi, J., and Kim, J. (2018). Flexible, transparent, sensitive, and crosstalk-free capacitive tactile sensor array based on graphene electrodes and air dielectric. *Advanced Electronic Materials* 4: 1700427.

80 Eller, H. and Denoth, A. (1996). A capacitive soil moisture sensor. *Journal of Hydrology* 185: 137–146.

6

Chemical Sensors

6.1 Introduction

6.1.1 Overview

The chemical sensors are devices that detect and measure the presence of specific chemicals or analytes selectively in a given chemical environment and provide quantitative or qualitative information about the concentration, identity, or presence of the target chemical/analyte through the variation in electrical, optical, colorimetric, acoustic, and other measurable responses. In solid-state chemical sensors, the chemical interaction with the target chemical analyte takes place in the bulk or at the surface of the active sensing material which interact selectively with the target analyte, and reversibly transforms chemical responses into electrical signals depending on its concentration, composition, and chemical properties, producing observable outcomes. These active sensor materials include polymers, metal oxides, composites, nanomaterials, biomaterials, biopolymers, and ceramics, and have strong affinity for corresponding target chemical/analytes. These sensors can detect a wide range of target chemicals, including gases, liquids, and solids, when they chemically come in contact with the sensor material. The chemical interaction between the active sensing material and the target analyte takes place through various mechanisms, such as adsorption, covalent and noncovalent molecular binding, electronic band bending, variation in bandgap, and charge transfer, leading to change in material property. Under exposure to the target analyte, when the active sensing material is electrically connected to a measuring device, the variations in the electrical property of the sensing material are captured. Thus, depending on the observable change in electrical response (in terms of measurable parameters like resistance, capacitance, voltage, and frequency), the concentration of the target analyte is measured through predetermined correlation between them. Thus, the chemical sensor serves as a transducer that converts

Solid-State Sensors, First Edition. Ambarish Paul, Mitradip Bhattacharjee, and Ravinder Dahiya.
© 2024 The Institute of Electrical and Electronics Engineers, Inc.
Published 2024 by John Wiley & Sons, Inc.

variations in chemical information into measurable electrical output signals which are processed by the analyzer to generate output sensory signals. The sensor output signals are often connected to sampling and signal processing units which are capable of reducing electronic noise and eliminating spurious data to provide accurate electronic sensory feedback. Solid-state chemical sensors have revolutionized the field of chemical sensing, offering a more reliable and accurate way of detecting and measuring chemical substances in real time. Chemical sensors are used in a variety of applications, including environmental monitoring, medical diagnostics, food safety, and industrial process control. They offer a highly sensitive and selective way of detecting and measuring chemicals, and can provide real-time monitoring of chemical processes.

There are several types of chemical sensors, each with its own unique sensing mechanism and application. Following are some common types of solid-state chemical sensors:

1) Gas sensors are used to detect and measure the concentration of specific gases in the environment. These sensors can detect a wide range of gases, including carbon monoxide, nitrogen dioxide, and methane. Biosensors can detect wide range of biological moieties, such as enzymes, antibodies, and antigens, and also detect and measure the presence of specific biomolecules like glucose and dopamine.

2) Optical sensors use laser or light source to detect and measure the presence of specific chemicals. These sensors work by measuring changes in the absorption, transmission, or reflection of light that occurs when a chemical interacts with the sensor material.

3) Mass sensors detect and measure the change in mass of the host molecule after chemically binding with the target guest molecule. The change in frequency of a vibrating sensor or resonator is measured when it interacts with the target chemical.

4) Thermal sensors detect and measure the presence of specific chemicals by measuring the change in temperature that occurs when the chemical reacts with the sensor material.

Each type of chemical sensor has its own strengths and weaknesses, and is best suited for specific applications. By understanding the different types of chemical sensors available, researchers and engineers can choose the right sensor for their application and design sensors that are more sensitive, selective, and reliable.

The solid-state chemical gas sensors are widely used worldwide due to their miniaturization, portability, cost-effectiveness, reliability, and possessing high sensitivity, selectivity, and stability. The conductive polymer sensors, ceramic sensors, and surface acoustic wave (SAW) sensors are widely used for gas sensing. Solid-state chemical gas sensors typically rely on one of several sensing mechanisms to detect and measure the presence of gases in the environment. Common gas sensing mechanisms include:

1) In catalytic combustion mechanism, the target gas is oxidized on a heated catalytic surface, which changes the resistance or voltage output.
2) Molecular absorption mechanism takes place in metal oxide semiconductors (MOS), where the target gas molecules adsorb onto the surface of the metal oxide film and change its electrical conductivity of the MOS film. Again, in SAW sensors, when gas molecules are adsorbed onto the surface of the piezo-electric material, they change the velocity of the acoustic wave which can be detected as a change in the electrical signal.
3) In ionization, a high-energy UV or electrical source ionizes gas molecules, creating ions that can be detected by a collector electrode. The amount of ionization is proportional to the gas concentration.

Chemical biosensors integrate biological recognition elements with transducers to detect and quantify a specific analyte. Chemical biosensors use biological elements (host analytes), such as enzymes, antibodies, functionalized nanomaterial, cells, biomaterial, cell aptamers, or a combination of each of them in combination with a transducer to detect and measure the presence of specific guest analytes, such as glucose, hormones, or bacteria. The host element in a biosensor recognizes the target guest analyte to generate chemical response through a specific and selective chemical reaction. The host sensing material is bound to the transducer by covalent or noncovalent chemical bonds and produces electrical signals in response to external chemical stimuli that is proportional to the amount of guest analyte present. This signal is then converted by the transducer into a measurable output, such as an electrical or optical signal. The chemical sensor for a guest analyte detection is designed using suitable host–guest chemistry, which is fast and preferentially operates at room temperature and pressure ambience. The sensor designed for repetitive and long-term operation requires appropriate host–guest chemistry that is temporary and lasts only during the time of exposure to external stimuli. However, for the design of disposable chemical sensor, suitable host–guest chemistry with fast, stable, permanent chemical response is desired. The transducers like the chemiresistors, field-effect transistors (FETs), and the pH electrodes convert chemical energy to electrical signals which after signal conditioning is displayed as change in voltage and capacitance in an analog circuit and triggers a detection circuit in case of digital output signal. Various biosensors include enzymatic, immunological, microbial, DNA, and aptamer-based biosensors. The scheme of chemical biosensor working is portrayed in Figure 6.1a. Chemical biosensors have many applications in medicine, environmental monitoring, food safety, and biotechnology. They offer a highly sensitive and selective way of

(a)

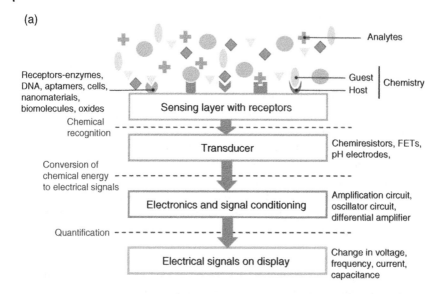

Analytes

Receptors-enzymes,
DNA, aptamers, cells,
nanomaterials,
biomolecules, oxides

Guest
Host

Chemistry

Sensing layer with receptors

Chemical
recognition

Transducer

Chemiresistors, FETs,
pH electrodes,

Conversion of
chemical energy
to electrical signals

Electronics and signal conditioning

Amplification circuit,
oscillator circuit,
differential amplifier

Quantification

Electrical signals on display

Change in voltage,
frequency, current,
capacitance

(b)

Figure 6.1 (a) Schematic representation of working biosensor. (b) Artist impression of functionalized nanowire-based chemical sensor exposed to various chemical analytes.

detecting and measuring analytes, and can provide real-time monitoring of biological processes. Figure 6.1b illustrates selective sensing of an analyte a using a functionalised nanowire, under exposure to various chemical analytes.

6.1.2 Global Limelight

The competency of chemical sensing technology has grown steadily over the years and depends on the emergence of new materials and measurement techniques, and driven by the rapid rise in various industries which require chemical sensors to monitor various industrial processes. Solid-state chemical sensors are portable, reliable, and most suitable for long-term use especially in continuous monitoring of chemical environment. Thus, solid-state chemical sensors have gained significant limelight due to its possible applications in environmental monitoring, industrial process monitoring, national defense and security, healthcare industry, public security, gas detection, and onsite emergency disposal due to its low cost of fabrication, small size, portability, large dynamic range, high sensitivity, and continuous detection. Hence, solid-state chemical sensors continue to the effective modern-day technology which rule over the global market in terms of demand and usage. Transparency Market Research states that the global chemical sensors market was valued at US$18.87 billion in 2018 and is anticipated to reach US$29.27 billion by 2027. Analysts predict that the global market will expand at a CAGR of 5.0% from 2019 to 2027.

6.1.3 Evolution of Chemical Sensors

The need for chemical sensors was felt when workers working in underground mines, tunnels, shafts, and other environments with confined spaces face health hazards from flammable oxygen-deficient atmosphere; and atmosphere with a high concentration of contaminants in poorly ventilated spaces. Before the advent of modern electronic sensors, early detection methods relied on mice and birds where coal miners would bring canaries down to the tunnels with them as an early detection system against life-threatening gases such as CO_2, CO, and CH_4. These odorless and colorless gases are equally deadly to both humans and canaries. However, since canaries are much more susceptible to the gas, and reacts more quickly and visibly than humans do, the caged canaries would stop chirping, thus alerting miners to the presence of the poisonous gas (Figure 6.2a). This technique to detect hazardous gases was first proposed by John Haldane (Figure 6.2b), a Scottish physiologist who is well-known for his discoveries of the physiology of breathing and the effect of different gases on humans. The first gas detector commonly known as the flame safety lamp (Figure 6.2c) was invented by Sir Humphry Davy (Figure 6.2d) in 1815 to detect the presence of CH_4 in underground mines. The lamp consisted of an oil flame which burnt with a specific height in fresh air. The flames height varied depending on the presence of methane (higher) or the

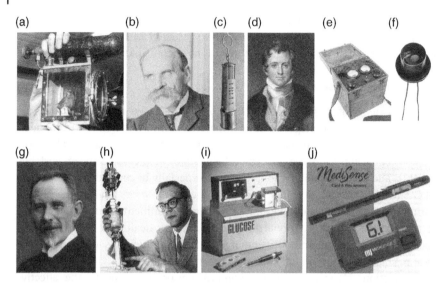

Figure 6.2 (a) Caged canaries as primitive gas sensors. (b) Portrait of Scottish physiologist John Haldane. (c) First gas detector known as the flame safety lamp invented by (d) Sir Humphry Davy. (e) Catalytic lower explosive limit (LEL) combustible gas sensor developed by Dr. Oliver W. Johnson. (f) Catalytic pellistors. (g) Søren Peder Lauritz Sørensen who introduced concept of hydrogen ion concentration and pH. (h) Leland Clark with his invention of a method of defoaming in a bubble oxygenator (1952). (i) Clark-type glucose biosensor, a commercial product of Yellow Springs Instruments (YSI). (j) First electrochemical blood glucose monitor, launched in 1987 by ExacTech by MediSense, Inc. *Sources:* (a) Fionn Taylor, (b) Wikimedia Commons, (c) Julo/Wikimedia Commons, (d) Unknown author/Wikimedia Commons, (e) H. Davey, (g) Julie Laurberg/Wikimedia Commons/Public domain, (h) Adapted from Severinghaus [1]/with permission of American Society of Anesthesiologists, (i) YSI, Inc., (j) Tom Robbins/Pinfold Technologies Ltd.

lack of oxygen (lower). If flammable gases were present, the flame burned very high with a blue tint. In 1926, Dr. Oliver W. Johnson created the first catalytic lower explosive limit (LEL) combustible gas sensor (Figure 6.2e) to be used in the combustible gas indicator. The device was labeled Model A and was developed to detect combustible mixtures in air to help prevent explosions in fuel storage tanks. The first practical "electric vapor indicator" meter labeled as Model B was developed in 1927. The world's first gas detection company, Johnson–Williams Instruments (or J-W Instruments) was formed in 1928 by Dr Oliver Johnson and Phil Williams. J-W Instruments is recognized as the first electronics company in Silicon Valley and later came out with Model C. The mid-1960s introduced gas sensors in the form of catalytic pellistors (Figure 6.2f). This sensor is made of two platinum wires positioned within a ceramic mass. Pellistors are useful for sensing a wide range of flammable gases and toxic vapors. MOS sensors were introduced in the 1992 by Sberveglieri et al. [2] where they demonstrated NO, NO_2, and H_2 sensors using SnO_2 thin film. Although earlier gas detectors could only detect a single gas, modern technology can detect several gases at a given time, or even a combination.

The foundation for the development of biosensors was laid by M. Cremer in 1906 when he demonstrated that the concentration of an acid in a liquid is proportional to the electric potential between parts of the fluid located on opposite sides of a glass membrane. But it was only in 1909 that the concept of hydrogen ion concentration or pH was introduced to the world by Søren Peder Lauritz Sørensen (Figure 6.2g). This was experimentally demonstrated in 1922 by W.S. Hughes. In 1952, American scientist L.C. Clark (Figure 6.2h) invented the first biosensor for cardiac surgery in animals. In 1954, he conceived a Pt electrode with a reference electrode inside a polyethylene (PE) membrane to determine blood oxygen level and was later known as the Clark electrode (Figure 6.2i). The Clark electrode emerged as the pivotal device that allowed real-time monitoring of patients' blood oxygen level and have made surgery safer and more successful for millions around the world. He is regarded as the "Father of Biosensors." The demonstration of an amperometric enzyme electrode for the detection of glucose by Clark in 1962 was followed by the discovery of the first potentiometric biosensor by Guilbault and Montalvo, Jr. in 1969 to detect urea by immobilizing urease on ammonia electrodes. The first commercially successful glucose biosensor using Clark's technology was developed by the Yellow Springs Instrument Company in 1975 as Model 23A YSI glucose analyzer for the direct measurement of glucose, which was based on the amperometric detection of hydrogen peroxide. This analyzer was of high cost due to the expensive platinum electrode and thus lacked extensive usage. In 1985, Dr. I. Lauks developed the I-Stat point-of-care blood analysis system. It comprises single-use cartridge containing an array of electrochemical sensors capable of measuring blood oxygen, CO, and pH. The device measures electrolytes such as potassium, sodium, chloride, calcium ions, glucose, urea, nitrogen, and hematocrit in few minutes with only a few drops of whole blood. The first electrochemical blood glucose monitor was launched in 1987 by ExacTech by Medisense, Inc. for self-monitoring of diabetic patients (Figure 6.2j). This pen-sized device used glucose dehydrogenase pyrroloquinolinequinone (GDH-PQQ) and a ferrocene derivative. The commercial success led to a revolution in the health care of patients with diabetes.

The reports of semiconductor materials exhibiting electrical conductivities in response to ambient gases and vapors were published earlier around 1965. In 1985, Wohltjen and Snow [3] demonstrated using copper phthalocyanine that its resistivity decreased in the presence of ammonia vapor at room temperature. They coined the term "chemiresistors" for the device. Although MOSFET was invented in 1959 by Egyptian engineer M.M. Atalla and Korean Engineer D. Kahng, it was in 1985 when Dutch engineer P. Bergveld realized that MOSFET can be adapted as sensing devices for chemical and biological applications [4]. Bergveld invented the chemical FET (ChemFET), which is a special type of ion-sensitive FET (ISFET), and used for the detections of ions and chemical analytes. Soon after, different variants of the ChemFET in the form of carbon nanotube FET [5], nanowire (NW) FET [6], nanoribbon FET [6], and the graphene FET [7] were chronologically introduced. With the advancements in microelectronics, ChemFET array was also developed in 2002 [8].

6.1.4 Requirements for Chemical Sensors

A chemical sensor operates by detecting and rapidly responding to the presence of analyte of a given concentration at the interface between the sensor and the medium. A prerequisite of any good chemical sensor lies in its selectivity, stability, and sensitivity in addition to the parameters like the response time and the limit of detection that brings in added advantages to the sensor. The parameters that play an important role in the operations of the chemical sensors are given as the following:

6.1.4.1 Selectivity
The selectivity of a chemical sensor is the most important qualification parameter that determines the success of the device. A chemical sensor with high selectivity effectively discriminates the detection of the target analyte from other interfering analytes in the medium. High selectivity in chemical sensors can be achieved by selecting a unique molecular recognition chemical reaction between the receptor (host) and the target analyte (guest) through which qualitative and quantitative discrimination can be achieved.

6.1.4.2 Stability
The chemical kinematics between the receptor and the target analyte as well as the consistency in electronic output signal over time determines the stability of the chemical sensor. A sensor with high stability provides low repeatability error, high signal-to-noise ratio, and high resolution. High stability may be achieved by selecting the suitable receptor for a given target analyte, fabrication considerations, and using a signal conditioning electronic circuit. Since the limit of detection of a given sensor depends on the signal-to-noise ratio of the output signal, fluctuations in output signals may affect the lowest concentration of analyte concentration that can be detected.

6.1.4.3 Sensitivity

The sensitivity of the sensor determines the effectiveness through which it can detect (produce measurable change in output signal) any variation in concentration of the analyte in the medium and forms the most important parameter in sensor characterization. It is desirable for a sensor to possess a linear calibration curve; the sensitivity is constant. For a nonlinear calibration curve, the linear portion of the curve may be selected as the dynamic range of the sensor.

6.1.4.4 Response Time

This property determines the time it takes for the biosensor to generate a signal or response following the interaction of the biological receptor with the target analyte. The response time is dependent on the chemical kinematics and stoichiometry of the receptor and the analyte. The sensor is said to be fast when the response time of the sensor is observed to be low. The response time of a good sensor must not be more than a few seconds.

6.1.4.5 Limit of Detection

A sensor with low limit of detection is desirable for trace detection of poisonous gases, and analytes such as heavy metal ions in ground water, and certain biomolecules such as the troponin II in blood for early detection of heart attack. The limit of detection can be drastically lowered by the use to nanomaterials due to its high surface area and increasing the concentration of the receptor molecule in the matrix. Increasing the signal-to-noise ratio of the output signal also lowers the limit of detection.

The ideal sensor of this kind is portable and responds selectively and instantaneously to the guest analyte upon exposure. However, in spite of rapid advances in fabrication techniques and with the advent of new materials, such ideal sensors are far from reality. These sensors not only suffer from chemical interference and delay in electrical response, but also usually possess complex device architecture and invite optimization processes for successful and effective operation.

6.2 Materials for Chemical Sensing

6.2.1 Metal Oxides

The metal elements are able to form a large diversity of oxide compounds which can form a vast number of structural geometries with varied electronic structures that exhibit metallic, semiconductor, or insulator behavior. In technological applications, oxides are used in the fabrication of microelectronic circuits, sensors, piezoelectric devices, fuel cells, coatings for the passivation of surfaces against corrosion, and as catalysts. Studies on various metal oxides indicated that the response of the metal oxide on exposure to different gases depends on temperature, humidity, chemical components, surface modification, and microstructures of the metal oxide layers. The metal oxide that are widely used for gas sensing applications are Cr_2O_3, Mn_2O_3, Co_3O_4, NiO, CuO, SrO, In_2O_3, WO_3, TiO_2, V_2O_3, Fe_2O_3, GeO_2, Nb_2O_5, MoO_3, Ta_2O_5, La_2O_3, CeO_2, and Nd_2O_3. The use of metal oxide for sensing applications is attributed to its electronic properties. The conductometric semiconducting metal oxide-based sensors have attracted much attention due to their low cost and flexibility in production, simplicity of their use, and large number of detectable fluids application. In addition, the detection can be performed by numerous operating principles including change of capacitance, work function, mass, and optical characteristics.

6.2.1.1 Types of Metal Oxides

Metal oxides selected for chemical sensors can be determined from their electronic structure. The metal oxides have been divided into two of the following categories depending on its electronic structure:

1) Transition metal oxides (Fe_2O_3, NiO, Cr_2O_3, etc.)
2) Nontransition metal oxides, which include:
   a) pretransition metal oxides (Al_2O_3, etc.)
   b) posttransition metal oxides (ZnO, SnO_2, etc.)

In the transitional metal oxides, the valence band is the oxygen 2p band and the conduction band has a metallic character and attributed to the d orbitals which is the lowest energy unoccupied orbitals of the conduction band. The d bands can be spit by ligand field effect and are partially occupied by d electrons. Since the energy difference between cationic d_n configuration and either d_{n+1} or d_{n-1} configurations are small, the transitional metal oxide are more responsive to chemical environment relative to the pretransitional metal oxides as described next. However, transitional metal oxides with d_0 and d_{10} electronic configurations are widely used as conductometric sensor applications. The d_0 configuration is found in TiO_2, V_2O_5, and WO_3, whereas the d_{10} electronic configuration is found in posttransitional metal oxides such as ZnO and SnO_2.

In pretransitional metal oxide, each ion has a closed shell electronic configuration. The highest energy filled level of the material is formed from the highest energy occupied orbital of the oxide anion, whereas the lowest energy empty orbitals level of the conduction band is formed from the lowest energy empty orbitals of the metal cation. As a result of this, all bands are either completely full or empty. Thus, pretransition metal oxides (MgO, etc.) are expected to be quite inert, because they have large bandgaps which render them unsuitable for sensor applications. Neither electrons nor holes can easily be formed. They are not preferred as chemical sensor materials due to their difficulties in electrical conductivity measurements.

Post-transition metals like the Tl(I), Pb(II), and Bi(III) are dominated by the group oxidation state N and a lower N-2 oxidation state, where the metal cations are preferentially associated with the formation of noncentrosymmetric coordination environments providing unique physicochemical properties in oxides containing lone pairs. These oxides are used as photocatalysts ($BiVO_4$), ferroelectrics ($PbTiO_3$), multiferroics ($BiFeO_3$), and p-type semiconductors (SnO).

6.2.1.2 Chemical Sensing Mechanism

Both n-type and p-type semiconducting metal oxides are widely used in various chemical-sensing applications especially in gas and volatile organic compounds (VOC) sensing. The p-type metal oxide (p-type MOX) semiconductor includes CuO, NiO, Co_3O_4, and Cr_2O_3 [9], while that of In_2O_3, SnO_2, ZnO, and ITO forms n-type metal oxide [10]. The p-type MOX thin films exhibit several advantages over n-type MOX, including a higher catalytic effect, low humidity dependence, and improved recovery speed, whereas n-type semiconductor offers low temperature sensing. The gases that are generally sensed using metal oxide include both reducing gases, such as volatile organic compounds (VOCs), H_2, and NH_3, and oxidizing gases, such as CO_2, NO_2, and O_3. Here, we describe the mechanism of chemical sensing using n-type metal oxide semiconducting materials where depletion regions are smaller than grain size.

Prior to the exposure to a target gas, the O_2 molecules of the atmosphere are adsorbed on the surface of metal oxides. They would extract electrons from the conduction band Ec of the metal oxide and trap the electrons at the surface in the form of O^- moiety. This will lead to a band bending and an electron-depleted region. When the metal oxide was exposed to the target gas/analyte, chemical reaction occurs between the O^- moiety and the reducing gases like the CO, where the gas is itself oxidized but releases the electrons at the surface of the material which produces reverse the band bending, resulting in an increased conductivity. O^- is believed to be dominant at the operating temperature of 300–450°C, which is the working temperature for most metal oxide gas sensors. Figure 6.3a–c schematically depicts the structural and band model of conductive mechanism upon exposure to a target gas. When gas sensors exposure to the reference gas with CO, CO is oxidized by O^- and released electrons to the bulk materials. Together with the decrease of the number of surface O^-, the thickness of space-charge layer decreases. Then, the Schottky barrier between two grains is lowered and it would be easy for electrons to conduct in sensing layers through different grains as illustrated in Figure 6.3c.

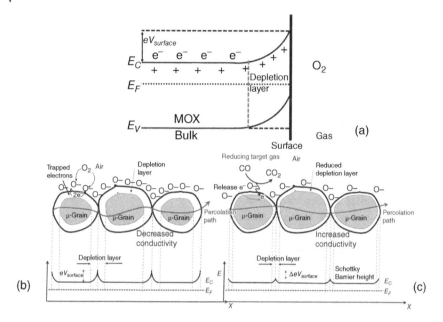

Figure 6.3 (a) The band model of conductive mechanism upon exposure to a target gas. Schematic representation showing the (b) adsorption of O_2 at the surface of metal oxide and (c) reaction between the metal oxide active sensor material and the target gas.

6.2.1.3 Metal Oxide Nanoparticles and Films as Sensor Materials

Owing to the quantum confinement in nanoparticles (NPs), they exhibit unique physical and chemical properties which are different from their bulk counterparts. The limited size and a high density of corner or edge surface sites of the metal oxide NPs influence three important groups of basic properties in any material (i) morphology, (ii) electronic behavior, and (iii) chemical reactivity. The first one is associated with the variation in structural characteristics, namely the lattice symmetry and cell parameters. As the particle size decreases, the increasing number of atoms at the surface and at the interface generates stress/strain in the crystal lattice and thus produces structural perturbations. Again, the NPs interacting with the substrate may induce structural perturbations or phases not observable in bulk state of the oxide.

The quantum confinement which arises from the presence of discrete, atom-like electronic state, produces variation in electronic properties in oxide NPs. The electronic effects of quantum confinement are related to the energy shift of exciton levels and the bandgap of the oxides. The structural and the electronic properties cumulatively affect the physical and chemical properties of the oxide NPs. In their bulk state, many oxides have wide bandgaps and a low reactivity. A decrease in the average size of an oxide NP widens the bandgap further and has a strong influence in the conductivity and chemical reactivity. The metal oxide NP can adopt varied morphologies based on the humidity, temperature, gas flow, and other ambient parameters during synthesis. Different morphologies develop unique edges, corners, ridges, and facets on the surface of the NP as a result of which the average surface area increases. This increase in the effective surface area increases the chemical reactivity of the oxide NP over bulk material, thus making the sensors more sensitive. In addition, the dangling bonds at the edges and facets of the NP acts as sites for chemical activity, thereby improving the sensitivity and the response time even more.

The two-dimensional (2D) oxide films have structural modifications in terms of rearrangement or reconstruction of geometries, and electronic modifications in the form of the presence of mid-gap states. In the case of nanostructured oxides, surface properties are strongly modified with respect to 2D infinite surfaces, producing solids with unprecedented sorption or acid/base characteristics. Furthermore, the presence of under coordinated atoms (like corners or edges) or O vacancies in an oxide NP should produce specific geometrical arrangements as well as occupied electronic states located above the valence band of the corresponding bulk material enhancing in this way the chemical activity of the system.

The key features responsible for the application of metal oxides as absorbents or catalysts are (i) the coordination environment of surface atoms, (ii) the redox properties, and (iii) the oxidation state at surface layers. Oxides having only s or p electrons in their valence orbitals tend to be more effective for acid/base catalysis, while those having d or f outer electrons find a wider range of uses.

6.2.2 Honeycomb Structured Materials

Honeycomb structures are found in beehives (Figure 6.4a) and in other naturally occurring structures. These structures are so well adopted and deployed by nature

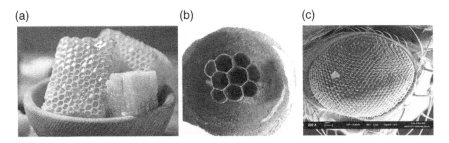

Figure 6.4 Naturally occurring hexagonal honeycomb structure in (a) beehives, (b) wasp nest, and (c) SEM image of the *Drosophila* compound eye. (a) *Source:* Africa Studio/Adobe Stock.

that they are often formed under numerous circumstances such as in honeycomb films of amphiphilic polymers formed at water–air interfaces, nest of wasp (Figure 6.4b), basalt rocks at the Gaint's causeway, Northern Ireland formed due to volcanic activity, compound eye of an insect (Figure 6.4c), and so on. In 1999, Professor Thomas C. Hales provided mathematical proof for the advantage of what he termed "honeycomb conjecture." He demonstrated that regular hexagons are the best way to divide a space into equal parts with minimal structural support such that they enclose more space using the same amount of wall material. The space efficiency is not the only benefit of the honeycomb structures. The hexagons fill the space in an offset arrangement with six short walls around each void giving structures a high compression strength. It also allows to dissipate heat well. Thus, efficiency, strength, and controlled heat loss pathway of the honeycomb structures have made them unique. The honeycomb structure is the basic building block of various materials such as: (i) stacked honeycomb structure assembles into three-dimensional (3D) graphite, (ii) 2D structure form graphene, boron nitride, WS2 monolayer, (iii) rolled honeycomb structure gives rise to one-dimensional carbon nanotubes (CNTs), and (iv) wrapped honeycomb structure produces zero-dimensional fullerenes. Graphene CNT and boron nitride are widely used in sensing application and are discussed next.

6.2.2.1 Graphene

Graphene has become one of the hottest topics in the fields of materials science, physics, chemistry, and nanotechnology [11]. This allotrope of carbon comprises periodic arrangements of six-atom rings to form a honeycomb network and can be visualized as aromatic planar sheet. During the past several years, various methods for producing graphene have been developed, such as micromechanical exfoliation, chemical vapor deposition, epitaxial growth, and chemically synthesis. Each carbon atom in the graphene honeycomb lattice is connected to its three nearest neighbors by strong in-plane covalent bonds. Each C atom in graphene forms a σ bond with three neighboring carbon atoms, having an average interlayer distance of 1.42 Å (Figure 6.5a). The C atom in graphene undergoes sp^2 hybridization between one 2s and two 2p orbitals resulting in sp^2 hybridized orbital. Graphene stability is attributed to its tightly packed C atoms and sp^2 orbital hybridization. The σ bonds in all carbon allotropes are responsible for the mechanical strength of that material. Graphene material has remarkable properties such as high stiffness and breaking strength. Graphene exhibits high planar surface (calculated value, $2630\,m^2/g$), Young's modulus of $E = 1.0\,TPa$, and intrinsic strength of 130 GPa, which makes graphene the strongest material ever measured.

In addition to three σ bonds connecting the neighboring three C atoms, the fourth valence electron occupies the 2pz orbital, which is aligned perpendicularly to the plane of graphene sheet and does not interact with the in-plane s electrons.

(a)

(b)

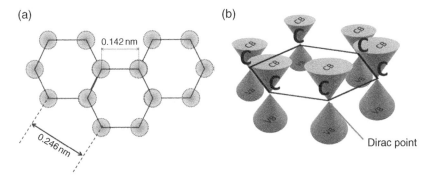

0.142 nm

0.246 nm

Dirac point

Figure 6.5 (a) Schematic representation of hexagonal lattice of graphene showing C–C bond distance of 0.142 nm and lattice constant of 0.246 nm. (b) Band structure of graphene near Fermi level.

The 2pz orbitals from neighboring atoms overlap resulting in delocalized π (occupied or valence) and $\pi*$ (unoccupied or conduction) bands. These π bonds are generally half filled and hybridize together to form π and $\pi*$ bands. The π bonds arising from the overlap of pz orbitals are responsible for most of graphene's notable electronic properties. The most intriguing features of graphene are its exceptional electronic quality and transport properties of individual graphene crystallites. Due to its 2D nature, graphene is theoretically a zero-bandgap semiconductor with excellent room temperature electrical conductivity. Moreover, the pronounced carrier mobility is well maintained at the highest electron or hole concentrations of up to 10^{13} cm^{-2} in ambipolar field-effect devices which suggest a ballistic transport on graphene crystal lattice.

High electron mobility in graphene is the most interesting property and has lured many researchers to explore its application in various devices. Graphene offers conductivity of 10^6 S/m and a sheet resistance of 31 Ω/sq and is attributed to its ultrahigh mobility of 2×10^5 cm^2/V s which is almost 140 times the mobility in silicon. This is because the π electrons in graphene lattice are free to move within the lattice, thereby exhibiting high electrical conductivity even at room temperature. Beside this, graphene is a typical semimetal where there is a small overlap between its conduction band and valence band. Thus, the electrons at the top of the valence band could flow into the bottom of the conduction band with lower energy. Even at the temperature of absolute zero, a certain concentration of electrons is already in the conduction band, while a certain concentration of holes is in the valence band. The valence band and conduction band of graphene exhibit cone-like structures which intercept at the Dirac point as shown in Figure 6.5b. However, the high electron mobility could not be utilized due to the lack of bandgap in the materials. The reactivity of graphene is very low as the extensive delocalization of π-bonding systems due to the strong interaction of all pz orbitals makes the 2D structure more inert. Although the strong hydrophobic nature of the 2D material is suitable for electronic applications, it is not appropriate for biosensor applications which require water medium for interaction. However, in graphene nanoribbons, the valence and conduction band are separated by bandgap, and thus, exhibits semiconductive behavior. Graphene has been well established as a promising material for high-performance next-generation FETs and many other advanced electronic devices.

Graphene is preferred over other allotropes of carbon for chemical sensor application because of the following reasons: (i) 2D structure with one-atom thickness allowing complete exposure to all the C atoms facilitating enhanced adsorption of gas molecules and (ii) high signal-to-noise ratio due to its high crystal quality and low resistance (typically few hundred ohms). Monolayered graphene has shown promise to detect gases down to single molecular level. This gas sensing mechanism of graphene is attributed to the adsorption and desorption behavior of gas molecules on graphene surface. The response of the graphene surface is based on the changes in the charge carrier concentrations due to p-type or n-type doping resulting from the adsorption of gas molecules. The use of graphene FET in gas sensing shows promise due to its integration capability in electronic devices as graphene is emerging as a potential candidate in field-effect devices with very high carrier mobility.

6.2.2.2 Carbon Nanotubes

In 1991, Sumio Iijima of NEC corporation' Fundamental Research Laboratory, Japan, was investigating the material on the walls of the arc discharge chamber under the electron microscope for fullerenes when he stumbled upon something different while he accidentally extracted some solid material that grew on the tips of carbon arc discharge electrodes under the same C_{60} formation conditions. Iijima found that the solids materials consisted of tiny tubes made up of numerous concentric "graphene" cylinders, which he in his report termed as the "helical microtubules of graphitic carbon." These cylinders usually consist of dome-shaped end-caps where the curvatures of the end-caps are brought about by the presence of pentagonal rings considered as topological defects in the otherwise hexagonal structure of the lattice. First, the pentagon gives a convex curvature to the edge of the tube which helps in closing the tube. Second, the pentagonal ring induces defects on the end-caps of the tube making the end-caps chemically active, and a chemical moiety is likely to be anchored at the defect site of the cap. The scientists were so fascinated by the unique properties of these tubules that they termed them CNTs. The uniqueness in the properties in terms of electrical, mechanical, thermal, field emission, and even acoustic behavior were so unending after 30 years of its discovery that it is still been explored for newer applications.

A typical CNT can be considered as a unique 1D hollow nanostructure resembling a quantum wire, and having a high aspect ratio (length-to-diameter ratio), where the building block consists of an all-carbon [12]. High-resolution images confirmed that these multiwalled CNTs (MWNTs) are domeless on either sides and that the spacing between adjacent layers are about 0.34 nm, close to the spacing observed between sheets of graphite. In 1996, Smalley produced high-purity SWNTs by laser vaporization of carbon impregnated with cobalt and nickel. The CNTs are uniquely identified using the chiral vectors C in terms of integers n and m corresponding to graphite vectors a_1 and a_2 as $C = n\,a_1 + m\,a_2$. The CNTs of type (n, n), $(n, 0)$, and (n, m) are called armchair, zigzag, and chiral nanotubes respectively, and depends on the rolling direction of the graphene sheet (Figure 6.6). The chiral angle Θ is defined as the angle between the vectors Ch and a_1, with $0 \leq \Theta \leq 30°$ because of the hexagonal symmetry of the honeycomb lattice. The electronic behavior of CNT depends on the rolling direction. The metallic and the semiconducting properties of the CNT are determined by the rule: $n - m = 3i$. The (n, m) CNT is metallic when i is an integer, whereas it shows semiconducting behavior when i is not an integer. Thus, in a given sample one third of the CNTs can be expected to show metallic behavior.

The unique electronic, chemical, mechanical, and optical properties is attributed to the electron confinement at the surface of CNT due to the partial rehybridization of the C atom from sp^2 to sp^3 hybridization state accounting for its curvature. The C atoms in graphene sheet have orbital on either side of

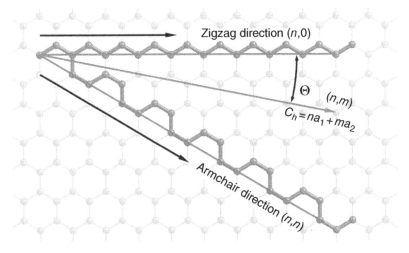

Figure 6.6 Carbon nanotube configurations with the chiral vector C and unit vectors **a** and **b**.

the plane. Because of the curvature of the CNT, the electron clouds are no longer uniformly distributed as in graphene sheet, but forms an asymmetric distribution of lobes inside and outside the cylindrical sheet of CNT. Due to the distorted electron cloud, a rich electron conjugation is formed on the outer surface of the CNT, which makes it potentially applicable for chemical sensors. The chemical reactivity of the nanotube surface also depends on its natural and strain-induced curvature. Electron-donating and electron-withdrawing chemical moiety can result in electron transaction to or from the electrons of CNT. The challenge behind the commercial use of CNT lies in the proper segregation and isolation of individual CNTs. Pristine CNTs are inert to most chemicals and need to be grafted with surface functional groups to increase their chemical reactivity and add new properties.

6.2.2.3 Other 2D Materials

Two-dimensional materials possess a great interest in sensing due to their high mobility, strong mechanical nature, and low electrical noise losses. Graphene, silicene, germanene, stanene, bismuthene, molybdenum disulfide (MoS_2), and hexagonal boron nitride (h-BN) (Figure 6.7a) are some of the 2D materials that have been used for sensing applications [13]. However, some 2D materials like graphene is not preferred due to its zero-bandgap [14] and low sensitivity, and MoS_2 (Figure 6.7b) for its low mobility [15]. In addition, thermal lattice fluctuations also result in unstable configuration of 2D materials and thus increased signal-to-noise ratio [16]. Thus, researchers are in a hunt for new kinds of 2D materials that may overcome all these shortcomings. Boron nitride exists in a variety of crystalline modifications and among them the cubic and hexagonal forms are the most common. The hexagonal form has properties similar to graphite, and the cubic form has properties similar to diamond. Hexagonal boron nitride (h-BN) (Figure 6.7a) has the

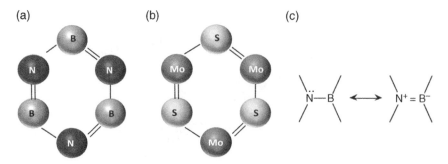

Figure 6.7 Schematic representation of (a) boron nitride (h-BN) and (b) molybdenum disulfide (MoS_2). (c) Dissimilar electronegativities of B and N, thereby providing an effective ionic character.

hexagonal Bernal structure resembling the honeycomb lattice with equal number of boron (B) and nitrogen (N) atoms, unlike graphene [17, 18]. The h-BN has AB-stacked structure, with B and N atoms above each other and vice versa. The B and the N atoms in h-BN have dissimilar electronegativities, thereby providing an effective ionic character as shown in Figure 6.7c. Since the structure of 2D boron nitride resembles monolayer graphene, it is often called the "white graphene" or borophene. Two-dimensional borophene possesses an anisotropic behavior that is different from the other 2D material graphene. This discovery of borophene has lured researchers for exploring newer avenues for more practical applications in different fields. The 2D h-BN has both armchair and zigzag edge-terminated structures, with each layer attached to another by weak van der Waals forces. The h-BN being isostructural with graphene has a high surface area, good stability, high thermal conductivity, high bandgap, fine mechanical stability, and good carrier mobility. The 2D materials with large band like the h-BN and the MoS2 (Figure 6.7b) have wide applications due to their chemical and thermal stability. Researchers demonstrated that these nitride-based materials are used for sensing applications [19]. Boron nitride is used for the sensing of CO_2 gas [20], hazardous methyl mercury [21], and dichlorosilane [22]. Higher electronic stability in borophene-based sensors facilitates large-scale commercial use.

6.2.3 Biopolymers

Biopolymers are polymers produced from natural sources. These can either be chemically synthesized from biological materials or biosynthesized by living organisms. The biopolymers may be derived from living organisms like plants and microbes or can be fossil based like the poly(alkylene dicarboxylate)s. Although the biopolymers are generally degradable, some may be nondegradable which incorporates the biobased as well the fossil-based polymers. The biopolymers find vast application in various industries ranging from food industries to manufacturing, packaging, and biomedical engineering. The demand for biopolymers arises due to its characteristics like abundance, biocompatibility, and unique properties like nontoxicity. The use of biopolymers and biomaterials in biosensors are highly preferred over other inorganic materials for surveillance and monitoring devices as they often interact with human consumable food and drinks and living environment to provide valuable information. During the last decade, the incorporation of biopolymers/biomaterials into devices has allowed significant advances in the direction of improved device performance. The biopolymers are classified on the basis of various factors such as type, origin, and monomeric units as illustrated in Figure 6.8.

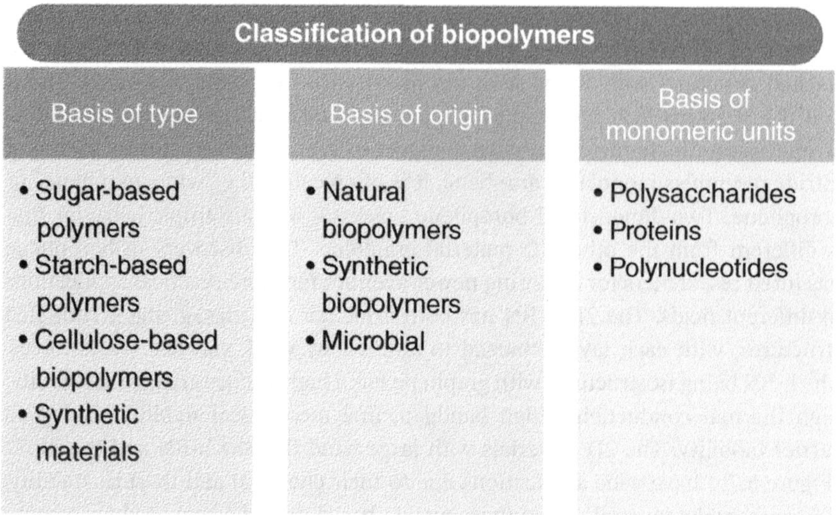

Figure 6.8 Classification of biopolymers in terms of type, origin, and monomeric units.

6.2.3.1 On the Basis of Type

Biopolymers are categorized on the basis of the type of the material and incorporate the following:

6.2.3.1.1 Sugar-Based Polymers

Polysaccharides are major classes of biomolecules which are long-chained carbohydrate molecules and composed of several smaller monosaccharides linked by glyosidic bond. These complex biomacromolecules serve as the source of energy for the animal cells and form the structural unit in plant cell. The main properties of the polysaccharides are: (i) complex and high-molecular weight carbohydrate molecules consisting of hydrogen, carbon, and oxygen with $H : O::2 : 1$, (ii) they are hydrophobic in nature and thus are insoluble in water, (iii) can be extracted in the form of white powder. The sugar-based polymer systems has huge implications on both the physiological and biological properties imparted by the saccharide units and are unique from synthetic polymers discussed next. The features which set the sugar-based biopolymers aside from other biopolymers include the ability of glycopolymers to preferentially target various cell types and tissues through receptor interactions, exhibit bioadhesion for prolonged residence time, and be the ability for rapid recognition and union internalized by malignant cells. The stimuli-sensitive properties of saccharide-modified polymers allow to assist drug release under given conditions [23]. The inherent pH- and temperature-sensitive properties and the enzyme-cleavable properties of polysaccharides are utilized for targeted bioactive delivery. Researchers from the University of Birmingham, UK, and Duke University, US, have created a new family of polymers [24, 25] from sustainable sugar-based starting materials. They used isoidide and isomannide as building blocks to create two new polymers, one which is stretchable like rubber and another which is tough but ductile, like most commercial plastics. The isoidide-based polymer showed a stiffness and malleability similar to common plastics, and a strength that is similar to high-grade engineering plastics such as Nylon-6. The isomannide-based polymer not only had similar strength and toughness, but also showed high elasticity, recovering its shape after deformation.

6.2.3.1.2 Starch-Based Polymers

Starch is a naturally occurring polysaccharide consisting of anhydroglucose units linked together primarily through glucosidic bonds [26]. It is a promising biopolymer for different application due to its inherent biodegradability, overwhelming abundance, and annual renewability. Starches offer a very attractive low-cost base for new biodegradable polymers due to their low material cost and ability to be processed with conventional plastic processing equipment. Starches can be isolated from grains such as corn, rice, and wheat, or from tubers such as potato and cassava (tapioca). Previous studies have showed that starch is a heterogeneous material containing two kinds of microstructures: linear and branched. Linear structure is called as amylose and branched structure is called as amylopectin. Starch has been considered an attractive raw material for polymer applications for over 200 years. However, since starch-based material suffer from lower mechanical strength and moisture sensitivity, various blends and composites have been developed in the last two decades for suitable applications. Ma et al. [27] developed an entirely starch-based hydrogel for flexible electronics including strain-sensitive batteries and self-powered (SP) wearable sensors. The biodegradable hydrogel was prepared using high-amylose starch, $CaCl_2$, and glycerol, where the preparation method is green and facile. The device can be used to detect human activities involving small strain such as wrist pulse and throat vibration, for which the signals are strong, clear, and stable. Zeng et al. [28] developed a facile one-step strategy to fabricate transparent, highly flexible, and multifunctional starch/polyacrylamide double-network hydrogels based on natural renewable starch. The resultant hydrogels exhibit fast self-adhesive ability and present high flexibility attributing to the double network consisting of cross-linked starch and polyacrylamide. The hydrogels can be assembled as transparent, self-adhesive, flexible, highly sensitive, and multifunctional strain/pressure and humidity sensors for accurate healthcare monitoring. The hydrogel-based sensor shows ultrahigh sensitivity to humidity (35–97% relative humidity [RH]).

6.2.3.1.3 Cellulose-Based Biopolymers Cellulose is a linear organic polysaccharide widely found in nature as the cotton, wood, cereal, fiber, and bacteria. Cellulose comprised hundreds to thousands of ringed glucose monosaccharide units with the chemical formula $(C_6H_{10}O_5)n$ and are linked together through glucosidic bonds [29]. The linear configuration of cellulose chains is formed due to the chemical interactions between hydroxyl group and oxygen of adjacent ring molecules, which stabilize the linkages and form the parallel linear nanofibers. The cellulose chains are formed by amorphous-like (disordered) and crystalline (highly ordered) regions which are used to form the cellulose nanocrystals. Although cellulose is hydrophilic substance insoluble in water and most organic solvents, it is a chiral molecule with biodegradable property. Thus, cellulose is often converted to its cellulose esters to increase solubility. The most common cellulose esters are cellulose acetate, cellulose acetate propionate (CAP), and cellulose acetate butyrate (CAB). Cellulose as the most dominating natural polymer has attracted more and more interest, especially in the field of medicine such as advanced medical diagnosis [30, 31]. It has also been widely applied in the flexible [32] and transparent bioelectronics [33] due to its robust mechanical property, biocompatibility, and biodegradability, as well as high surface area, low thermal expansion, and good flexibility. Kim et al. [34] developed a label-free cellulose surface-enhanced Raman spectroscopy (SERS) biosensor chip with pH-functionalized, gold NP (AuNP)-enhanced localized surface plasmon resonance (LSPR) effects for identification of subarachnoid hemorrhage (SAH)-induced cerebral vasospasm and hydrocephalus caused by cerebrospinal fluid (CSF). Arakawa et al. [35] developed a mouth guard (MG) glucose sensor with electrodes coated with cellulose acetate (CA) membrane as an interference rejection membrane to measure glucose in saliva.

6.2.3.1.4 Synthetic Fossil-Based Biopolymers Synthetic nonbiodegradable polymers termed as "plastic" are widely used because of its low cost, versatility, and durability. The term nonbiodegradable refers to the polymers that do not break down to a natural, environmentally safe condition over time by biological processes. The durability of these synthetic and nonbiodegradable polymers are high as these polymers are unaffected by, and an uncommon target for, bacteria. If we oversee their environmental impacts, synthetic plastic materials are extremely useful in the packaging sector, due their useful characteristics, such as transparency, strength ability, flexibility, thermal performance, permeability, and simple sterilization methods, all of which making them appropriate for the food packaging sector. Fossil-based nonbiodegradable biopolymers (i.e. ethylene vinyl alcohol [EVOH], polypropylene [PP], PE, polyurethane [PU], poly(ethylene terephthalate)

[PET], polystyrene [PS], expanded PS, polyamides [PA], and poly(vinyl chloride) [PVC]) are in huge demand for excellent mechanical and physical characteristics.

However, researchers around the world have recently shifted their focus on the development of environmentally benign polymers to reduce the ill effects of synthetic and nonbiodegradable plastic leakage on the ecosystems. Strict norms and regulations against nondegradable chemical pollution causing conventional synthetic polymers play a driving force behind the development of renewable, biodegradable, sustainable, and environmentally benign materials. The synthetic biopolymers are degradable polymers and possess huge advantages over fossil-based nondegradable polymers by means of cost-effectiveness, ecofriendliness, and user-friendly materials. Biodegradable polymers are not necessarily made from renewable resources in order to be completely biodegradable. Polymers such as polybutyrate adipate terephthalate (PBAT), polybutylene succinate (PBS), polycaprolactone (PCL), and polyvinyl alcohol (PVOH, PVA) are synthetic biodegradable polymers as their structure contains chemical groups that can be easily broken down by the action of microorganisms. The use of synthetic biodegradable biopolymers in gas sensors, medical, tissue engineering, military, and environmental applications suggest a paradigm shift in environmental awareness. Han et al. [36] developed an ammonia (NH_3) gas sensor based on organic FET (OFET) where poly(vinyl alcohol) (PVA) was used as the gas accumulation layer which improved the sensitivity of the device. The interaction between NH_3 and PVA is stronger under the influence of the applied electrical field on the hydroxyl dipoles. The bias-induced reorientation of the hydroxyl dipoles can modulate the influence of NH_3 on the trapping (absorption) and detrapping (desorption) processes of charge carriers at the pentacene channel and the PVA interface. Again, Dhiman et al. [37] fabricated a highly specific capacitive nanosensor for nonenzymatic glucose detection. The authors performed capacitance measurements with polyvinyl alcohol-capped copper oxide (PVA-CuO) thin films on indium tin oxide (ITO)-coated glass using Arduino UNO for glucose detection. In addition, PVA was also used for making wearable devices for human activity monitoring [38, 39]. The PVA was also used in medical assignments as it is a liver tissue mimicking material in terms of needle–tissue interaction [40]. PBAT commercially known as Ecoflex is widely used in stretchable devices and in the development of strain sensors [41]. Both PVA and the Ecoflex are biodegradable fossil-based biopolymers. Some examples of biopolymers are listed in Figure 6.9.

6.2.3.2 On the Basis of Origin

6.2.3.2.1 Natural Biopolymers Extracellular cell matrix is an abundant source of natural chitosan biopolymers and widely available in sources as in forest products, grasses, tunicates, crustacean, and stalks (Figure 6.10). Biopolymers have

Figure 6.9 Classification of synthetic biopolymers into biodegradable and nonbiodegradable fossil and natural polymers.

Natural biopolymers

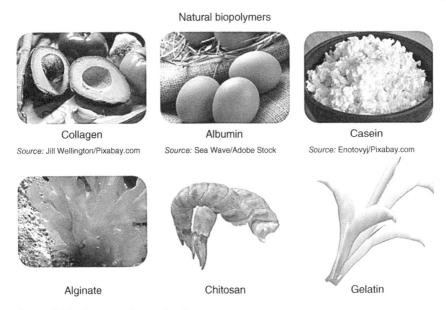

Figure 6.10 Sources of natural polymers.

advantages of being biocompatible as well as biodegradable besides their abundance, renewability, and cost-effectiveness. The use of biopolymers in devices has attracted interest in the areas of physics, chemistry, materials, and microelectronics in the past few years due to numerous potential applications. The applications of natural biopolymer include the field of advanced biomedical materials (tissue engineering, artificial bones, or gene therapy) and extend to the electronic devices, in the form of FET-based electronic switches, gates, storage devices, biosensors, and biologic transistors. Alternative sources of natural biopolymer-like bacterial or fungal cellular components, soy protein, and fish skin gelatin are continuously being explored for the isolation. Natural biopolymers like starch, cellulose, chitosan, carrageenan, gelatin, alginates, and their derivatives have proved themselves to be promising alternatives to synthetic fossil-based polymers.

Collagen contains special types of amino acids called glycine, proline, hydroxyproline, and arginine. Since these amino acids have a regular arrangement in each of the three chains of these collagen subunits, the collagen has a triple helix-like structure. The sequence often follows the pattern Gly-Pro-X or Gly-X-Hyp, where X may be any of various other amino acid residues. The collagen molecule is approximately 300 -nm long and 1.5 nm in diameter. Each fiber of collagen contains thousands of individual collagen molecules that are bound together by crosslinking and staggered covalent bonds. The tensile strength of collagen depends on the formation of covalent intermolecular cross-links between the individual protein subunits. There are different types of collagen fibers of which some are more flexible and form meshworks, while some contain more disulfide bonds which allow them to form long and stable microfibrils. It is known that collagen degradation is promoted by oxidative stress.

Gelatin is a natural biopolymer which can be extracted from seaweeds, and skin, tendons, ligaments, and bones of animals. It is formed by the collection of peptides and protein produced by partial hydrolysis of collagen which is also a natural polymer. The amino acid content of hydrolyzed collagen is the same as collagen. Hydrolyzed collagen contains 19 amino acids, predominantly glycine (Gly) 26–34%, proline (Pro) 10–18%, and hydroxyproline (Hyp) 7–15%, which together represent around 50% of the total amino acid content. Other amino acids include alanine, arginine, aspartic acid, and glutamic acid. During hydrolysis, some of the bonds between- and within-component protein fibrils are broken into smaller peptides, depending on the physical and chemical methods of denaturation, to form gelatin. Thus, the chemical composition of gelatin, in many aspects, is similar to that of its parent collagen. Gelatin is soluble in polar solvents like hot water, glycerol, and acetic acid, but insoluble in organic solvents like alcohol. Gelatin absorbs 5–10 times its weight in water to form a gel. Its main function is to look after the connective tissues such as bones, tendons, and cartilage. It is widely used in various industries due to its properties of biodegradability and biocompatibility. Its main function is to look after the connective tissues such as bones, tendons, and cartilage.

Chitosan (CS) is one of the conducting biopolymer that has much attention in the research for sensor applications. The recognition of chitosan as a promising material for sensor application is attributed to its abundant availability, low in cost, and biodegradable, and having excellent film forming ability and nontoxic properties, making its suitability in biomolecular sensors, gas sensors, and volatile organic compound sensors. Researchers [42] fabricated interdigitated electrodes which were functionalized with layer-by-layer (LbL) films of hyaluronan (HA) and chitosan (CS) to detect prostatic tumor cells (PC3 line). Chitosan which is obtained from modified seafood waste such as crab and shrimp is a unique material that is appropriate for microdevices due to its potential to be selective because of high density of amine groups, which impart active binding sites. Other natural biopolymers such as starch (Section 6.2.3.1.2) and cellulose (Section 6.2.3.1.3) are discussed earlier.

6.2.3.2.2 Synthetic Biopolymers
We have discussed the fossil-based synthetic biopolymers in Section 6.3.2.1.4, where we discussed various nonbiodegradable and biodegradable subcategories. In this section, we shift our focus toward the nonfossil-based counterpart of synthetic biopolymers which are generally renewable and

degradable materials like polylactic acid (PLA). The PLA is a widely used alternative to fossil fuel-based biopolymer and reduces their derogatory effects on the environment. PLA or polylactide is an environmentally sustainable, compostable, and biodegradable biopolymer obtained from renewable feedstock.

PLA is obtained from the monomer lactic acid (LA) also known as "2-hydroxy propanoic acid." PLA exists in three forms: poly-L-lactic acid (PLLA), poly-D-lactic acid d-PLA (PDLA), and a racemic mixture of D,Ll-PLA (PDLLA). Of these, PLLA and PDLLA have shown promise in bioengineering. The advantage for the widespread applications of PLA is (a) the production of PLA involves the utilization of CO_2, (b) offers enhanced processability for fiber spinning, injection molding, etc., (c) since PLA itself is not toxic or noncarcinogenic and neither are its degradation products, it is a biocompatible biopolymer, and (d) provides good physiomechanical characteristics such as high tensile and elastic strength. Its biocompatibility can be attributed to the fact that PLA hydrolyses into α-hydroxy acid after implantation into the human body, gets assimilated into the citric acid cycle. In addition to these properties of the PLA, it is well suited for biomedical and bioengineering applications because of its biodegradability. Hydrolytic degradation of PLA occurs in the body through deesterification to generate lactic acid in the body, and is excreted by tricarboxylic acid cycle. Researchers [43] have developed 3D-printed protective mask which can provide sufficient barrier protection against particles of a size corresponding to microorganisms including viruses *Staphylococcus epidermidis*, *Escherichia coli*, *Candida albicans*, and SARS-CoV-2 by decontamination using ethanol and sodium hypochlorite-containing disinfectant. Despite the various advantages of PLA as a biomaterial, some disadvantages have raised concerns about the biocompatibility of this polymer. The PLA and polylactic acid-*co*-glycolic acid (PLGA) may release acidic degradation products which can cause inflammatory response. The molecular weight PLLA can take more than five years to be completely resorbed in the body. The PLLA and PLGA can cause platelet adhesion and coagulation in the body. Applications of PLA also include food packaging material, textiles, and recently also as engineering plastics.

6.2.3.2.3 Biopolymers from Microbial Source Many microorganisms can generate different biopolymers which incorporate polysaccharides, glycogen, cellulose, PA, and polyhydroxyalkanoates based biopolymers. Several bacteria, yeast, and fungi are known to produce polysaccharides through microbial fermentation. These polysaccharides derived from bacteria are widely used as ingredients in modern food industry and offered controlled and modifiable properties, and can be utilized as thickeners/viscosifiers, gelling agents, encapsulation and film-making agents, or stabilizers. Recently, some of these biopolymers have gained special interest owing to their immune-stimulating/therapeutic properties and may lead to the formation of novel functional foods and nutraceuticals.

The microbial polysaccharide xanthan gum derived from *Xanthomonas campestris* is a water-soluble biopolymer which has a trisaccharide side chain and appears to be rigid, linear, and of low flexibility. Xanthan is extensively used in many food applications because of its unique rheological properties. Curdlan is a high-molecular-weight water-insoluble extracellular microbial polysaccharide and is derived from bacteria *Alcaligenes faecalis*. Curdlan has utility as a food additive in its ability to form an elastic gel. Curdlan forms a heat-set gel at both relatively high and low temperatures or on neutralization or dialysis of alkaline solution of curdlan. Pullulan, levan, and elsinan are extracellular microbial polysaccharides that are edible and biodegradable. Pullulan is a natural, water-soluble polysaccharide with excellent film-forming and binding properties and is produced from polymorphic fungus *Aureobasidium pullulans*. Pullulan is seen as an ingredient in cosmetics and beauty products, specifically antiaging products, because of its ability to provide an instant skin-tightening effect as it adheres to the skin. Pullulan is used in food and is a natural prebiotic. It provides few calories and promotes the growth of beneficial bifidobacteria in the intestine. Pullulan films cast from aqueous solutions are clear, odorless, and tasteless, and have good oxygen barriers. Pullulan coatings have been used successfully as oxygen barriers to prolong food shelf life. Levan is a naturally occurring fructan present in many plants and microorganism species. It is produced as exopolysaccharides (EPS) in the extracellular matrix of bacteria from various genera, such as *Acetobacter*, *Aerobacter*, *Azotobacer*, *Bacillus*, *Corynebacterium*, *Erwinia*, *Gluconobacter*, *Mycobacterium*, *Pseudomonas*, *Streptococcus*, and *Zymomonas*. Levan can be used in a wide range of skin care applications for body or face and also used as edible coating materials for foods and pharmaceuticals due to their low oxygen permeability properties. Dextran is a complex branched glucan (polysaccharide derived from the condensation of glucose), originally derived from wine. It is synthesized by the action of the bacterium *Leuconostoc mesenteroides* on sucrose. Glycogen is secreted from the capsular polysaccharides that are associated with the surface of the cell and result in the development of the matrix of the biofilm. Alginates that are produced by *P. aeruginosa* are responsible for the protection of the cells from phagocytosis. Glycogen is found in many bacteria, and usually accumulates in environmental conditions that limit growth and also offer excess carbon supply. Glycogen is a multibranched polysaccharide of glucose and is stored in fungi like the *Neurospora crassa* and bacteria like *Lactobacillus acidophilus*. Glycogen is secreted from the capsular polysaccharides that are associated with the surface of the cell and result in the development of the matrix of the biofilm.

Escherichia coli produces phosphoethanolamine cellulose which mediate strong connections between proteinaceous curli fibers in complex biofilms and provide resistance under high-shear conditions. Hyaluronate produced by *Streptococcus pyrogenes* and *Bacillus cereus* helps in mimicking hyaluronate being present within the connective tissues, thus protecting them from phagocytosis. The alginate and the hyaluronate-based biopolymers are utilized in the development of artificial extracellular matrix in the form of 3D scaffolds. PA or poly(amino acid) chains such as poly(γ-D-glutamic acid) (γ-PGA) and poly(ε-L-lysine) (ε-PL) or the intracellular cyanophycin (a copolymer of L-aspartic acid and L-arginine) can be produced from bacteria. Many nonpathogenic polyamide-producing bacteria, such as *Bacillus licheniformis*, *Bacillus megaterium*, and most cyanobacteria, are known to produce polyamide-based materials. PA are highly charged and can be polyanionic (γ-PGA) or polycationic (ε-PL). PA are biodegradable, nontoxic, and renewable and are used as replacements for chemically synthesized polymers in industrial applications. The γ-PGA can be used as a flocculant to replace synthetic flocculants in wastewater treatment, while the ε-PL has antibacterial properties as it disrupts membrane integrity, and its cross-linked form was used in antimicrobial coatings. Polyhydroxyalkanoates (PHA) are a large family of biopolyesters produced by many bacteria for carbon and energy storage. The monomeric structure of PHA gives flexible properties such as brittleness, elasticity, and stickiness to the biopolymer. PHAs can be produced by bacteria of diverse types – wild type, pure/mixed bacteria, or metabolically engineered, and includes genera *Azotobacter*, *Bacillus*, *Pseudomonas*, *Burkholderia*, *Halomonas*, and *Aeromonas*. The PHA can also be produced from bacteria *Rhodobacter sphaeroides*, *Ralstonia eutropha*, *Wautersia eutropha*, *Corynebacterium lutamicum*, and *E. coli*.

6.2.3.3 On the Basis of Monomeric Units

The biopolymer can be classified in terms of the chemical composition of the monomeric units. The biopolymers may be composed of chain of branched or linear arrangement of numerous monomeric units which can be composed of monosaccharides, amino acids, or nucleic acid molecules. The biopolymers are grouped as polysaccharides, proteins, and polynucleotides when the monomeric units are monosaccharides, amino acids, or nucleic acid molecules, respectively, as described below.

6.2.3.3.1 Polysaccharides Sucrose is a disaccharide biopolymer and composed of two monomeric units of glucose and fructose which are bonded together by a glycosidic linkage. Sucrose is a simple carbohydrate with limited reactivity that is used as a transport and energy storage molecule in most plants and as an energy source for animals. Raffinose is a trisaccharide composed of galactose, glucose, and fructose. They are a family of soluble sucrose derivatives that constitute an important form of transported carbon in some plants (e.g. pumpkin).

Starch and cellulose are polysaccharides consisting of numerous glucose units as described in Sections 6.2.3.1.2 and 6.2.3.1.3, respectively.

6.2.3.3.2 Proteins Proteins are large biomolecules that comprise one or more long chains of amino acid. A linear chain of amino acid residue is called a polypeptide. A protein contains at least one long polypeptide. Short polypeptides, containing less than 20–30 residues, are generally not considered to be proteins and are called peptides. The individual amino acid residues are bonded together by peptide bonds and adjacent amino acid residues. Silk fibroin (SF) ($\approx$75 wt%) and sericin ($\approx$25 wt%) are the two major proteins of silk [44]. SF is a natural fibrous protein composed of a heavy and a light chain that are connected together by a disulfide bond and a glycoprotein (P25) noncovalently linked to these chains. Silks are natural fibrous proteins existing in the glands of some arthropods, such as silkworms, spiders, scorpions, mites, and bees. In recent years, silk from *Bombyx mori* (silkworm) has been widely studied due to its high yield [45], excellent tensile strength (0.5~1.3 GPa) [46] and toughness ($6 \times 10^4 \sim 16 \times 10^4$ J/kg) [47], predominant biocompatibility, tunable biodegradability, and ease of processing. These predominant characterizations give silk proteins versatile applications in biological fields, including tissue engineering, wound healing, and drug delivery.

SF was used in the implantable sensor as a stiff support that enables insertion in mouse brain tissues [48], memresistive device [49], and flexible electronics [50], and biosensor for the detection of nitric oxide (NO) at nanomolar levels [51] as well as biotriboelectric generator generating a high voltage with large surface area [52]. Therefore, advances in SF-based flexible electronics would open new avenues [53] for employing biomaterials in the design and integration of high-performance and high-biocompatibility electronics for future applications in biosensors, wearable sensor, E-skins, and biomedical diagnosis. Collagen [54] is another example of biopolymer with monomeric amino acid chain and discussed in Section 6.2.3.2.1.

6.2.3.3.3 Polynucleotides Nucleotides are organic molecules consisting of a nucleoside and a phosphate and serve as monomeric units of the nucleic acid polymers: deoxyribonucleic acid (DNA) and ribonucleic acid (RNA). Each monomeric nucleotide unit is made up of three parts, namely, a phosphate group, a furanose sugar moiety, and one of the four possible nitrogenous bases (nucleoside): adenine (A), guanine (G), cytosine (C), or thymine (T). The acidity of the nucleic acid results from the distribution of a lone negative charge among the two pendant oxygen atoms in the PO^{-4} group that constitutes the "rigid

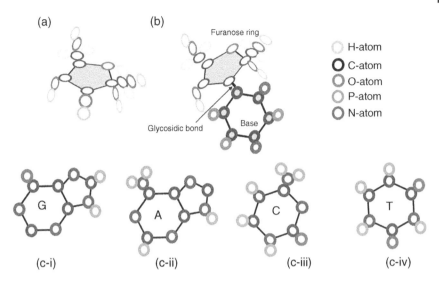

Figure 6.11 (a) Molecular structure of sugar furanose ring (b) representation of glycosidic bond between furanose sugar ring sugar and one of the bases. Molecular structure of purine bases (c-i) Guanine (G) and (c-ii) Adenine (A) and Pyramidine bases (c-iii) Cytosine (C) and (c-iv) Thymine (T).

backbone" of the polynucleotide entity. The furanose moiety consists of a five-membered ring with exocyclic alcohol and hydroxyl groups attached to it. In order to avoid short contacts between the pendant hydrogen (H) atoms of the $HO-CH_2-C^{ring}-H$ entity, the furanose ring is puckered into an energetically favorable conformation as shown in Figure 6.11a. Each base is linked to the sugar through a covalent glycosidic bond to form a nucleoside (Figure 6.11b). The phosphate group bonds with the sugar covalently to form a mononucleotide and also increases conformational possibilities. Mononucleotides are linked to each other through phosphodiester bonds to form a polynucleotide. The planar nitrogenous bases can be composed of either purines or pyramidines which, however, have no internal flexibility. Guanine and adenine are the purines (Figure 6.11c – i, ii), whereas the cytosine and thymine form the pyramidine class (Figure 6.11c – iii, iv). The plane of the base is almost perpendicular to the plane of the sugar ring and has conformational possibilities in the plane of the nitrogenous bases.

DNA is a polymer with deoxyribonucleotides as its repeating or monomeric units. Double-stranded DNA (Figure 6.12a) consists of two single strands of polynucleotide twirled together, connected by pairs of complementary nitrogenous

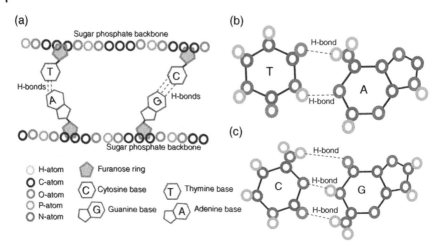

Figure 6.12 (a) Schematic representation of two single strands of polynucleotides, connected by pairs of complementary nitrogenous bases through H- bonds forming a ladder with rung arrangement. Complementary base pairs of (b) A–T connected by two H-bonds and (c) G–C connected by three H-bonds.

bases forming a ladder with rung arrangement with the two strands arranged anti-parallel to each other. (A and T) and (G and C) form base pairs, respectively, due to energetically favorable hydrogen bonding schemes, as the complementary base pairs are of the same size (Figure 6.12b). Each purine base pairs up with its respective complementary base through hydrogen bond (H bond) formed between the electronegative oxygen (O) atom of the purine or the pyramidine base and the H atom linked to the nitrogen (N) atom of the bases. Such arrangements of base pairing minimize the energy of intertwirled structure. Hence, the complementary bases are defined with respect to the H-bonding scheme between the bases. Any cross-pairing would lead to distortion in the double helix structure and is not energetically favorable. The furanose–phosphate chain of the polynucleotide backbone forms the side of the ladder, where the base pairs are placed perpendicular to the axis. These aromatic purines or pyramidine rings consist of pi-lobes extending from their atoms. Since the pi-electron lobes of the nucleotide bases extend above and below the rungs of the ladder, the lobes overlap with the pi-electron counterparts from neighboring rungs. As the aromatic rings of the nucleotide bases in a DNA molecule are positioned nearly perpendicular to the helical axis of the strands, the faces of the aromatic rings are arranged parallel to each other, allowing the bases to participate in aromatic interactions. Thus, the pi-orbital of the aromatic rings of a pair of complementary bases are stacked over each other resulting in pi-stacking interactions. These interactions are mostly van der Waals and electrostatic interactions, but collectively are strong forces due to the sum of all pi-stacking interactions within DNA. In this way, net stabilizing energy is provided to the molecule. Such bonds are termed as noncovalent bonds and are individually weaker than covalent bonds.

6.2.4 Functionalization

Extraordinary electronic properties of certain conductive materials such as CNTs, graphene, Au, Ag NWs, and NPs have lured attention of the research community for possible applications as biosensing and gas sensing applications. However, due to various shortcomings of different materials has restricted their potential application in wider regime. Materials like the pristine CNT and monolayered graphene are hydrophobic which limits their application in biosensors and also is less sensitive to different gases and humidity. On the other hand, sensors developed with Au, Ag NWs, and NPs in its pristine form lacks selectivity, specificity, and suffers from low responsivity for certain chemical moieties and analytes that have less chemical affinities to them. These shortcomings can be overcome by functionalizing the materials which forms the active sensing material by specific chemical moieties which have strong affinities for the target analytes that are desired to be detected. The functional chemical moieties are termed as the host or receptors and the target analytes are termed as the guest. Functionalization is the process of adding new functions, features, capabilities, or properties to a material by changing the surface chemistry of the material. Thus, chemical surface functionalization in its strictest sense implies the attachment of chemical functional groups to surfaces. Surface modification refers to the modification of the surface of a material by bringing physical, chemical, or biological characteristics different from the ones originally found on the surface of a material. Surface chemical functionalization refers to a method of changing the structure and state of the surface of a material by chemical reaction or chemisorption between the surface of the material and the treatment agent to achieve the purpose of surface modification (Figure 6.13). The effectiveness of the sensor in terms of sensitivity depends on the strength of the host–guest chemistry, whereas the selectivity of the sensor is governed by the uniqueness of the chemical interaction between the host and the guest. Functionalization also sometimes affects the sensor dynamics. Functionalization is a fundamental technique used in chemistry, materials science, biological engineering, textile engineering, and nanotechnology for different applications including the development of smart sensors. There are two main approaches for the surface functionalization: (i) covalent functionalization and

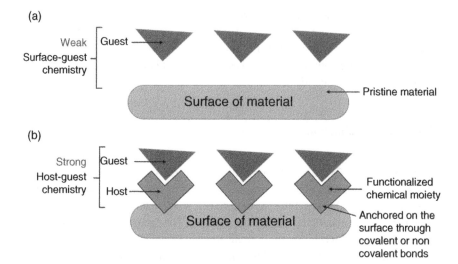

Figure 6.13 Schematic representation of sensing a guest target molecule using (a) unfunctionalized material and (b) functionalized pristine material.

(ii) noncovalent functionalization, depending on the types of linkages of the functional entities onto the surface of the material.

6.2.4.1 Covalent Functionalization

Covalent bond occurs when a functional chemical moiety is linked to the surface of the material through covalent bonds. The attachment of functional group to the surface of the material takes place predominantly at the edges and at the corners where there are defects in the crystallographic structure or at the sites of the dangling bonds where reactive sites are available. The introduction of functional groups via a covalent approach can be used to tune the electronic structure of the material, improving solubility and stability, and opening of bandgap. Different covalent functionalization methods have been developed by the researchers, including the formation of covalent bonds between the surface and functional groups such as organic molecules, biomolecules, and polymers. Various materials can be covalently functionalized directly with organic molecules such as epoxide, carboxylic acid groups, amino groups, hydroxyl groups. Furthermore, highly reactive intermediates like nitriles, carbenes, and aryl diazonium salts can be functionalized directly on the surfaces of CNT, graphene, and GO. Upon functionalization, the sp^2 hybridization of the CNT and graphene could be altered to sp^3 using chemical reaction. Some functional groups on material improve the solubility of the nanomaterials such as CNT and graphene.

The advantages of covalent functionalization are:

1) Significant improvement in solubility and processability
2) Covalent functionalization is chemically and thermally more stable
3) The electronic and chemical properties of the material and its derivatives can be easily and effectively tailored using the covalent functionalization approach

However, there are various shortcomings for this approach as well:

1) Introduces defect in the crystal structure or lattice
2) Alters the intrinsic electronic properties of the material. For example, the sidewalls themselves contain defect sites such as pentagon–heptagon pairs called Stone–Walls defects, sp^3-hybrideized defects, and vacancies in the nanotube lattice. Direct covalent sidewall functionalization is associated with a change of hybridization from sp^2 to sp^3 and a simultaneous loss of p-conjugation system on the planar sp^2-hybridized layer. Sidewall covalent functionalization can be achieved through covalent interaction of nanotube by molecules with high chemical reactivity.

Functionalization of CNTs with various functional groups changes the electronic property, thus enhancing their selectivity, thereby increasing their response toward target gases [55]. The carboxylic group functionalized CNT showed enhanced response for benzene as compared to pristine CNTs as the interaction of COOH–CNTs with benzene gas shifts the Fermi level of the COOH–CNTs away from the valence band, resulting in an increase of resistance of the COOH–CNTs [56]. The development of a bandgap through chemical doping is a powerful method for the use of graphene in nanoelectronic devices [57, 58]. Through controlled covalent functionalization, researchers have developed graphene-based FET devices for biosensing applications [59]. Due to the rich presence of carboxyl and hydroxyl groups on GO, it is widely proffered for the attachment of other functional group to form wide range of graphene derivatives through the covalent attachment of organic groups on its surface. Such graphene derivatives have no additional structural perturbations as compared to GO as the functional groups are linked through oxygen atoms. However, covalent functionalization of graphene causes several shortcomings including:

1) mechanical damage to hexagonal framework of the CNT/graphene leading to degradation in electronic properties and electrical conductivity and
2) use of strong chemicals in the form of concentrated acids or strong oxidants causes harm to the environment if not disposed suitably.

Noncovalent functionalization technique is gaining research attention as it retains the excellent electrical, electronic, and mechanical properties of CNT/graphene.

6.2.4.2 Noncovalent Functionalization

Noncovalent functionalization are generally found in π-conjugated systems which have copious presence p-orbitals on the surface and are best suited to indulge in π–π stacking interaction with the receptor functional chemical moiety. Here the functional groups are attached on the surface of the π-conjugated systems such as CNT/graphene through van der Waals π–π stacking interactions. The different functional moieties used for noncovalent functionalization are: small aromatic molecules, long-chain polymer, surfactants, and biopolymers. Since the functional moieties attached to the surface of the CNT/graphene via π–π stacking interaction, the π-conjugated system of the CNT/graphene electronic structure of the sidewalls are retained and thus does not affect the hexagonal framework structure of CNT/graphene. The noncovalent functionalized CNT/graphene shows enhanced electrical properties as the pristine electronic properties of CNT/graphene is retained providing ballistic transport behavior in devices. The noncovalent functionalization is an alternative method for tuning the interfacial properties of CNT/graphene. The CNT/graphene can be noncovalently functionalized by aromatic compounds, surfactants, and polymers, employing π–π stacking between the functional group and the CNT/graphene surface or hydrophobic interaction.

The advantages of noncovalent functionalization include:

1) The major purpose of noncovalent functionalization is to enhance the bioaffinity of hydrophobic surface to establish effective electrical communication between the receptors and the transducer in sensor platforms.
2) The noncovalent functionalization approach retains the desired electronic properties of π-conjugated systems (CNT/graphene).
3) Improves the solubility of hydrophobic material in aqueous solution.
4) Functionalization of CNTs with high-affinity molecules, such as surfactants, conjugated aromatic molecules, and polymers, are widely used to enhance the adsorption of target analytes on their surfaces.

Noncovalent functionalization of single-stranded deoxyribonucleic acid (ssDNA) on the surface of CNT has been well studied and demonstrated by researchers [60]. This not only facilitates good and stable solubility in aqueous solution, but is also used in the development of different environmental and molecular sensors owing to the copious availability of delocalized electrons on

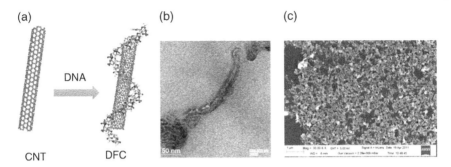

Figure 6.14 (a) Schematic representation and (b) TEM image of DNA functionalized CNT (DFC) and (c) SEM image showing the thick covering of ssDNA on CNT. (a) *Source:* Paul et al. [12]/Taylor & Francis. (b) *Source:* Paul et al. [62].

the phosphate backbone of the ssDNA [61]. The ssDNA is noncovalently functionalized on the surface of the CNT through $\pi-\pi$ stacking interaction between the aromatic nucleobases, guanine, thymine, adenine, cytosine, and the sidewalls of the CNT. A long strand of ssDNA due to its intrinsic helical structure wraps around the surface of CNT through $\pi-\pi$ stacking interaction due to its torsional moment (Figure 6.14a, b). The effectiveness of a DNA functionalized CNT as a sensor material depends on the extent of coverage of ssDNA on the surface of CNT (Figure 6.14c).

DFC hold great promise as molecular probes and sensors targeted for moieties that do not interact or interact weakly with unmodified CNTs. DFC have been widely used for the detection of glucose [63, 64] peroxide [65, 66], pesticides [67], proteins [68], and ions [69]. DFC-FET has also been used in sensing isopropanol (IPA) [70]. Gao and coworkers [71] detected trace levels of Hg(II) ions in solution using DFC by monitoring the decrease in CD spectrograph intensity. Recently, Su and coworkers [72] developed a resistive device with metal-decorated DNA functionalized CNTs – which can sense various gases including H_2, H_2S, NH_3, and NO_2. The Pt, Pd, and Au NPs were used due to its high binding affinity toward specific analytes which enhances the selectivity and sensitivity to a particular target molecule. Paul et al. [62, 73, 74] used ssDNA functionalized CNT (DFC) network to develop FET-based humidity sensor utilizing the excellent charge transfer property of electron-rich phosphate backbone of the DNA with water molecules. Tu and coworkers [75] developed a porphyrin-modified ITO electrode for the detection of chlorite. The porphyrin used was noncovalently functionalized with graphene nanoplatelets. Graphene nanoribbons functionalized with water-soluble iron(III) mesotetrakis (*N*-methylpyridinum-4-yl) porphyrin (FeTMPyP) via $\pi-\pi$ noncovalent interactions was used as a biosensor for the detection of glucose in human serum [76]. Graphene was stabilized using various organic biopolymers to enhance its dispersion in aqueous solution [77, 78].

6.2.5 Biocomposites

The success of polymer and nanocomposites as sensors is largely attributed to easy fabrication process and enhanced sensor performance in terms of reproducibility and sensitivity for target analyte. A biocomposite is a material composed of two or more distinct constituent materials involving a naturally derived material which generally forms the biopolymer matrix, and reinforcing materials (like fibers, silica, nanomaterials-CNT, graphene, NPs) which form the filler in the composite. The biopolymers used in biocomposites have intrinsic biocompatibility, low immunogenicity, nontoxicity, and biodegradable properties, and can be used for various clinical applications. These attractive properties of biopolymers have encouraged their use with reinforcing components such as mineral particles or natural fibers for the development of biopolymer matrix composite commonly termed as biocomposites. The biocomposite so developed is found to offer enhanced sensor performance to a target analyte as compared to either the matrix or the fillers. The biocomposite materials are widely used in promising applications such as in the biomedical research field, wound dressings, wound healing, tissue engineering, drug delivery, and medical implants. Researchers have been exploring varied biocomposites which are aimed at targeted detection of analyte. New research projects are targeting the development of highly specific biopolymer composite receptors and new transducer platforms for developing electrical noses (e-noses) for wide range applications in industry, environmental monitoring, disease monitoring, defense, and public safety. Different sensors have been developed with blended biopolymer films, assembled monolayers of biopolymers, NP-reinforced biocomposites film, biopolymers hybridized with conducting organic polymers, and conductive fiber-reinforced biopolymers like CNT–biopolymer composites, graphene flakes based composites, and tested for various gases and vapors. The biocomposites are being increasingly recognized as better alternatives to traditional polymer composites due to its ability to get degraded by environmental agents, and hence find specific applications in health and wearable industries. Again the bionanocomposites have lured much attention of the researchers as it creates a new avenue for fascinating interdisciplinary research field that brings together materials science, biology, and nanotechnology.

Nanocomposites are a new class of materials where one of its components, generally the fillers, constitutes nanomaterials like NPs, NWs, nanotubes, nanofibers, and graphene flakes and derivatives. The matrix in the nanocomposites may be a biopolymer or fossil-based nondegradable polymer. Polymer nanocomposites are advanced composite materials which have received much attention in research for the development of different sensors. CNTs and graphene are found to enhance the sensing properties of polymer-based and oxide-based gas sensors significantly. CNT/graphene–polymer nanocomposites have received much attention because of the remarkable conductive properties' of CNT, where the polymer matrix serves as the active material for the sensing of target analyte. The highly conductive nanocomposite produces variation in electronic output signal in response to target analyte. The thin-film nanocomposites in addition offer excellent flexibility and are suitable for development of wearable sensors. Both conductive and semiconducting polymers can also be used as active sensing material and employs (i) charge transfer capability to and from the target analyte and (ii) swelling behavior under adsorption of target analyte. When insulating polymer is used in preparation of the nanocomposite-based electronic sensor, it must be ensured that the concentration of the conductive filler must be above the percolation threshold for the flow electric current across the dimension of the nanocomposite.

Cellulose fiber-reinforced composites attracted our attention since the 1970s. To improve the material properties of cellulose for electronic and sensor applications, cellulose matrix-based nanocomposites with metal oxide fillers were investigated and found to possess enhanced chemical stability, electrical conductivity, and photosensitivity than its individual constituents. Mahadeva and Kim [79] developed a glucose biosensor using glucose oxidase (GOx)/immobilized cellulose–SnO_2 hybrid nanocomposite, where porous GOx was linked to the cellulose–SnO_2 hybrid nanocomposite through covalent bonding between GOx and SnO_2. Silver/starch nanocomposites [80] and ZnO/Au NP nanocomposites [81] were used for several biomedical applications. More bionanocomposites were widely used by researchers for sensor applications [82–84]. Kafi and coworkers [85] developed a biodegradable chitosan biostrip with mesoporous–chitosan–graphene oxide (m-Chit-GO) composite (Figure 6.15) as the sensing electrode and graphene interconnects for the detection of dopamine in physiological fluid. The work demonstrates the effectiveness of mesoporous electrodes for the detection of DA at low detection limit of 10 pM. Farea et al. used polypyrrole–graphene oxide PPy/GO composite using *in situ* polymerization for the detection of carbon monoxide at room temperature [86]. Although 1D nanostructure of metal oxides such as ZnO, SnO_2, and Cu_2O NWs or nanorods (NRs) has been widely explored for sensing applications, their low conductivities limit their application as sensors. The conductivity and hence the sensing performance were improved by blending the metal oxide nanostructures with graphene sheets [87].

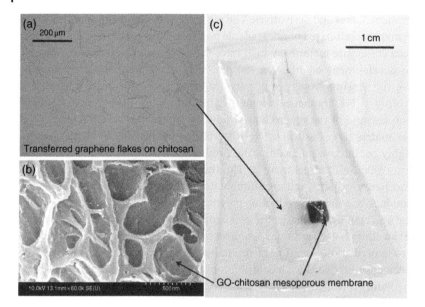

Figure 6.15 (a) Microscopic image showing crumpled morphology of graphene flake network- transferred on chitosan substrate to form the conductive graphene based electrical interconnect of the biostrip. (b) SEM image showing the mesoporous morphology of Chitosan GO electrode. (c) Optical image of chitosan based biostrip with m-Chit-GO as the active sensing material.

6.3 Architectures in Chemical Sensors

6.3.1 Chemiresistors

Chemiresistors are devices based on the change in resistance of the active sensor material in response to selective binding of target chemical analytes on the sensor material through ionic, covalent, or stacking interactions. A simple chemiresistive electronic circuit consists of a pair of electrodes fabricated on a rigid or flexible substrate and are bridged by a conductive material having finite resistance (Figure 6.16). The conductive material is usually in the form of thin film or resistive network of functionalized or unfunctionalized NWs/tubes, fibers, conductive polymers, and nanocomposites. The selectivity of the chemiresistors are achieved through desired functionalization of NWs, nanotubes, and fibers with appropriate chemical moieties, utilization of oxide semiconductor depending on its material

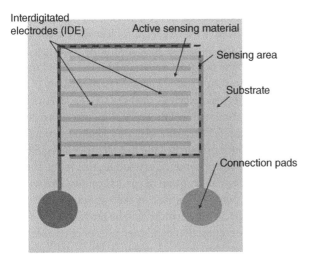

Figure 6.16 Active sensing material deposited on interdigitated electrode (IDE).

properties, and the addition of reinforcing active sensory NPs and materials in a conductive polymer matrix for nanocomposite thin films. The sensitivity of the chemiresistors depends on the electron affinity and the amount of charge transfer between the host and the guest/active sensing material. The conditions required for chemical response and electron affinity of the reacting chemical moieties decide the fast response of the chemiresistors. Thus, some gas sensors which use metal oxides as the active sensing material often require high temperature to initiate the chemical reaction. Chemiresistors finds application as various gas sensing applications and in the detection of various chemical moieties and biomolecules. The chemiresistors operate by the host–guest chemistry where the target chemical analyte considered as the guest, selectively reacts with the active sensor material termed as the host, producing a derivative or a complex compound which modifies the resistance of the sensing circuit. The chemical interaction between the host and the guest decides the increase or decrease in the resistance of the sensing circuit depending on the electron transfer mechanism between the host and the guest.

When the metal oxide film is exposed to ambient atmosphere, the O_2 molecules are adsorbed on the surface of metal oxides. The atmospheric oxygen molecules extract electrons from the conduction band Ec of the metal oxide and trap them to form negatively charged O moieties at the surface. This results in band bending and the formation of electron-depleted region or the space charge region of the length of the band bending region. The reaction of these negatively charged oxygen species, adsorbed on the oxide surface, with target gas molecules or a competitive adsorption and replacement of the adsorbed oxygen anions by target gas molecules decreases or increases the band bending, depending on the type of oxide and nature of the target gas, thereby resulting in the change in conductivity of the oxide layer. For n-type oxides, reducing gases such as the SO_2, CO, H_2, NH_3, H_2S, and C_2H_5OH (donors) increases the conductivity of the oxide film, whereas oxidizing gases such as O_2, NO_2, CO (acceptors) decreases and correspondingly converse for p-type materials. The chemiresistance property of the metal oxides is regulated by the width of the space charge region formed on the crystallites due to the transfer of electrons during the adsorption and desorption of gas molecules. The width of the space charge region acts as a potential barrier in the conduction process between the grains and induces changes in the Fermi level. The process of creation and annihilation of electrons from the conduction band increases or decreases the band bending, and hence Fermi level modulation as a result of which the conductivity decreases or increases, respectively.

CNT with covalently linked functional groups provides larger surface area of adsorption for the target gas molecules and thus facilitates increased electron transfer with the functional group. The transfer of free electron to the conduction band of oxide groups (carboxylic and amide) in the functionalized CNTs changes the hole concentration. Such a process creates hole–electron recombination and reduces the number of holes in the tube. As a result, the resistance of CNTs increased as well as their sensitivity upon exposure to the benzene vapors. Similarly, when organic molecules are noncovalently attached on the graphene surface, its extended aromatic character and its electronic properties are perturbed. When the target analyte reacts with the chemical moiety functionalized on the graphene, the target molecules perturbs the electronic landscape of the functional moiety. This functional moiety which is linked to the surface of graphene through π–π stacking interactions changes the electronic distribution on the surface of graphene. This electronic variation on graphene surface leads to increase or decrease in resistance of the material depending on the chemical properties of the chemical moiety and the target analyte. A change in electrical resistance of graphene may also occur through formation of a charge transfer complex by interaction with chemical target acting as electron donor or acceptor [88, 89]. Conductive polymers have been widely investigated in chemical sensors due to the high sensitivity on electrical changes when exposed to diverse types of gases or liquids such as alcohols, ethers, halocarbons, ammonia, NO_2, and CO_2 with low detection limit and potential to operate at or near room temperature [90–92]. This is mainly attributed to the π-conjugated structure of conducting polymer chains. Increase/decrease of polaron and/or bipolaron densities inside the bandgap of the polymer through interaction with chemical target molecules results in electrical changes. Conductive polymers and chemiresistive sensor material are often used in fabrics for wearable applications. Thin films of conductive polymers, including polypyrrole or polyaniline, were coated on polyester or nylon threads, followed by weaving into fabric mesh.

The chemiresistor-type sensor offers the advantage of a simple device architecture and easy signal processing compared to gas-sensitive capacitors or transistors. Two kinds of conductive polymers are in use: intrinsically conductive and extrinsically conductive polymers. For the first class, the polymer itself is conductive, while for the second class a conductive filler material is introduced into an insulating polymer to render it conductive. While the latter show lower sensitivity, they outperform intrinsically conductive polymer in long-term stability and stability to moisture or oxygen.

6.3.2 ChemFET

Among various potentiometric techniques, sensing based on FETs has attracted considerable attention because of its potential for miniaturization, parallel sensing, fast response time, and seamless integration with electronic manufacturing processes. The presence of a biomolecule or a chemical analyte and its concentration along with its chemical behavior with the host receptor material can be determined using the chemical FET (ChemFET) when the chemical information of the target (guest) molecule is transduced into electrical signal through the field-effect phenomenon of the transistors.

The working of the ChemFET is associated with the charge transfer mechanism between the host (receptor) and the guest (target) analytes. The mechanism of charge transfer is studied and understood by analyzing the transfer characteristics of ChemFET under different analyte concentrations. A ChemFET consists of source and drain electrodes which are bridged by a semiconducting material and forms the channel of the FET. The electric field modulation in the channel contributing to the field effect is achieved by insulating the channel by a thin film dielectric material over which the gate electrode is deposited. An ideal FET should have completely insulated gate dielectric to ensure reduced leakage current. Oxide semiconductor materials, CNTs, graphene, and organic materials like the pentacene are often used as the channel material, while SiO_2 and biomaterials like cellulose, chitosan, polyimide, and other dielectric materials are used as gate dielectric. For sensing, the semiconducting channel must be chemically responsive to exposure of target (guest) analyte or the channel material is functionalized with host receptor materials for the utilization of the host–guest chemistry for chemical sensing. The chemical responsiveness of the receptor analyte for the target analytes is attributed to the charge transferred between the target analyte and the receptor molecules of or chemically bond to the channel.

To understand the charge transfer mechanism between the target and the receptor, the ChemFET is held at a fixed source drain voltage V_{DS} and the gate voltage VGS sweep is performed between the source and the gate elelctrodes. If the current flowing between the source and the drain is measured to be IDS, then the transfer characteristics IDS–VGS plots are constructed for different concentrations of the target analyte. The shift in threshold voltages of respective transfer characteristics plot for different target analyte concentration is indicative of the charge transfer mechanism between the analyte and the receptor.

The shift in threshold voltage V_{th} under exposure to increased concentration of target analyte may be explained by the virtue of varied surface coverage of the receptor by the target molecules which lead to the transfer of charge. Since the charge in the receptor channel is directly proportional to the gate voltage V_g near V_{th}, we can express this as:

$$en = C\left(V_g - V_{th}^0\right) \tag{6.1}$$

where C is the capacitance per unit length between the channel and the gate electrode, n is the number of electrons per unit length on the channel surface, and V_{th}^0 is the threshold voltage before target analyte exposure. In the presence of target analyte molecules covering the surface of the receptor channel, the equation is modified to:

$$en + \frac{ea\theta d}{a} = C\left(V_g - V_{th}^{\text{target}}\right) \tag{6.2}$$

where α is the number of electrons transferred per target analyte molecule, a is the area that a target analyte molecule occupy on the surface of the receptor molecule, θ is the amount of surface coverage of target analyte molecule on receptor, d is the nanotube diameter constituting the channel, and V_{th}^{target} is the threshold voltage after target analyte exposure. The shift in threshold voltage ΔV_{th} can be expressed as:

$$|\Delta V_{th}| = \left|\left(V_{th}^{\text{target}} - V_{th}^0\right)\right| = \left|\frac{ea\theta d}{aC}\right| \tag{6.3}$$

The shift in threshold voltage ΔV_{th} is directly proportional to the surface coverage of target analyte on the receptor channel which in turn increases with increase in target analyte concentration. The direction of the threshold voltage shift to the positive or negative side of the transfer characteristics plot provides information about the electron transfer mechanism. The shift in threshold voltage toward the negative gate voltage side of the ID-VG plot signify transfer of electrons from the analyte molecule into the channel of the ChemFET, whereas the shift in threshold voltage in the positive side signify transfer of electrons from the receptor channel to the target analyte or the addition of holes to the ChemFET channel.

Saturation in V_{th} may also occur at high analyte concentration because of complete surface coverage θ of the target molecules on the receptor channel. It is well-known that the channel conductance of the nanotube FET depends exponentially with the charge in the channel or the threshold voltage V_{th} by the relation [93],

$$G = G_0 e^{-e(V_g - V_{th})/K_B T} \tag{6.4}$$

where $G_0(=2e^2/h$, h = Planck's constant) and G represents quantum conductance at absolute zero and at temperature T (in Kelvin), respectively, e is the electronic charge, K_B denotes Boltzman constant, and V_g and V_{th} represents the gate voltage and threshold voltage at which the ChemFET turns on, respectively. With the increase in target analyte concentration, transfer of charges between the receptor molecules and the target analyte molecules takes place, which brings about variation in the threshold voltage of the ChemFET. Thus, the channel conductance of the ChemFET varies exponentially with increase in target analyte concentration. Because of the exponential variation of conductance, ChemFET is extremely sensitive to the charge in the channel and thus promotes sensitive detection of target analytes.

6.4 Applications

6.4.1 Gas Sensors

CNTs have shown the promise to be an ideal material for sensing applications [94, 95] because of their hollow center, nanometer size, and large surface area. Since all the C atoms of the CNT are exposed to the external environment, any weak chemical change in local neighborhood can drastically affect the electrical conductivity of CNT. Moreover, since the carrier distribution of CNT is weakly dependent on the temperature variations due to the van Hove singularities located at the band edges, the CNT-based devices show insignificant shift in device performance with change in temperature.

Since the discovery of semiconducting single-walled CNTs (SWNT) based FET by Tans et al. [96], it has become the subject of much interest in the scientific community, not only in the field of electronics but also finds wide applications as biosensors [97–101]. The difficulty in achieving uniform dispersion of CNT in solvent has held back the widespread use of CNT networks as sensors. The problem was solved by original research contributions from Paul et al. [73, 102] where DNA-wrapped CNT network was used to develop resistive and FET-type devices for environmental monitoring and chemical sensing.

The resistance and the DOS of a single CNT or their network show significant change when exposed to certain gases. The first gas sensor developed with individual SWNT was demonstrated by Kong and coworkers [5] where they reported that the threshold voltage (V_{th}) of nanotube NTFETs shifts toward the higher and lower gate voltage (V_g) with increasing concentrations of NO_2 and NH_3, respectively. This behavior indicates that charge transfer due to interactions with the absorbates is the principal mechanism in changing the conductivity of CNT under the influence of various gases. The charge transfer property of the chemical moiety can be obtained from the Hammett parameter [103], which is related to the electron donating and withdrawing properties on the adsorbed molecules. Electron donating properties of various organic molecules immobilized on the surface of CNT by noncovalent bonds were studied by obtaining the Hammett parameter [104]. However, variation in electrical properties in CNT-based devices may also occur as a result of Schottky barrier modification at the site of contact between the nanotubes and the electrodes. Heinze and coworkers [105] studied the adsorption of gases both on the nanotube surface and onto the metal contacts by suitable passivation.

CNT-based sensors showed excellent electrical response to NO_2, N_2, NH_3 [106, 107], ethanol [108], and other volatile organic compounds. Kuzmych and coworkers [109] developed a gaseous NO sensor with poly(ethylene imine) (PEI)-coated CNT having a lower detection limit of 5 ppb. The device was tested for cross-sensitivity against CO_2 and O_2 for possible application to detect exhaled human breath components.

6.4.2 Environmental Sensors

Environmental monitoring and protection, together with prevention from air-borne and water-borne diseases for continuous enhancement of people's well-being are the main application areas of the chemical sensors. A variety of chemicals sensors are widely used in environmental protection and monitoring applications and are required to adhere to environmental standards as specified by the relevant authority. This also incorporates the requirement for a complex monitoring strategy covering the transport of toxic chemicals from source to human exposure and the influence of regulatory requirements. The environmental monitoring incorporates a wide variety of toxins, however, the most common occupational exposures include biotoxin monitoring (blood, urine, breath), local monitoring of gases, vapors, and aerosols, soil/groundwater monitoring, and dermal exposure in the form of UV rays monitoring. The toxins can be measured at the emission source where the toxins are released in the transport media into which the toxin is incorporated *en route* to human exposure, at the point of human exposure, and through possible consequences of exposure. The toxin concentration is typically greater, at the source than after dispersal in a transport medium. The chemical sensor design is greatly influenced by the regulatory requirements at the point of human exposure and the threshold for environmental alert is issued based on the limit of exposure of a given toxin within a specified time.

The environmental sensors, especially for the detection of water-soluble toxins, soil pollutants, and poisonous gases must be designed with highly active sensing materials to achieve high sensitivity and low detection limits for trace detection. Moreover, the sensors are designed for small size and weight to offer portability, low cost to provide commercialization, and easily fabricable to facilitate mass production. However, the emphasis of sensor design lies on its performance in terms of accuracy, fast response time for real-time monitoring, and low power consumption. Some commonly used environmental sensors for different applications are discussed here.

6.4.2.1 Pollutants/Aerosols Sensors

In recent years, ambient air has been severely contaminated by aerosols and some gas pollutants such as NO_2, sulfur dioxide SO_2 which can induce adverse impacts on human health. The exposure to air pollutants may also cause diseases such as ischemic heart disease (IHD), cerebrovascular disease (stroke), chronic obstructive pulmonary disease (COPD), lung cancer (LC), and acute lower respiratory infection (ALRI) in a long run. Thus, concentrations of pollutant gases and aerosols must be monitored to protect the environment. In the field of toxic pollutant gas sensing, researchers like Vetter et al. [110] fabricated a p-type semiconducting cobalt oxide (Co_3O_4)-based carbon monoxide (CO) gas detector which can sense CO at low concentrations at 473 K, Zhang et al. [111] fabricated a stable H_2S sensors with ZnO–carbon nanofibers prepared from 30.34 wt% carbon by the electrospinning route and showed excellent detection limit of 40–20 ppm, and Liu et al. [112] synthesized copper oxide (CuO) film on porous zinc oxide (ZnO) by the electrodeposition process for the development of NOx sensor at room temperature. High-temperature metal oxide detectors were also widely used for the detection of pollutants in air. Ionescu et al. [113] used a novel nanocrystalline WO_3-based sensor produced by advanced gas deposition for the detection of ethanol and H_2S, mixed with air or NO_2. The device operated in a dynamic mode at 150–250°C with square voltage pulses applied to its heating element. The identification and quantification of the gases were achieved by pattern recognition of data obtained from the fast Fourier and discrete wavelet transforms of the sensor output signals.

Chlorine is one of the toxic gases that are widely used in variety of applications in manufacturing, industrial, domestic, and health care. Chlorine gas hazards can occur during manufacture, use, and transportation (Figure 6.17). Recent evidences of the explosion of a chlorine gas container during transport in the port of Aqaba, Jordan, shocked the world when nearly 13 died and 360 workers were injured due to its carcinogenic hazards. This explosion has increased environmental hazards in the neighboring areas. Researchers have used various metallic oxides including porous tin dioxide (SnO_2) structures [114], indium

Figure 6.17 Representative image of explosion of a chlorine gas container during transport.

oxide $(In_2O_3)/SnO_2$ heterojunction microstructures [115], indium oxide nanosheets [116], polypyrrole (PPy) composite films modified with ZnO NWs [117], and silicone rubber with fluorescent porphyrin H2TPP [118] for the development of chlorine detectors which offer detection limits in ppm levels to detect gas leakages and for environmental monitoring. Organic isocyanates belong to the more important basic substances for producing pesticide, herbicides, fuel cells, batteries, plastics, and foam materials. Methyl isocyanide (MIC) because of its high toxicity, the air at the working places where MIC is produced and further treated, must be carefully monitored. However, MIC gas leakage in an industrial plant in Bhopal, India took away 3787 lives, left 574,366 permanently disabled, and exposed 500,000 victims to toxic MIC gas in 1984. The colorimetric MIC gas detectors [119] that were prevalent then and involved benzidine reagent or tetra base reagent producing a blue color indicating the presence of the isocyanide group could not be deployed leading to vast casualties. Later researchers [120] identified indicators for MIC which detected the presence of organic isocyanates in air. The method of invention included exposing test gas to *bis*-4-(dimethylamino)-phenyl-methylenimine-hydrochloride in the presence of a catalyst and on a carrier with a subsequent exposure of the carrier to a developer for producing a coloration of the carrier which is indicative of the presence of organic isocyanates. Alternative chemistry was also explored for easy and fast detection of MIC [121].

6.4.2.2 Water Quality Monitoring Sensors

Water pollution is caused due to toxic metal ions anthropogenic activities and harmful industrial waste. Heavy metal pollution has emerged due to anthropogenic activity caused to mining the metal, smelting, foundries, and other industries that are metal based and leaching of metals from different sources such as landfills, waste dumps, excretion, livestock and chicken manure, runoffs, automobiles, and road works. Heavy metal ions like $Hg(II)$, $Cd(II)$, $Pb(II)$, and $Ar(II)$, when present in soil or in ground water beyond its allowable permissible limits are dangerous to the humans as well as to the aquatic flora and fauna. Since the heavy metal ions are not degraded in the environment, they are prone to enter in the food chain and accumulate inside the body, where these metal ions can be converted to more toxic forms or can directly interfere with metabolic processes. This heavy metal toxicity produces various disorders and damage due to oxidative stress and even causes genetic modifications. The toxic effect of heavy metal ions on human depends on their concentration in drinking water, exposure duration, individual susceptibility, and the presence of other contaminants. To mitigate the risk of heavy metal exposure, proper water treatment and monitoring are essential. Regular testing of drinking water sources can identify potential contamination, allowing for appropriate remediation measures to be taken to ensure the water is safe for consumption. Environment researchers have devised various heavy metal detectors that can sense trace levels of toxic metal ions. Paul et al. [102] developed a heavy metal ion resistive sensor with DNA-functionalized CNT as the active material which can selectively detect $Hg(II)$, $Cd(II)$, and $Pb(II)$ ions in traces of 5 pM. Chen et al. [122] reported FET-based $Hg(II)$ ion sensors based on thermally reduced graphene oxide (rGO) with thioglycolic acid (TGA)-functionalized gold NPs(Au NPs) (or rGO/TGA-AuNP hybrid structures) in aqueous solutions. The sensor offered a lower detection limit of 25 nM. Mcghee et al. [123] designed fluorescent, colorimetric, SERS, electrochemical, and electrochemiluminscent sensors for the detection of metal ions in the environment and imaging them in living cells. The sensor uses DNAzymes as the active material for the detection. The DNAzymes are an emerging class of metal ion-dependent enzymes which are selectively reactive toward almost any metal ion. These DNAzymes are functionalized with fluorophores, NPs, and other imaging agents and incorporated into sensors for the detection of metal ions in environmental samples and for imaging metal ions in living cells. Synhaivska et al. [124] fabricated silicon nanoribbon (SiNR) ion-sensitive FET ISFET devices modified with a Gly–Gly–His peptide for the detection of copper ions in a large concentration range in water medium. The sensor works on the principle of chelate chemistry where specific binding of copper ions causes a conformational change of the ligand and a deprotonation of secondary amine groups. The invention of solid-state heavy metal sensors has greatly revolutionized heavy metal detection through easy portable and user-friendly techniques.

Water pollution can also occur due to the degradation of plastics into microplastics and may expose humans and other living organisms to contaminated drinking water. Microplastics enter freshwater systems through surface runoffs, treated wastewater, industrial effluent, contaminated plastic debris, and atmospheric deposition. Since microplastics adsorb numerous contaminants such as heavy metals, chlorinated and aromatic chemicals, and possibly persistent organic pollutants due to their hydrophobic nature, the presence of microplastics in fresh water or soil may enter the food chain thereby producing carcinogenic effects affecting the endocrine and reproductive functions of organisms. Researchers have explored various techniques such as agglomeration into biological flocs, combination of oxidation and fluorescent staining, polycarbonate filters, as well as filtration with 0.45 μm filter paper for the removal of microplastics and hence heavy metals from water [125–128]. Alternative membrane-based technology as proposed by Nkosi et al. [129] used PVDF membrane modified with carbon nano-onions (CNOs) to remove high-density PE, poly(1,4-butylene terephthalate), Nylon 12, and cellulose from water resulted from hotels, houses, schools, industries, and commercial businesses surrounding the treatment plant. High concentrations of toxic heavy metals such as arsenic, copper and zinc were attached to the microplastics.

6.4.2.3 Humidity Detectors

Water content detection is important for our daily life. For environmental detection, the RH is an index for comfortable living atmosphere. Researchers have used various fabrication techniques (printed [130], drop casting [62], and micropatterned [73]), used various active sensing materials (nanomaterials, biomaterials, metal oxides, polymers, and composites) with printed [131], photolithography etched [132], and plated [133] electrodes on flexible [134], stretchable [135], and rigid [136] substrates, and explored different principles (capacitive [137], optical [138], chemical, and piezoelectric [139]) for the development of humidity sensors. In this section, we focus on the devices which produce electronic variation in the output signal in response to reversible and irreversible interactions between the active sensing materials and water molecules. The effect of humidity on the transfer characteristics of SWNT FET was first observed by Kim et al. [140], where they observed hysteresis in SWNT FET in the presence of water molecules in the environment. Paul et al. [62, 73] used the DNA-functionalized CNT to develop humidity sensor in the RH range 30–90% and offered fast response time of 4 s. Humidity sensors were also constructed using nanocomposite of SWNT and silicone-containing polyelectrolyte [141], CNT-poly(dimethyldiallylamonium chloride) [142] and CNT/Nafion composite [143]. Jeong et al. [144] developed humidity sensors with gravure printed Ag electrodes and drop-casted TiO_2 nano-flowers as a sensing layer for achieving high sensitivity between 20% and 95% RH. Kim and Gong [131] used screen novel photocured copolymer (MEPAB/CMDAB/MMA) for humidity sensing. The sensor offered high reliability with resistance to water, showing minimal change in behavior when soaked in water for up to 60 minutes. All the humidity sensors discussed here incorporates the vivid advancement in the field of environmental monitoring and extends their scope for future Internet of things (IoT) applications where environmental data may be accessed remotely.

6.4.2.4 UV Radiation Exposure Monitoring

Although ultraviolet (UV) radiation has a numerous beneficial effect on human health (vitamin D3 and endorphins synthesis), excess exposure to UV rays often may causes damage to skin (burns, cancer, premature aging) and eyes (cataracts, degeneration of the macula lutea). Thus, UV dosimeters are designed and developed to monitor the UV exposure on human. The researchers have utilized either various UV-responsive organic materials [145] or employed natural biological monitors such as bacterial spores [146], vegetative bacteria [147], and DNA fragments of bacteriophages [148]. These organic dyes used as UV-responsive materials are used in solution medium [149, 150] or use living organisms that serves as natural indicators for UV exposures and does not involve electronic quantified observables for measurement and thus excluded from this discussion. However, colorimetric UV dosimeters, which is pale yellow/colorless in the absence of UV light, and turns red upon exposure to UV light, were developed by researchers [151]. The dosimeter utilized a mixture of tetrazolium dye and neotetrazolium chloride (NTC) dissolved in polyvinyl alcohol (PVA) to form a film as the active sensing material. The UV dosimeter exhibits a cosine-like dependence on irradiance angle and response of the dosimeter is temperature independent over the range 20–40°C. Flat UV dosimeters with foils made from poly(vinyl alcohol) (PVA) and either 10,12-pentacosadiynoic acid (PDA) or a representative of tetrazolium salts – 2,3,5-triphenyl tetrazolium chloride (TTC) and encapsulated by parylene C – were used for the development of water-resistant dosimeters for under water usage [152]. For dermal exposure monitoring, Sou et al. [153] invented Schottky barrier structure-based UV detectors which used ZnMgS as the UV-responsive active material. The detector composed on stacked layers of n-Zn MgS : Al as the first electrode layer, an active layer of Zn–MgS formed on at least a part of the first electrode, a second transparent conductive electrode layer formed on the upper surface of active layer, and a conducting material formed on first electrode as an Ohmic contact. The composition of the Mg was reported to be less than 30%.

6.4.3 Biomolecule Sensors

Glucose oxidase-functionalized NT FET can be used as a glucose sensor in liquid environment [154]. The decrease in pH of the solution lowered the conductance of the CNT. The shift in threshold voltage to more negative values implied that the charged groups on the glucose oxidase become less negatively charged on decreasing the pH. The sensing mechanism was attributed to the dissolution of the double layer near the surface of CNT which decreased the capacitance of the liquid gate. Lerner et al. [155] devised a scalable and noninvasive glucose sensor with pyrene-1-boronic acid functionalized CNT-based FET which responds to human saliva in the range of 1 μM–100 mM and has a minimum detectable limit of 300 nM.

6.4.4 Food Quality Monitoring

There has been a gradual increase in demands for rapid, highly sensitive, and very selective sensors for food safety and environmental monitoring [156]. Various detection platforms for indicators like humidity, gases, and toxic metals have arisen as very effective and promising tool for quality assessment of food. Effective food quality monitoring demands accurate chemical data in no time with cost-effectiveness.

6.4.4.1 Relative Humidity Monitoring

The food industry often requires stable RH sensor with full-range measurements and high measurement accuracy for different applications. The equilibrium RH (ERH) of a food product is defined as RH of the air surrounding the food that is in equilibrium with its environment. When the equilibrium is obtained, the ERH (in percent) is equal to the water activity multiplied by 100, i.e. ERH (%) = $aw \times 100$. The ERH of a product indicates its tendency either to absorb or to lose moisture, depending on the RH of the environment. Products with a higher ERH than the RH of the environment will tend to dry out, but those with a lower ERH than the RH of the environment will absorb moisture during storage. Any such changes in the moisture content during the storage will lead to major changes in the sensory quality of the product. The choice of correct packaging is vital to minimize moisture transfer between the product and the environment. The RH sensors are used to detect chemical deterioration of foods and beverages, stability and shelf life of confectionery products, like wafers, waffles, and adjuncts, measure the RH in the preparation of dry cured meat products in meat drying rooms, and monitoring of postharvest storage of cereals. Some notable RH sensors are discussed in Section 6.4.2.3.

6.4.4.2 Gas Monitoring

Storage of cereals requires environment composed of carbon dioxide, nitrogen, and oxygen. Increased carbon dioxide concentrations and lowered oxygen concentrations can have adverse effect on the growth and survival of insects, whereas increasing nitrogen and oxygen concentration are often used to control the RH of the environment, thereby facilitating storage and preservation of crops and avoid risk of spoilage. However, the chemical composition of atmospheric gases in the cereal storage containers and bins varies with the presence of molds and insects in the containers. Respiration of molds and insects can lead to elevated carbon dioxide levels and stored seeds and cereals produce small amounts of not just carbon dioxide, but also highly toxic carbon monoxide gas. Thus, detection of carbon dioxide serves as an effective indicator for early detection of spoilage than monitoring other conditions in the storage area such as temperature. The composition of different gases is kept at their optimum ranges by food sensor technology to avoid unnecessary wastage of food. For remote monitoring, rapid-response IoT sensors with good sensitivity are needed. Today, many perishable food products such as produce and meat are preserved in packages that help secure them from environmental impacts such as light, oxygen, moisture, microorganisms, as well as contaminants. Carbon dioxide (CO_2) is one of the most common byproducts for food to perish which is the outcome of raising microorganism activity. A CO_2 sensor integrated within the food package could effectively monitor the food freshness and quality of the item up until it gets to the end consumer. Researchers [157, 158] have developed electronic nose for ethylene for nondestructive monitoring the fruit ripeness. In 2012, Esser et al. [159] succeeded in making an economical gas sensor, with the sensitivity of much less compared to one component per million, for monitoring ethylene (C_2H_4) gas released by fruit which can suggest the ripening process [160].

6.4.4.3 Temperature Monitoring

Food supply chains (FSC) incorporate various interdependent processes and sequences which take food from producer to consumer. Quality of food product forms an important part of the food chain as it is consumed by human beings. FSC encapsulates mainly nature of product which is perishable and very sensitive to fluctuations of temperature and humidity. Since the food industry across globe is pivoted against quality assurance and product safety, the FSC is highly prioritized in this trade. Perishable nature of food product requires coordination with temperature assessment for better quality of food to be delivered to the customer. Collaboration of FSC with temperature trends broadens its scope and makes supply chains intelligent. Thus, temperature sensors are a vital part of the food quality monitoring in the food chain. Caselli et al. [161] designed a temperature sensor in 65-nm CMOS technology aimed at food quality monitoring in IoT and RFID contexts. The sensor occupies an area of 0.26 mm^2, with an inaccuracy of $\pm0.37°$C over the temperature range of -20 to $80°$C, and offered a power consumption lower than 15.5 μW. Xiao et al. [162] proposed a flexible battery-free wireless electronic system (FBES) for food monitoring by the laser direct scribing on commercial PE terephthalate/polyimide/copper (PET/PI/Cu) film. The FBES could acquire the temperature, weighing, and pH sensor data, and transmit to the wireless reader by radiofrequency identification/near-field communication (RFID/NFC) with 13.56 MHz. The multiple sensors like temperature, humidity, and gas sensors can be integrated together through communication technologies ZigBee, Wi-Fi, radiofrequency identification (RFID) for IoT implementation of food monitoring.

6.4.4.4 Presence of Toxic Metals

Guan et al. [163] used microfluidic paper-based analytical devices fabricated by atom stamp printing (ASP) technique for the detection of different concentrations of Cu^{2+} via a colorimetric method and achieved a Cu^{2+} detection limit of down to 1 mg/l. The authors claimed that the developed paper-based solid–liquid extraction device (PSED) platform allows the detection of heavy metal ions more cost-effectively and easily and offered fast response time at the point of care, and it has great promise for applications in food safety in resource-limited areas.

6.4.5 Water Quality Management in Public Pools

Water quality is a health concern in all public pools and hot tubs as they are often polluted by human waste (urine, sweat). Although sterile, urine reacts with pool disinfectants like chlorine, to form dangerous byproducts including trichloramine, which can pose health hazards. The human wastes react with chlorine to create byproducts that decrease the chlorine content in water and if sufficiently low may even fail to kill the bacteria such as *E. coli* present in the pool water. Thus, scientists have devised new ways to detect the presence of urine in pool for appropriate water quality monitoring. Blackstock et al. [164] have identified a possible urinary marker acesulfame potassium (ACE) for the detection of urine in pool. ACE is an artificial sweetener that is often used in processed foods and is widely consumed by many. The ACE being chemically stable, it passes through the digestive tract and into consumers' urine. The researchers developed a rapid, high-throughput mass spectrometric technique to test more than 250 water samples from 31 actively used pools and hot tubs in two Canadian cities, and more than

90 samples of clean tap water to validate the detection and quantification of ACE in water. The researchers estimated that swimmers released more than 7 gallons of urine in a 110,000-gallon pool in one instance.

6.4.6 Health Monitoring

Human health monitoring devices have been realized to provide real-time measurement of physiological status of the body, for example body temperature [165, 166], heart rate [167], pulse oxygenation [168], respiration rate [169], blood pressure [170], blood glucose [171], electrocardiogram signal [172], electromyogram signal [173], and electroencephalograph signal [174]. Thus, there is a burst of research activities and a growing demand for smart devices in the form of wearable sensors and implantable devices for human health monitoring. Solid-state chemical sensors are widely employed to measure the concentrations of different bioanalytes and biomarkers like glucose, lactate, and pH in body fluids. Modern-day wearable and implantable sensors can successfully detect and measure various signals or analytes with high specificity and sensitivity [175, 176]. The use of biochemical sensors as wearable health monitoring device is in high demand due to its high specificity, rapidity, portability, low price, and power consumption. The working principle for most biochemical sensors includes the employment of receptors such as enzyme, antibody, and DNA as recognition elements for the detection of target analytes to generate physiochemical signal output [177–180]. The transducer transforms the physiochemical output into electrical signals which is measured and quantified.

Researchers have developed various chemical sensors that employ biofluids such as sweat, saliva, tears, and interstitial fluid (ISF) as the noninvasive medium of detection of bioanalytes. Wearable biochemical sensors detect bioanalytes noninvasively and offer continuous, real-time, routine monitoring of biomarkers, and thus play an important role in self-management of chronic diseases and for monitoring abnormal and unforeseen situations. Noninvasive glucose sensing is an attractive method for diabetes control and management and is carried out by employing interstitial fluid and sweat as the sensing medium. Wang et al. [181] developed a wearable, tattoo-based noninvasive amperometric glucose detection device on a flexible substrate. The iontophoretic and glucose sensing electrodes were fabricated on a conformal and wearable platform using screen printing, and the device was aimed for single-use glucose measurements [182, 183]. This tattoo-based interstitial fluid glucose detection device holds promise for continuous noninvasive glucose monitoring. Although the device shows good sensitivity and selectivity for interstitial glucose concentration, the results may vary for different patients and with the amount of sweat execrated. Despite the success in the noninvasive glucose detection, much need to be improved in terms of long-term

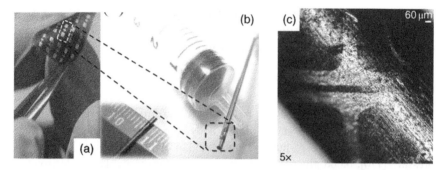

Figure 6.18 (a) Fabricated array of sensors on GO–chitosan substrate and (b) a sensor wrapped around the tip of syringe needle with diameter of 1 mm. (c) Microscopic images of GO/chitosan chemiresistive sensor-on-probe (SoP) conforming to the needle surface.

use, sensor integration with wireless electronics, and established in terms of examination with large populations of both healthy and individuals with diabetes. Vilouras et al. [184] reported a graphene oxide (GO)-chitosan-based chemiresistor for *in situ* label-free detection of serotonin (Figure 6.18). This ultraflexible sensor consists of Au electrodes, separated by a gap length of 60 μm, patterned on 2-μm thick GO–chitosan film which acts as the active sensing material. The device is easily conformable on a clinical needle of radius of curvature $r = 500$ μm and was demonstrated as sensor-on-probe (SoP) platform for easy and fast detection of serotonin. The sensors were characterized with synthetic blood equivalent (Dulbecco's modified Eagle's medium [DMEM]) and showed good electrical stability over a dynamic range of 2 μM–2 mM. With its biocompatible, biodegradable, and eco-friendly properties, the GO–chitosan-based sensors could find application in rapid point-of-care (POC) determination of diseases such as carcinoid syndrome. Despite the rapid progress in wearable biochemical sensors technology in the last decade, commercially successful applications aimed at detecting biomarkers, and bioanalytes is in the infancy. Noninvasive wearable biochemical sensors are a huge technological advancement that would revolutionize real-time measurements and must be a priority in the development of future diagnosis platforms.

6.4.7 Defense and Security

In addition to biological and nuclear warfare, the chemical warfare (CW) is also a brutal mass destruction weapon created by mankind. The main advantages of chemical warfare agents (CWA) over the biological and nuclear counterparts are inexpensive and relatively easy production techniques and its brutal capability to create physiological damage to humans silently. Thus, they are widely used by small terrorist groups to create mass casualties with small quantities. The effectiveness of CWA depends on several factors, including the toxicity of the compound, its volatility and concentration, the route of exposure, the duration of the exposure, and the environmental conditions. The CWA detectors are not only required by security establishments to curb terrorism and other criminal activities against mankind, but also frequently used by medical personnel's during emergency rescue operations [185]. Thus, identification of CWA is equally important as medical treatment of chemical casualties at the site of CWA exposure. The CWA are physiologically classified in terms of harm caused to humans as (a) nerve agents, (b) vesicant (blistering agents), (c) blood agents, (d) choking agents, (e) Riot control agents, (f) psychomimetic agents, and (g) toxins. Among the listed CWAs, the nerve agents are the most toxic substances and can cause death within a few minutes to a few hours as it attacks the central nervous system of the humans. Tabun, sarin, soman, and cyclosarin are some of the nerve agents that are known to cause brutal damage to humans.

Single-walled CNTs with diameter of 15–30 nm have been demonstrated as effective sensors for nerve gas agents sarin and soman. Fennell et al. [186, 187] developed a chemiresistive CWA sensor using single-walled CNTs (SWCNTs) wrapped with poly(3,4-ethylenedioxythiophene) (PEDOT) derivatives where they demonstrated that a pendant hexafluoroisopropanol group on the polymer enhanced the sensitivity of the device toward dimethyl methylphosphonate (DMMP), a precursor of nerve agent sarin. The experiment was conducted with different derivatives of PEDOT/SWNT composites as the active sensor material in both nitrogen and air environments at DMMP concentrations as low as 5 and 11 ppm, respectively. This chemiresistive detection of DMMP is manifested by the variation in conductance of PEDOT/SWNT and consistent with the transduction mechanism of modulation of SWNT conductance by charge transfer between DMMP and the sensor material. A network of films 1–2 microns thick on a PET substrate can detect traces of chemical agent vapors with a sensitivity of 25 ppm. Strong sensors responses were obtained that were not affected by environmental conditions such as air quality and humidity. Single-stranded DNA along with single-walled CNTs FETs have been used to detect chemical warfare agents. These sensors have shown high sensitivity and stability up to 50 cycles of operation [97]. Detection of V-type nerve agent has been experimentally demonstrated using CNTs. The detection is based on enzyme-catalyzed hydrolysis of nerve agents and amperometric detection of thiol containing hydrolysis product that is performed at the CNT-modified screen-printed electrode. The sensitivity demonstrated for such sensors is 258 ppb [188]. Carbon black composite sensing arrays have been used to detect explosives and chemical warfare agent such as sarin and soman [189]. Aptamers are single-stranded DNA that can bind target molecules with high affinity and specificity. Such DNA aptamers are biocompatible, biodegradable, and highly stable even at elevated temperatures and dry conditions, especially when conjugated with NPs. Cyanobacteria toxins are the most encountered toxins in water sources. After obtaining toxin-specific aptamers, DNA-functionalized gold NPs will be used to obtain rapid, autonomous, cost-effective, and easily interpretable sensors for soldiers in the field. The sensors will be made into a test strip format that can be easily used by soldiers or other users without a chemistry background. Importantly, the same technology will be applicable to detect a broad range of chemical and biological agents without an extensive support system.

6.5 Conclusions

The solid-state chemical sensors have established their high significance in the global market, owing to the increasing demand for rapid, compact, accurate, and portable diagnostic sensing systems. Moreover, nanotechnology has dramatically changed the operating characteristics of chemical sensors and is gaining prominence in various smart applications. The solid-state chemical sensors are mostly preferred over other types of genres of sensors due to (i) the lack of sampling with its inherent errors; (ii) no sample pretreatment (such as dilution steps with their errors); and (iii) real-time analysis. Most sensing heads can be fabricated at a cost that makes them cheap enough to be disposed of after use. With the advent of advanced microfabrication tools and the emergence of nanomaterials, the chemical sensors have been highly miniaturized with highly efficient smart architectures, bestowed with enhanced sensor performance in terms of sensitivity and reduced active sensor area, and widened the range of application in health care and wearable devices with the use of flexible substrates and printed electronics. Due to their small size, sensors have high spatial flexibility and can be transported easily. Chemical sensors are devices that can convert a chemical signal into an analytic one. This device includes a chemical or biological recognition feature coupled with a transduction system. The chemical signal is created through a selective interaction between the active sensing material placed in the sensor and a target analyte. The goal of sensing technology is to revolutionize how we measure key parameters related to diagnostics, monitoring of the environment, safety, and protection. Since the mechanism of signal transduction emerges from the properties of the materials, the choice of materials in their suitable form is pivotal for the efficiency and operation of the sensor. Different classes of materials including inorganic, organic, and hybrid nanomaterials (quantum dots, metallic, and semiconducting NPs, carbon nanomaterials), 2D materials, metal–organic frameworks, supramolecular systems (macrocyclic compounds, cavitand molecules), smart soft materials (stimuli-responsive hydrogels and polymers), molecularly imprinted polymers, and DNA-based sensors are widely used in the development of chemical sensors for targeted application. The chemical sensing using metallic oxides takes place by variations in the space charge due to the extractions or injection of electrons by acceptor or donor spaces. The detection of the target gas leads to an electronic transfer between the adsorbed molecules and the active surface, which results in the variation in electric conductivity. The use of nano-materials in chemical sensors and biosensors led to substantial improvements in device performance due to fascinating physicochemical properties that are absent in their bulk counterparts. Polymers either in the form of blends or com-posites have proved to be useful to make sensors that provide fast, reversible, and reproducible feedback. The use of biomaterials, such as synthetic biodegradable polymers, proteins, and polysaccharides, has revolutionized the field of sensors by addressing most challenges which also included nondegradability. Biopoly-mers such as chitosan, alginate, cellulose, pectin, gelatin, and acacia gum have been extensively utilized for the development of biosensors. Biopolymer-reinforced composites are widely used as better alternatives to traditional non-biodegradable materials due to their ability to get degraded by environmental agents. The materials like the bionanocomposites offer a fascinating interdisci-plinary area that brings together materials science, biology, and nanotechnology. Biopolymer-based sensors are less expensive and their production techniques are quite simple. The chemical sensors have expanded their field of applications from gas sensing, leak detection, healthcare monitoring, environmental and food quality monitoring, water quality monitoring, and defense to include space applications, engine emissions, wearable human breath, exercise monitoring, and detection of corona virus to assist the health services. With the growth of multisensory systems and remote monitoring of sensor parameters, the integra-tion of different sensors through the IoT is in high demand.

List of Abbreviations

2D	Two-dimensional
ACE	Acesulfame Potassium
CAB	Cellulose Acetate Butyrate
CAGR	Compound Annual Growth Rate
CAP	Cellulose Acetate Propionate
CHEM-FET	Chemical Field-effect Transistors
CNTs	Carbon Nanotubes
COPD	Chronic Obstructive Pulmonary Disease

CS	Chitosan
CW	Chemical Warfare
CWA	Chemical Warfare Agents
DAB	3,3'-Diaminobenzidine
DFC-FET	DNA-functionalized CNT Field-effect Transistors
DMEM	Dulbecco's Modified Eagle's Medium
DMMP	Dimethyl Methylphosphonate
DNA	Deoxyribonucleic Acid
EPS	Exopolysaccharides
ERH	Equilibrium Relative Humidity
EVOH	Ethylene Vinyl Alcohol
FBES	Flexible Battery-free Wireless Electronic System
FET	Field-effect Transistors
FSC	Food Supply Chains
GDH-PQQ	Glucose Dehydrogenase Pyrroloquinolinequinone
GO	Graphene Oxide
Gox	Glucose Oxidase
HA	Hyaluronan
IHD	Ischemic Heart Disease
IoT	Internet of Things
ISF	Interstitial Fluid
ISFET	Ion-Sensitive Field-effect Transistors
ITO	Indium Tin Oxide
LA	Lactic Acid
LC	Lung Cancer
LEL	Lower Explosive Limit
LSPR	Localized Surface Plasmon Resonance
M-chit-GO	Mesoporous–Chitosan–Graphene Oxide
MEPAB	[2-[(Methacryloyloxy)ethyl]dimethyl]propylammonium Bromide
MG	Mouth Guard
MIC	Methyl Isocyanide
MMA	Methyl Methacrylate
MOSFET	Metal Oxide Semiconductor Field-effect Transistors
NP	Nanoparticle
NRs	Nanorods
NTC	Neotetrazolium Chloride
NTFET	Nanotube Field-effect Transistors
NWs	Nanowires
OFET	Organic Field-effect Transistor

PA	Polyamides
PBAT	Polybutyrate Adipate Terephthalate
PBS	Polybutylene Succinate
PC3 line	Prostatic Tumor Cells
PCL	Polycaprolactone
PDA	Pentacosadiynoic Acid
PDLA	Poly-D-lactic Acid D-PLA
PEI	Polyethylene
PEI	Poly(ethylene Imine)
PHA	Polyhydroxyalkanoates
PLA	Polylactic Acid
PLGA	Polylactic Acid-*co*-glycolic Acid
PLLA	Poly-L-lactic Acid
POC	Point-of-care
PP	Polypropylene
PS	Polystyrene
PSED	Paper-based Solid–Liquid Extraction Device
PU	Polyurethane
PVC	Poly(vinyl chloride)
RH	Relative Humidity
RNA	Ribonucleic Acid
SARS-CoV-2	Severe Acute Respiratory Syndrome Coronavirus 2
SERS	Surface-enhanced Raman Spectroscopy
SF	Silk Fibroin
SiNR	Silicon Nanoribbon
SoP	Sensor-on-probe
ssDNA	Single-stranded Deoxyribonucleic Acid
SWNT	Single-walled Carbon Nanotubes
VOC	Volatile Organic Compounds
γ-PGA	Poly(γ-D-glutamic acid)
ε-PL	Poly(ε-L-lysine)

References

1 Severinghaus, J.W. (2002). The invention and development of blood gas analysis apparatus. *Anesthesiology* 97 (1): 253–256.
2 Sberveglieri, G., Faglia, G., Groppelli, S., and Nelli, P. (1992). Methods for the preparation of NO, NO_2 and H_2 sensors based on tin oxide thin films, grown by means of the r.f. magnetron sputtering technique. *Sensors and Actuators B: Chemical* 8 (1): 79–88.

3 Wohltjen, H., Barger, W.R., Snow, A.W., and Jarvis, N.L. (1985). A vapor-sensitive chemiresistor fabricated with planar microelectrodes and a Langmuir-Blodgett organic semiconductor film. *IEEE Transactions on Electron Devices* 32 (7): 1170–1174.

4 Bergveld, P. (1985). The impact of MOSFET-based sensors. *Sensors and Actuators* 8 (2): 109–127.

5 Kong, J., Franklin, N.R., Zhou, C. et al. (2000). Nanotube molecular wires as chemical sensors. *Science* 287 (5453): 622–625.

6 Yi, C., Qingqiao, W., Hongkun, P., and Charles, M.L. (2001). Nanowire nanosensors for highly sensitive and selective detection of biological and chemical species. *Science* 293 (5533): 1289–1292.

7 Schedin, F., Geim, A.K., Morozov, S.V. et al. (2007). Detection of individual gas molecules adsorbed on graphene. *Nature Materials* 6 (9): 652–655.

8 Polk, B.J. (2022). ChemFET arrays for chemical sensing microsystems. *2002 IEEE Sensors*, Orlando, FL, USA (12–14 June 2002).

9 Moumen, A., Kumarage, G.C.W., and Comini, E. (2022). P-Type metal oxide semiconductor thin films: synthesis and chemical sensor applications. *Sensors* 22 (4): 1359.

10 Tiwari, N., Nirmal, A., Kulkarni, M.R. et al. (2020). Enabling high performance n-type metal oxide semiconductors at low temperatures for thin film transistors. *Inorganic Chemistry Frontiers* 7 (9): 1822–1844.

11 Kumar, C.V. et al. (2017). *Chapter 1 – Discovery of graphene and beyond*. In: *Introduction to Graphene* (ed. J. Fedor), 1–15. Elsevier.

12 Paul, A. and Bhattacharya, B. (2010). DNA functionalized carbon nanotubes for nonbiological applications. *Materials and Manufacturing Processes* 25 (9): 891–908.

13 Bhimanapati, G.R. et al. (2016). Chapter Three – 2D Boron Nitride: Synthesis and Applications. In: *2D Materials* (ed. F. Iacopi, J. Boeck, and C. Jagadish), 101–147. Elsevier.

14 Sahu, S. and Rout, G.C. (2017). Band gap opening in graphene: a short theoretical study. *International Nano Letters* 7 (2): 81–89.

15 Cui, X., Lee, G.-H., Kim, Y.D. et al. (2015). Multi-terminal transport measurements of MoS2 using a van der Waals heterostructure device platform. *Nature Nanotechnology* 10 (6): 534–540.

16 Krishnamoorthy, A., Baradwaj, N., Nakano, A. et al. (2021). Lattice thermal transport in two-dimensional alloys and fractal heterostructures. *Scientific Reports* 11 (1): 1656.

17 Kumar, C.V. and Pattammattel, A. (2017). Chapter 4 – Inorganic analogues of graphene: Synthesis, characterization, and applications. In: *Introduction to Graphene* (ed. J. Fedor), 75–101. Amsterdam, Netherlands: Elsevier.

18 Kang, M.-H., Lee, D., Sung, J. et al. (2019). Chapter 4 - structure and chemistry of 2D materials. In: *Comprehensive Nanoscience and Nanotechnology*, 2e (ed. D.L. Andrews, R.H. Lipson, and T. Nann), 55–90. Oxford: Academic Press, Elsevier.

19 Khan, M.I., Aziz, S.H., Majid, A., and Rizwan, M. (2021). Computational study of borophene/boron nitride (B/BN) interface as a promising gas sensor for industrial affiliated gasses. *Physica E: Low-dimensional Systems and Nanostructures* 130: 114692.

20 Goel, N. and Kumar, M. (2021). Recent advances in ultrathin 2D hexagonal boron nitride based gas sensors. *Journal of Materials Chemistry C* 9 (5): 1537–1549.

21 Roondhe, B., Jha, P.K., and Ahuja, R. (2020). Haeckelite boron nitride as nano sensor for the detection of hazardous methyl mercury. *Applied Surface Science* 506: 144860.

22 Doust Mohammadi, M., Abdullah, H.Y., Louis, H., and Mathias, G.E. (2022). 2D boron nitride material as a sensor for H_2SiCl_2. *Computational and Theoretical Chemistry* 1213: 113742.

23 Zhang, Y., Chan, J.W., Moretti, A., and Uhrich, K.E. (2015). Designing polymers with sugar-based advantages for bioactive delivery applications. *Journal of Controlled Release* 219: 355–368.

24 Stubbs, C.J., Worch, J.C., Prydderch, H. et al. (2022). Sugar-based polymers with stereochemistry-dependent degradability and mechanical properties. *Journal of the American Chemical Society* 144 (3): 1243–1250.

25 Petersen, S.R., Prydderch, H., Worch, J.C. et al. (2022). Ultra-tough elastomers from stereochemistry-directed hydrogen bonding in isosorbide-based polymers. *Angewandte Chemie International Edition* 61 (17): e202115904.

26 Jiang, T., Duan, Q., Zhu, J. et al. (2020). Starch-based biodegradable materials: challenges and opportunities. *Advanced Industrial and Engineering Polymer Research* 3 (1): 8–18.

27 Ma, C., Xie, F., Wei, L. et al. (2022). All-starch-based hydrogel for flexible electronics: strain-sensitive batteries and self-powered sensors. *ACS Sustainable Chemistry & Engineering* 10 (20): 6724–6735.

28 Zeng, S., Zhang, J., Zu, G., and Huang, J. (2021). Transparent, flexible, and multifunctional starch-based double-network hydrogels as high-performance wearable electronics. *Carbohydrate Polymers* 267: 118198.

29 Nevell, T.P. and Zeronian, S.H. (1985). *Cellulose Chemistry and Its Applications*. Halsted Press, John Wiley.

30 Chang, M., Song, T., Liu, X. et al. (2020). Cellulose-based biosensor for bio-molecules detection in medical diagnosis: a mini-review. *Current Medicinal Chemistry* 27 (28): 4593–4612.

31 Ratajczak, K. and Stobiecka, M. (2020). High-performance modified cellulose paper-based biosensors for medical diagnostics and early cancer screening: a concise review. *Carbohydrate Polymers* 229: 115463.

32 Kim, J.-H., Mun, S., Ko, H.-U. et al. (2014). Disposable chemical sensors and biosensors made on cellulose paper. *Nanotechnology* 25 (9): 092001.

33 Kamel, S. and Khattab, T.A. (2020). Recent advances in cellulose-based biosensors for medical diagnosis. *Biosensors* 10 (6): 67.

34 Kim, W., Lee, S.H., and Ahn, Y.J. (2018). A label-free cellulose SERS biosensor chip with improvement of nanoparticle-enhanced LSPR effects for early diagnosis of subarachnoid hemorrhage-induced complications. *Biosensors and Bioelectronics* 111: 59–65.

35 Arakawa, T., Tomoto, K., and Nitta, H. (2020). A wearable cellulose acetate-coated mouthguard biosensor for in vivo salivary glucose measurement. *Analytical Chemistry* 92 (18): 12201–12207.

36 Han, S., Zhuang, X., Jiang, Y. et al. (2017). Poly(vinyl alcohol) as a gas accumulation layer for an organic field-effect transistor ammonia sensor. *Sensors and Actuators B: Chemical* 243: 1248–1254.

37 Dhiman, T.K., Poddar, M., Lakshmi, G.B.V.S. et al. (2021). Non-enzymatic andÂ rapid detection of glucose on PVA-CuO thin film using ARDUINO UNO based capacitance measurement unit. *Biomedical Microdevices* 23 (3): 36.

38 Dong, X., Tong, S., Dai, K. et al. (2022). Preparation of PVA/PAM/Ag strain sensor via compound gelation. *Journal of Applied Polymer Science* 139 (14): 51883.

39 Peng, Y., Yan, B., Li, Y. et al. (2020). Antifreeze and moisturizing high conductivity PEDOT/PVA hydrogels for wearable motion sensor. *Journal of Materials Science* 55 (3): 1280–1291.

40 de Jong, T.L., Pluymen, L.H., van Gerwen, D.J. et al. (2017). PVA matches human liver in needle-tissue interaction. *Journal of the Mechanical Behavior of Biomedical Materials* 69: 223–228.

41 Zhou, J., Long, X., Huang, J. et al. (2022). Multiscale and hierarchical wrinkle enhanced graphene/ecoflex sensors integrated with human-machine interfaces and cloud-platform. *NPJ Flexible Electronics* 6 (1): 55.

42 Rocha Neto, J.O.B.M., Soares, A.C., Bataglioli, R.r.A. et al. (2020). Polysaccharide multilayer films in sensors for detecting prostate tumor cells based on hyaluronan-CD44 interactions. *Cells* 9 (6): 1563.

43 Vaňková, E., KaÅiparovÃi, P., Khun, J. et al. (2020). Polylactic acid as a suitable material for 3D printing of protective masks in times of COVID-19 pandemic. *PeerJ* 8: e10259–e10259.

44 Zhang, X., Tsukada, M., Morikawa, H. et al. (2011). Production of silk sericin/silk fibroin blend nanofibers. *Nanoscale Research Letters* 6 (1): 1–8.

45 Tang, X., Ye, X., Wang, X. et al. (2021). High mechanical property silk produced by transgenic silkworms expressing the spidroins PySp1 and ASG1. *Scientific Reports* 11 (1): 20980.

46 Shao, Z. and Vollrath, F. (2002). Surprising strength of silkworm silk. *Nature* 418 (6899): 741–741.

47 Andersson, M., Johansson, J., and Rising, A. (2016). Silk spinning in silkworms and spiders. *International Journal of Molecular Sciences* 17 (8): 1290.

48 Fernández-García, L., MarÃ-BuyÃ, N., Barios, J.A. et al. (2016). Safety and tolerability of silk fibroin hydrogels implanted into the mouse brain. *Acta Biomaterialia* 45: 262–275.

49 Yong, J., Hassan, B., Liang, Y. et al. (2017). A silk fibroin bio-transient solution processable memristor. *Scientific Reports* 7 (1): 14731.

50 Wen, D.-L., Sun, D.-H., Huang, P. et al. (2021). Recent progress in silk fibroin-based flexible electronics. *Microsystems & Nanoengineering* 7 (1): 35.

51 Musameh, M.M., Dunn, C.J., Uddin, M.H. et al. (2017). Silk provides a new avenue for third generation biosensors: sensitive, selective and stable electrochemical detection of nitric oxide. *Biosensors and Bioelectronics* 103: 26–31.

52 Zhang, X.-S., Brugger, J.R., and Kim, B. (2016). A silk-fibroin-based transparent triboelectric generator suitable for autonomous sensor network. *Nano Energy* 20: 37–47.

53 Prakash, N.J., Mane, P.P., George, S.M., and Kandasubramanian, B. (2021). Silk fibroin as an immobilization matrix for sensing applications. *ACS Biomaterials Science & Engineering* 7: 2015–2042.

54 Banerjee, P., Lenz, D., Robinson, J.P. et al. (2008). A novel and simple cell-based detection system with a collagen-encapsulated B-lymphocyte cell line as a biosensor for rapid detection of pathogens and toxins. *Laboratory Investigation* 88 (2): 196–206.

55 Norizan, M.N., Moklis, M.H., Ngah Demon, S.Z. et al. (2020). Carbon nanotubes: functionalisation and their application in chemical sensors. *RSC Advances* 10 (71): 43704–43732.

56 Janudin, N., Abdullah, N., Wan Yunus, W.M.Z. et al. (2018). Effect of functionalized carbon nanotubes in the detection of benzene at room temperature. *Journal of Nanotechnology* 2018: 2107898.

57 Liu, H., Liu, Y., and Zhu, D. (2010). Chemical doping of graphene. *Journal of Materials Chemistry* 21 (10): 3335–3345.

58 Denis, P.A. (2010). Band gap opening of monolayer and bilayer graphene doped with aluminium, silicon, phosphorus, and sulfur. *Chemical Physics Letters* 492 (4): 251–257.

59 Mishyn, V., Rodrigues, T., Leroux, Y.R. et al. (2021). Controlled covalent functionalization of a graphene-channel of a field effect transistor as an ideal platform for (bio)sensing applications. *Nanoscale Horizons* 6 (10): 819–829.

60 Zheng, M., Jagota, A., Semke, E.D. et al. (2003). DNA-assisted dispersion and separation of carbon nanotubes. *Nature Materials* 2 (May): 338–342.

61 Strano, M.S., Zheng, M., Jagota, A. et al. (2004). Understanding the nature of the DNA-assisted separation of single-walled carbon nanotubes using fluorescence and Raman spectroscopy. *Nano Letters* 4 (4): 543–550.

62 Paul, A., Pramanick, B., Bhattacharya, B., and Bhattacharyya, T.K. (2013). Deoxyribonucleic acid functionalized carbon nanotube network as humidity sensors. *IEEE Sensors Journal* 13 (5): 1806–1816.

63 Karachevtsev, V.A., Glamazda, A.Y., Leontiev, V.S. et al. (2007). Glucose sensing based on NIR fluorescence of DNA-wrapped single-walled carbon nanotubes. *Chemical Physics Letters* 435 (1-3): 104–108.

64 Xu, Y., Pehrsson, P.E., Chen, L. et al. (2007). Double-stranded DNA single-walled carbon nanotube hybrids for optical hydrogen peroxide and glucose sensing. *The Journal of Physical Chemistry C* 111 (24): 8638–8643.

65 Liang, Z., Lao, R., Wang, J. et al. (2007). Solubilization of single-walled carbon nanotubes with single-stranded DNA generated from asymmetric PCR. *International Journal of Molecular Sciences* 8: 705–713.

66 Xu, Y., Pehrsson, P.E., Chen, L., and Zhao, W. (2008). Controllable redox reaction of chemically purified DNA-single walled carbon nanotube hybrids with hydrogen peroxide. *Journal of the American Chemical Society* 130 (31): 10054–10055.

67 Viswanathan, S., Radecka, H., and Radecki, J. (2009). Electrochemical biosensor for pesticides based on acetylcholinesterase immobilized on polyaniline deposited on vertically assembled carbon nanotubes wrapped with ssDNA. *Biosensors and Bioelectronics* 24 (9): 2772–2777.

68 Wu, Z., Zhen, Z., Jiang, J.-H. et al. (2009). Terminal protection of small-molecule-linked DNA for sensitive electrochemical detection of protein binding via selective carbon nanotube assembly. *Journal of the American Chemical Society* 131 (34): 12325–12332.

69 Jin, H., Jeng, E.S., Heller, D.A. et al. (2007). Divalent ion and thermally induced DNA conformational polymorphism on single-walled carbon nanotubes. *Macromolecules* 40 (18): 6731–6739.

70 Chen, C.-L., Yang, C.-F., Agarwal, V. et al. (2010). DNA-decorated carbon-nanotube-based chemical sensors on complementary metal oxide semiconductor circuitry. *Nanotechnology* 21 (9): 095504. (1–8).

71 Gao, X., Xing, G., Yang, Y. et al. (2008). Detection of trace Hg^{2+} via induced circular dichroism of DNA wrapped around single-walled carbon nanotubes. *Journal of the American Chemical Society* 130: 9190–9191.

72 Su, H.C., Zhang, M., Bosze, W. et al. (2013). Metal nanoparticles and DNA co-functionalized single-walled carbon nanotube gas sensors. *Nanotechnology* 24: 505502–505512.

73 Paul, A., Bhattacharya, B., and Bhattacharyya, T.K. (2014). Fabrication and performance of solution-based micropatterned DNA functionalized carbon nanotube network as humidity sensors. *IEEE Transactions on Nanotechnology* 13 (2): 335–342.

74 Paul, A., Pramanick, B., Bhattacharya, B., and Bhattacharyya, T.K. (2013). DNA functionalized carbon nanotube network as humidity sensors. *IEEE Sensors Journal* 13 (5): 1806–1816.

75 Tu, W., Lei, J., Zhang, S., and Ju, H. (2010). Characterization, direct electrochemistry, and amperometric biosensing of graphene by noncovalent functionalization with picket-fence porphyrin. *Chemistry* 16 (35): 10771–10777.

76 Zhang, S., S. Tang, J. Lei, et al., Functionalization of graphene nanoribbons with porphyrin for electrocatalysis and amperometric biosensing. *Journal of Electroanalytical Chemistry*, 2011. 656(1–2): p. 285–288.

77 Qi, X., Pu, K.Y., Li, H. et al. (2010). Amphiphilic graphene composites. *Angewandte Chemie International Edition* 49 (49): 9426–9429.

78 Kerscher, B., Appel, A.-K., Thomann, R., and Muȋ^lhaupt, R. (2013). Treelike polymeric ionic liquids grafted onto graphene nanosheets. *Macromolecules* 46 (11): 4395–4402.

79 Mahadeva, S.K. and Kim, J. (2011). Conductometric glucose biosensor made with cellulose and tin oxide hybrid nanocomposite. *Sensors and Actuators B: Chemical* 157: 177–182.

80 Tariq, A., Bhawani, S.A., Nisar, M. et al. (2021). 13 – Starch-based nanocomposites for gene delivery. In: *Polysaccharide-Based Nanocomposites for Gene Delivery and Tissue Engineering* (ed. S.A. Bhawani, Z. Karim, and M. Jawaid), 263–277. Woodhead Publishing.

81 Liu, Y., Zhong, M., Shan, G. et al. (2008). Biocompatible ZnO/Au nanocomposites for ultrasensitive DNA detection using resonance Raman scattering. *The Journal of Physical Chemistry B* 112 (20): 6484–6489.

82 Xie, F., E. Pollet, P.J. Halley, and L. Avérous, Starch-based nano-biocomposites. *Progress in Polymer Science*, 2013. 38(10): p. 1590–1628.

83 Bai, H., Liang, Z., Wang, D. et al. (2021). Biopolymer nanocomposites with customized mechanical property and exceptionally antibacterial performance. *Composites Science and Technology* 199: 108338.

84 Sumrith, N., Rangappa, S.M., Dangtungee, R. et al. (2019). Biopolymers-based nanocomposites: properties and applications. In: *Bio-based Polymers and Nanocomposites: Preparation, Processing, Properties & Performance* (ed. M.L. Sanyang and M. Jawaid), 255–272. Cham: Springer International Publishing.

85 Kafi, M.A., Paul, A., Vilouras, A., and Dahiya, R. (2020). Mesoporous chitosan based conformable and resorbable biostrip for dopamine detection. *Biosensors and Bioelectronics* 147: 111781.

86 Farea, M.A., Mohammed, H.Y., Sayyad, P.W. et al. (2021). Carbon monoxide sensor based on polypyrrole–graphene oxide composite: a cost-effective approach. *Applied Physics A* 127 (9): 681.

87 Khan, M., M.N. Tahir, S.F. Adil, et al., Graphene based metal and metal oxide nanocomposites: synthesis, properties and their applications. *Journal of Materials Chemistry A*, 2015. 3(37): p. 18753–18808.

88 Hu, T. and Gerber, I. (2013). Theoretical study of the interaction of electron donor and acceptor molecules with graphene. *The Journal of Physical Chemistry C* 117: 2411–2420.

89 Varghese, N., Ghosh, A., Voggu, R. et al. (2009). Selectivity in the interaction of electron donor and acceptor molecules with graphene and single-walled carbon nanotubes. *Journal of Physical Chemistry C* 113: 16855–16859.

90 Wang, Y., Liu, A., Han, Y., and Li, T. (2020). Sensors based on conductive polymers and their composites: a review. *Polymer International* 69 (1): 7–17.

91 Nambiar, S. and Yeow, J.T.W. (2011). Conductive polymer-based sensors for biomedical applications. *Biosensors and Bioelectronics* 26 (5): 1825–1832.

92 Liu, X., Zheng, W., Kumar, R. et al. (2022). Conducting polymer-based nanostructures for gas sensors. *Coordination Chemistry Reviews* 462: 214517.

93 Leyden, M.R., Schuman, C., Sharf, T. et al. (2010). Fabrication and characterization of carbon nanotube field-effect transistor biosensors. In: *Organic Semiconductors in Sensors and Bioelectronics: Proceedings of the SPIE*, vol. 7779, 77790H–77790H-11.

94 Mahar, B., C. Laslau, R. Yip, and Y. Sun, Development of carbon nanotube-based sensors—a review. *IEEE Sensors Journal*, 2007. 7(2): p. 266–284.

95 Li, J., Lu, Y., and Meyyappan, M. (2006). Nano chemical sensors with polymer-coated carbon nanotubes. *IEEE Sensors Journal* 6 (5): 1047–1051.

96 Tans, S.J., Devoret, M.H., Dai, H. et al. (1997). Individual single-wall carbon nanotubes as quantum wires. *Nature* 386 (6624): 474–477.

97 Staii, C., Alan, J., and Johnson, T. (2005). DNA-decorated carbon nanotubes for chemical sensing. *Nano Letters* 5 (9): 1774–1778.

98 Lee, Y.D., Cho, W.-S., Moon, S.-I. et al. (2006). Gas sensing properties of printed multiwalled carbon nanotubes using the field emission effect. *Chemical Physics Letters* 433 (1–3): 105–109.

99 Byon, H.R. and Choi, H.C. (2006). Network single-walled carbon nanotube-field effect transistors (SWNT-FETs) with increased Schottky contact area for highly sensitive biosensor applications. *Journal of the American Chemical Society* 128: 2188–2189.

100 Barone, P.W., Baik, S., Heller, D.A., and Strano, M.S. (2005). Near-infrared optical sensors based on single-walled carbon nanotubes. *Nature Materials* 4 (1): 86–92.

101 Bestel, I., Campins, N., Marchenko, A. et al. (2008). Two-dimensional self-assembly and complementary base-pairing between amphiphile nucleotides on graphite. *Journal of Colloid and Interface Science* 323 (2): 435–440.

102 Paul, A., Bhattacharya, B., and Bhattacharyya, T. (2015). Selective detection of Hg(II) over Cd(II) and Pb(II) ions by DNA functionalized CNT network. *IEEE Sensors Journal* 15 (5): 2774–2779.

103 Hammett, L.P. (1937). The effect of structure upon the reactions of organic compounds. Benzene derivatives. *Journal of the American Chemical Society* 59 (1): 96–103.

104 Star, A., Han, T.-R., Gabriel, J.-C.P. et al. (2003). Interaction of aromatic compounds with carbon nanotubes: correlation to the Hammett parameter of the substituent and measured carbon nanotube FET response. *Nano Letters* 3 (10): 1421–1423.

105 Heinze, S., Tersoff, J., Martel, R. et al. (2002). Carbon nanotubes as Schottky barrier transistors. *Physical Review Letters* 89 (10): 106801.

106 Valentini, L., Cantalini, C., Armentano, I. et al. (2004). Highly sensitive and selective sensors based on carbon nanotubes thin films for molecular detection. *Diamond and Related Materials* 13 (4-8): 1301–1305.

107 Li, J., Lu, Y., Ye, Q. et al. (2003). Carbon nanotube sensors for gas and organic vapor detection. *Nano Letters* 3 (7): 929–933.

108 Anoshkin, I.V., Nasibulin, A.G., Mudimela, P.R. et al. (2013). Single-walled carbon nanotube networks for ethanol vapor sensing applications. *Nano Research* 6 (2): 77–86.

109 Kuzmych, O. and Brett, L.A.A.S. (2007). Carbon nanotube sensors for exhaled breath components. *Nanotechnology* 18 (37): 375502.

110 Vetter, S., Haffer, S., Wagner, T., and Tiemann, M. (2015). Nanostructured Co3O4 as a CO gas sensor: temperature-dependent behavior. *Sensors and Actuators B: Chemical* 206: 133–138.

111 Zhang, J., Zhu, Z., Chen, C. et al. (2018). ZnO-carbon nanofibers for stable, high response, and selective H_2S sensors. *Nanotechnology* 29 (27): 275501.

112 Liu, N., Li, T.-t., Yu, H., and Xia, L. (2018). Fabrication of a disordered mesoporous ZnO matrix modified by CuO film as high-performance NO_x sensor. *ChemistrySelect* 3 (19): 5377–5385.

113 Ionescu, R., Hoel, A., Granqvist, C.G. et al. (2005). Ethanol and H_2S gas detection in air and in reducing and oxidising ambience: application of pattern recognition to analyse the output from temperature-modulated nanoparticulate WO3 gas sensors. *Sensors and Actuators B: Chemical* 104 (1): 124–131.

114 Zhang, W., Li, Q., Wang, C. et al. (2019). High sensitivity and selectivity chlorine gas sensors based on 3D open porous SnO_2 synthesized by solid-state method. *Ceramics International* 45 (16): 20566–20574.

115 Wang, D., Hu, P., Xu, J. et al. (2009). Fast response chlorine gas sensor based on mesoporous SnO_2. *Sensors and Actuators B: Chemical* 140 (2): 383–389.

116 Krajewski, A., Houshyar, S., Wang, L., and Padhye, R. (2022). Chlorine gas sensor with surface temperature control. *Sensors* 22 (12): 4643.

117 Joshi, A., Aswal, D.K., Gupta, S.K. et al. (2009). ZnO-nanowires modified polypyrrole films as highly selective and sensitive chlorine sensors. *Applied Physics Letters* 94 (10): 103115.

118 Baron, M.G., Narayanaswamy, R., and Thorpe, S.C. (1995). A kineto-optical method for the determination of chlorine gas. *Sensors and Actuators B: Chemical* 29 (1–3): 358–362.

119 Grabtree, E.V., E.J. Poziomek, and D.J. Hoy, Isocyanide indicator, in Google Patents, US Patent 3, 382, Editor. 1971, Office of the Secretary of the Army Washington, DC.

120 Westrup, B. and Marcoll, J. (1986). Indicator for determining organic isocyanates on a carrier. Google Patents, U.P. Office, Editor. USA: Dragerwerk Aktiengesellschaft.

121 Samuilov, A.Y., Nesterov, S.V., Balabanova, F.B. et al. (2014). Quantum-chemical study of isocyanate reactions with linear methanol associates: IX. Methyl isocyanate reaction with methanol-phenol complexes. *Russian Journal of Organic Chemistry* 50 (2): 155–159.

122 Chen, K., Lu, G., Chang, J. et al. (2012). Hg(II) ion detection using thermally reduced graphene oxide decorated with functionalized gold nanoparticles. *Analytical Chemistry* 84 (9): 4057–4062.

123 McGhee, C.E., Loh, K.Y., and Lu, Y. (2017). DNAzyme sensors for detection of metal ions in the environment and imaging them in living cells. *Current Opinion in Biotechnology* 45: 191–201.

124 Synhaivska, O., Mermoud, Y., Baghernejad, M. et al. (2019). Detection of Cu^{2+} ions with GGH peptide realized with Si-nanoribbon ISFET. *Sensors* 19 (18): 4022.

125 Napper, I.E. and Thompson, R.C. (2016). Release of synthetic microplastic plastic fibres from domestic washing machines: effects of fabric type and washing conditions. *Marine Pollution Bulletin* 112 (1–2): 39–45.

126 Raju, S., Carbery, M., Kuttykattil, A. et al. (2020). Improved methodology to determine the fate and transport of microplastics in a secondary wastewater treatment plant. *Water Research* 173: 115549.

127 Franco, A.A., Arellano, J.M., AlbendÃn, G. et al. (2021). Microplastic pollution in wastewater treatment plants in the city of Cádiz: abundance, removal efficiency and presence in receiving water body. *Science of the Total Environment* 776: 145795.

128 Yang, Z., Li, S., Ma, S. et al. (2021). Characteristics and removal efficiency of microplastics in sewage treatment plant of Xi'an City, northwest China. *Science of the Total Environment* 771: 145377.

129 Nkosi, S.D., Malinga, S.P., and Mabuba, N. (2022). Microplastics and heavy metals removal from fresh water and wastewater systems using a membrane. *Separations* 9 (7): 166.

130 Lim, D.-I., Cha, J.-R., and Gong, M.-S. (2013). Preparation of flexible resistive micro-humidity sensors and their humidity-sensing properties. *Sensors and Actuators B: Chemical* 183: 574–582.

131 Kim, M.-J. and Gong, M.-S. (2012). Water-resistive humidity sensor prepared by printing process using polyelectrolyte ink derived from new monomer. *Analyst* 137 (6): 1487–1494.

132 Aziz, S., Chang, D.E., Doh, Y.H. et al. (2015). Humidity sensor based on PEDOT: PSS and zinc stannate nano-composite. *Journal of Electronic Materials* 44 (10): 3992–3999.

133 Zhang, D., Chang, H., and Liu, R. (2016). Humidity-sensing properties of one-step hydrothermally synthesized tin dioxide-decorated graphene nanocomposite on polyimide substrate. *Journal of Electronic Materials* 45 (8): 4275–4281.

134 Zhang, R., Peng, B., and Yuan, Y. (2018). Flexible printed humidity sensor based on poly(3, 4-ethylenedioxythiophene)/reduced graphene oxide/Au nanoparticles with high performance. *Composites Science and Technology* 168: 118–125.

135 Komazaki, Y. and Uemura, S. (2019). Stretchable, printable, and tunable PDMS-CaCl2 microcomposite for capacitive humidity sensors on textiles. *Sensors and Actuators B: Chemical* 297: 126711.

136 Burman, D., Santra, S., Pramanik, P., and Guha, P.K. (2018). Pt decorated MoS2 nanoflakes for ultrasensitive resistive humidity sensor. *Nanotechnology* 29 (11): 115504.

137 Huang, T.-H., Chou, J.-C., Sun, T.-P., and Hsiung, S.-K. (2008). A device for skin moisture and environment humidity detection. *Sensors and Actuators B: Chemical* 134 (1): 206–212.

138 Wang, X., Yuan, W., and Sang, M.-H. (2015). An optical relative humidity sensor based on the enhanced Goos-Hanchen shift. *2015 Opto-Electronics and Communications Conference (OECC)*, Shanghai, China (28 June–2 July 2015). IEEE Xplore, pp. 1–3.

139 Sun, C., Shi, Q., Yazici, M.S. et al. (2018). Development of a highly sensitive humidity sensor based on a piezoelectric micromachined ultrasonic transducer array functionalized with graphene oxide thin film. *Sensors* 18 (12): 4352.

140 Kim, W., Javey, A., Vermesh, O. et al. (2003). Hysteresis caused by water molecules in carbon nanotube field-effect transistors. *Nano Letters* 3 (2): 193–198.

141 Li, Y., Yang, M.J., and Chen, Y. (2005). Nanocomposites of carbon nanotubes and silicone-containing polyelectrolyte as a candidate for construction of humidity sensor. *Journal of Materials Science* 40: 245–247.

142 Liu, L., Ye, X., Wu, K. et al. (2009). Humidity sensitivity of carbon nanotube and poly(dimethyldiallylammonium chloride) composite films. *IEEE Sensors Journal* 9 (10): 1308–1314.

143 Su, P.-G., Sun, Y.-L., and Lin, C.-C. (2006). A low humidity sensor made of quartz crystal microbalance coated with multi-walled carbon nanotubes/Nafion composite material films. *Sensors and Actuators B: Chemical* 115 (1): 338–343.

144 Jeong, H., Noh, Y., and Lee, D. (2019). Highly stable and sensitive resistive flexible humidity sensors by means of roll-to-roll printed electrodes and flower-like TiO_2 nanostructures. *Ceramics International* 45 (1): 985–992.

145 Chanishvili, A., Chilaya, G., Petriashvili, G. et al. (2005). Cholesteric liquid crystal mixtures sensitive to different ranges of solar UV irradiation. *Molecular Crystals and Liquid Crystals* 434 (1): 25/[353]–38/[366].

146 Munakata, N. (1989). Genotoxic action of sunlight upon *Bacillus subtilis* spores: monitoring studies at Tokyo, Japan. *Journal of Radiation Research* 30 (4): 338–351.

147 Karentz, D. and Lutze, L.H. (1990). Evaluation of biologically harmful ultraviolet radiation in Antarctica with a biological dosimeter designed for aquatic environments. *Limnology and Oceanography* 35 (3): 549–561.

148 Hegedus, M., Modos, K., RontÃ, G., and Fekete, A. (2003). Validation of phage T7 biological dosimeter by quantitative polymerase chain reaction using short and long segments of phage T7 DNA. *Photochemistry and Photobiology* 78 (3): 213–219.

149 Smith, G.J., Ultraviolet radiation actinometer, in Google Patents, U.P. Office, Editor. 1988, *Melbourne Grattan Street Parkville Victoria Australia a Body Politic and Corporate Established Under Melbourne University Act of State of Victoria*, University of New Zealand Government Property Corp A Statutory Body of New Zealand New Zealand Government Property Corp: USA.

150 Moroson, H. and Gregoriades, A. (1964). A sensitive chemical actinometer for ultra-violet radiation. *Nature* 204 (4959): 676–678.

151 Mills, A., Grosshans, P., and McFarlane, M. (2009). UV dosimeters based on neotetrazolium chloride. *Journal of Photochemistry and Photobiology A: Chemistry* 201 (2): 136–141.

152 Kozicki, M., Sąsiadek, E., Kadłubowski, S. et al. (2018). Flat foils as UV and ionising radiation dosimeters. *Journal of Photochemistry and Photobiology A: Chemistry* 351: 179–196.

153 Sou, I.K., Wu, M.C.W., Wong, K.S., and Wong, G.K.-L. (2003). ZnMgS-based UV detectors. In: (ed. Google patents, U.P. office). USA: Hong Kong University of Science and Technology HKUST.

154 Besteman, K., Lee, J.-O., Wiertz, F.G.M. et al. (2003). Enzyme-coated carbon nanotubes as single-molecule biosensors. *Nano Letters* 3 (6): 727–730.

155 Lerner, M.B., Kybert, N., Mendoza, R. et al. (2013). Scalable, non-invasive glucose sensor based on boronic acid functionalized carbon nanotube transistors. *Applied Physics Letters* 102 (18): 183113.

156 Loutfi, A., Coradeschi, S., Mani, G.K. et al. (2015). Electronic noses for food quality: a review. *Journal of Food Engineering* 144: 103–111.

157 Janssen, S., Schmitt, K., Blanke, M. et al. (2014). Ethylene detection in fruit supply chains. *Philosophical Transactions of the Royal Society A: Mathematical, Physical and Engineering Sciences* 372 (2017): 20130311.

158 Brezmes, J., Llobet, E., Vilanova, X. et al. (2000). Fruit ripeness monitoring using an electronic nose. *Sensors and Actuators B: Chemical* 69 (3): 223–229.

159 Esser, B., Schnorr, J.M., and Swager, T.M. (2012). Selective detection of ethylene gas using carbon nanotube-based devices: utility in determination of fruit ripeness. *Angewandte Chemie International Edition* 51 (23): 5752–5756.

160 Nie, W., Chen, Y., Zhang, H. et al. (2022). A novel colorimetric sensor array for real-time and on-site monitoring of meat freshness. *Analytical and Bioanalytical Chemistry* 414: 6017–6027.

161 Caselli, M., Ronchi, M., and Boni, A. (2021). An integrated low power temperature sensor for food monitoring applications. *2021 IEEE International Symposium on Circuits and Systems (ISCAS)*, Daegu, Korea (22–28 May 2021).

162 Xiao, X., Mu, B., Cao, G. et al. (2022). Flexible battery-free wireless electronic system for food monitoring. *Journal of Science: Advanced Materials and Devices* 7 (2): 100430.

163 Guan, Y. and Sun, B. (2020). Detection and extraction of heavy metal ions using paper-based analytical devices fabricated via atom stamp printing. *Microsystems & Nanoengineering* 6 (1): 14.

164 Jmaiff Blackstock, L.K., Wang, W., Vemula, S. et al. (2017). Sweetened swimming pools and hot tubs. *Environmental Science & Technology Letters* 4 (4): 149–153.

165 Yang, Y., Lin, Z.-H., Hou, T. et al. (2012). Nanowire-composite based flexible thermoelectric nanogenerators and self-powered temperature sensors. *Nano Research* 5 (12): 888–895.

166 Ghosh, S.K., T.K. Sinha, M. Xie, et al. (2020). Temperature–pressure hybrid sensing all-organic stretchable energy harvester. *ACS Applied Electronic Materials* 3 (1): 248–259.

167 Lin, Z., Chen, J., Li, X. et al. (2017). Triboelectric nanogenerator enabled body sensor network for self-powered human heart-rate monitoring. *ACS Nano* 11 (9): 8830–8837.

168 Lochner, C.M., Khan, Y., Pierre, A., and Arias, A.C. (2014). All-organic optoelectronic sensor for pulse oximetry. *Nature Communications* 5 (1): 1–7.

169 Choi, J.H. and Kim, D.K. (2009). A remote compact sensor for the real-time monitoring of human heartbeat and respiration rate. *IEEE Transactions on Biomedical Circuits and Systems* 3 (3): 181–188.

170 Meng, K., Chen, J., Li, X. et al. (2019). Flexible weaving constructed self-powered pressure sensor enabling continuous diagnosis of cardiovascular disease and measurement of cuffless blood pressure. *Advanced Functional Materials* 29 (5): 1806388.

171 Shichiri, M., Asakawa, N., Yamasaki, Y. et al. (1986). Telemetry glucose monitoring device with needle-type glucose sensor: a useful tool for blood glucose monitoring in diabetic individuals. *Diabetes Care* 9 (3): 298–301.

172 Muankid, A. and Ketcham, M. (2019). The real-time electrocardiogram signal monitoring system in wireless sensor network. *International Journal of Online & Biomedical Engineering* 15 (2).

173 Veer, K. (2016). Development of sensor system with measurement of surface electromyogram signal for clinical use. *Optik* 127 (1): 352–356.

174 Sullivan, T.J., Deiss, S.R., and Cauwenberghs, G. (2007). A low-noise, non-contact EEG/ECG sensor. *2007 IEEE Biomedical Circuits and Systems Conference*, Montreal, QC, Canada (27–30 November 2007). IEEE.

175 Gao, Y., Zhang, C., Yang, Y. et al. (2021). A high sensitive glucose sensor based on Ag nanodendrites/Cu mesh substrate via surface-enhanced Raman spectroscopy and electrochemical analysis. *Journal of Alloys and Compounds* 863: 158758.

176 Pervan, P., Svagusa, T., Prkacin, I. et al. (2017). Urine high sensitive Troponin I measuring in patients with hypertension. *Signa Vitae* 13 (3): 62–64.

177 Mano, T., Nagamine, K., Ichimura, Y. et al. (2018). Printed organic transistor-based enzyme sensor for continuous glucose monitoring in wearable healthcare applications. *ChemElectroChem* 5 (24): 3881–3886.

178 Lee, H., Hong, Y.J., Baik, S. et al. (2018). Enzyme-based glucose sensor: from invasive to wearable device. *Advanced Healthcare Materials* 7 (8): 1701150.

179 McGrath, M.J. and Scanaill, C.N. (2013). *Sensor Technologies: Healthcare, Wellness, and Environmental Applications*. Springer Nature.

180 Kazanskiy, N.L., Butt, M.A., and Khonina, S.N. (2022). Recent advances in wearable optical sensor automation powered by battery versus skin-like battery-free devices for personal healthcare – a review. *Nanomaterials* 12 (3): 334.

181 Bandodkar, A.J. and Wang, J. (2014). Non-invasive wearable electrochemical sensors: a review. *Trends in Biotechnology* 32 (7): 363–371.

182 Kim, J., Jeerapan, I., Imani, S. et al. (2016). Noninvasive alcohol monitoring using a wearable tattoo-based iontophoretic-biosensing system. *ACS Sensors* 1 (8): 1011–1019.

183 De Guzman, K. and Morrin, A. (2017). Screen-printed tattoo sensor towards the non-invasive assessment of the skin barrier. *Electroanalysis* 29 (1): 188–196.

184 Vilouras, A., Paul, A., Kafi, M., and Dahiya, R. (2018). Graphene oxide-chitosan based ultra-flexible electrochemical sensor for detection of serotonin. In: *2018 IEEE SENSORS*. New Delhi: IEEE.

185 Lãpez, P., R. Triviño, D. Calderón et al. (2017). Electronic nose prototype for explosive detection. *2017 CHILEAN Conference on Electrical, Electronics Engineering, Information and Communication Technologies (CHILECON)*, Pucon, Chile (18–20 October 2017).

186 Fennell, J.F. Jr., Hamaguchi, H., Yoon, B., and Swager, T.M. (2017). Chemiresistor devices for chemical warfare agent detection based on polymer wrapped single-walled carbon nanotubes. *Sensors* 17 (5): 982.

187 Fennell, J.F. Jr. (2017). *Polymer and Covalent Functionalization of Single Walled Carbon Nanotubes for Electronic Sensor Applications*. Massachusetts Institute of Technology.

188 Joshi, K.A., Prouza, M., Kum, M. et al. (2006). V-Type nerve agent detection using a carbon nanotube-based amperometric enzyme electrode. *Analytical Chemistry* 78 (1): 331–336.

189 Lewis, N.S. (2001). *Electronic Nose Chip Microsensors for Chemical Agent and Explosives Detection*. Pasadena, California: California Institute of Technology, Noyes Laboratory.

7

Optical Sensors

7.1 Introduction

Solid-state optical sensors majorly are of three types, such as a photodiode, photo-resistor/conductor, and photovoltaic device. In these cases, the optical devices either emit (i.e. LED, laser) or detect (i.e. photodiode, light-dependent resistor, solar cell) optical energy. However, several other types of optical sensors are based on different optical phenomena such as absorbance, surface plasmon resonance (SPR), scattering, spectrometry, dichroism, and luminescence. There are optical sensors that combine a few of these optical properties. For example, colorimetric sensors are mostly based on the reflection or transmission of light that involves a light source (laser or LED) and a photodetector (photodiode or LDR) for measurement purposes.

Optical sensing has emerged as one of the major sensing technologies in recent years. Several optical materials have emerged to enhance the performance of optical sensors and they have been used in a wide range of applications such as water quality monitoring, health care, toxic material, and microorganism detection.

Aside from the core solid-state devices, there are a few physical phenomena that also are used extensively for different optical sensing along with different solid-state devices. Hence, it is also important to discuss these in brief to provide some insight that will help in understanding optical sensing techniques in general.

Optical sensors are very important in different fields of sensing due to the high sensitivity, tunable optical properties, and ease of fabrication. Conventional electronic devices such as CMOS and FET-based sensor fabrication require high-end fabrication facilities that make it unsuitable for affordable device fabrications. On the other hand, in modern-day health care, environmental challenges require rapid and affordable systems to combat the problems. In this direction, solid-state optical sensors provide an affordable solution. The major advantage of using an optical sensor is the construction of the device. Mostly, the sensors are photodetector that is planner in nature and contains a film or track of optical sensing material. A sensing system that generates light requires

Solid-State Sensors, First Edition. Ambarish Paul, Mitradip Bhattacharjee, and Ravinder Dahiya.
© 2024 The Institute of Electrical and Electronics Engineers, Inc.
Published 2024 by John Wiley & Sons, Inc.

a photodetector for the sensing purpose. In some systems, the reflection, scattering, and absorption phenomenon of light is used and, in these cases, a light source and photodetector both are used for sensing purposes. This is an indirect process of sensing where the sensing mechanisms are not disturbed by the electrical readout arrangement. However, in most of the other sensors, such as chemical, capacitive, and resistive, the electrodes often create hindrance in sensing. Thus, optimization and proper fabrication of electrodes are one of the major steps in these sensors.

In this chapter, we discuss different optical phenomena in the context of solid-state sensors. A generalized optical sensing phenomenon is illustrated schematically in Figure 7.1. Furthermore, a description of different smart and

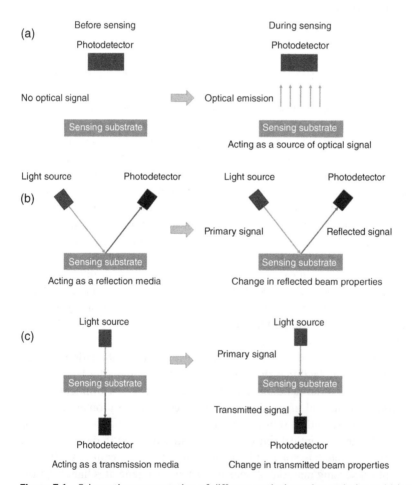

Figure 7.1 Schematic representation of different optical sensing techniques (a) based on the photodetector and light-emitting sensors, (b) based on a light source and photodetector in reflection mode, and (c) based on the light source and photodetector in transmission mode.

nanomaterials for solid-state sensor fabrications are given. The chapter also contains a description of fabrication techniques and various applications. Finally, the future of optical sensors along with existing limitations is described to conclude the discussion.

7.2 Classifications of Optical Properties

7.2.1 Absorbance

Optical absorbance [1] is the measure of light absorption when it passes through a material. Almost all the materials have optical absorption properties but the capacity of absorbing the light varies from material to material. Depending on the applications, people use different light-absorbing materials for fabricating different devices or sensors. Applications that require the high transmission of light through the material demand low light-absorbing materials and so on.

There is another property known as optical density that often gets confused with optical absorbance. Optical density is the amount of attenuation or loss of intensity when light passes through a material or optical component. However, optical absorption is only the absorption of light within the material or optical component.

Both absorbance and optical density of material are measured using spectroscopic techniques. Usually, UV-Vis spectroscopy is used to measure the absorbance and optical density of a material. In general, the range of a usual UV-Vis spectrophotometer is 200–800 nm of wavelength, which means that the spectrophotometer scans in the mentioned wavelength range to find out the highest absorption or transmission zone of the material.

In order to elaborate on the concept, one can say that optical density, generally termed OD, is the measure or ability to delay the transmission of light through the medium. It is a measure of the speed of light corresponding to a particular wavelength through a medium. Hence, the slower the speed of light through a specific medium, the higher the optical density of that medium.

Mathematically, absorbance or decadic absorbance of a material can be defined as: $A = \log_{10}(P_i/P_t) = -\log_{10}T$, where P_t is the transmitted power and P_i is the incident power. Here, the transmittance of the material can be defined as $T = (P_t/P_i)$. In the case of spectral absorbance, the frequency and wavelength-specific radiant powers $(P_{t,f}, P_{i,f})$ or $(P_{t,\lambda}, P_{i,\lambda})$ comes into consideration. Although absorption stands for the physical process of absorbing light, absorbance considers attenuation of the transmitted power caused by several other optical phenomena like reflection and scattering.

The optical absorbance is related to the molar absorption coefficient, optical path length, and the concentration of the attenuating material. Hence, mathematically, the absorbance can be defined as, $A = \log_{10}(P_i/P_t) = \varepsilon l c$, where ε is the molar absorption or attenuation coefficient or absorptivity, l is the optical path length, and c is the concentration of the attenuating material. The molar attenuation or absorption coefficient, i.e. the absorptivity, depends on the sample. It defines how strong the absorber of the sample is for a given optical wavelength. The concentration here defines how many moles of the sample are dissolved per liter of the solution, and the optical length is the length of the cuvette where the solution is kept, i.e. how long the light travels through the sample. It is clear from this equation that if one keeps the optical path length and absorptivity constant then the absorbance will change linearly with the concentration of the material. Hence, during absorption spectroscopy, it is important to use cuvettes having the same dimensions just to avoid variation due to the change in optical length. For designing a device for measuring the concentration of a material, one has to be careful about the optical length so that variations in the measurement due to the same can be avoided. Moreover, it is also important to select the correct wavelength of the light source that gives the highest absorptivity of the material under test for the absorption-based devices with a single optical source.

7.2.2 Reflectance

Optical reflection [2, 3] is a phenomenon that deepens significantly on the surface of an optical medium. Reflectance in this regard refers to the effectiveness of reflecting the incident radiant energy by an optical medium. This optical property and technique are widely used to characterize materials and the electronic properties of a thin film. The photoreflectance method measures the change in the reflectivity of material upon the incidence of an amplitude-modulated light beam. The reflectance or hemispherical reflectance can be defined as: $R = (P_r/P_i)$, where P_r is the reflected power and P_i is the incident power.

Optical reflectance largely depends on the surface morphology of a film. Broadly speaking, the smoother the surface, the higher will be the reflectance. In many cases, the reflected wave suffers a phase change however, it depends on different material combinations.

7.2.3 Light Scattering

Like other optical properties, scattering [4, 5] is another optical property that plays an important role. Many optical devices such as solar cells utilize this phenomenon to increase the efficiency of the devices. In the case of scattering, the particles that interact with light absorb and scatter or re-emit the light in different directions with different intensities. This phenomenon is different from reflection, where a light ray is deflected in a direction maintaining the laws of reflection. However, in scattering, radiation such as acoustic and electromagnetic is deviated in different directions due to the nonuniformity of the medium. In this view, a few cases of reflection can also experience scattering, and these are termed diffused reflection.

Scattering also takes place due to elastic and nonelastic collision of particles such as electron–electron and electron–proton. Based on the types of collision, the scattering can be classified into five different types: Rayleigh scattering, Mie scattering, Brillouin scattering, Raman scattering, X-ray scattering, and Compton scattering. Among these, Rayleigh scattering and Mie scattering are due to an elastic collision and the others originate due to inelastic collisions. The scatterings are mathematically differentiated using a dimensionless size parameter, $\alpha = (\pi D_p/\lambda)$. Where $\alpha \ll 1$ refers to Rayleigh scattering, $\alpha \approx 1$ refers to Mie scattering, and $\alpha \gg 1$ refers to a geometric optical domain where geometric optics plays the role. The scattering phenomenon is mostly explained using Maxwell's equations theoretically. But due to the lack of exact solutions for irregular geometries, computational strategies are being adopted for solving the equation by discretizing the domains under consideration. Figure 7.2 shows the schematic illustration of the physical phenomenon such as absorption, transmission, and reflection.

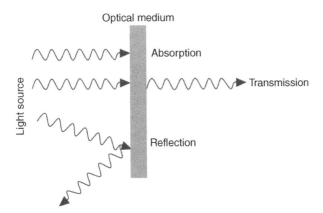

Figure 7.2 Schematic representation of different optical phenomena such as absorption, transmission, and reflection through an optical medium.

7.2.4 Luminescence

Luminescence [6] refers to the emission of light from an object when it is relatively cool and hence the emission does not result from heat. It is termed cold-body radiation. There is a different origin of this emission such as chemical reaction, atomic motion, crystal stress, or electric energy. Depending on the origin, luminescence can be classified into various types, such as bioluminescence, chemiluminescence, crystalloluminescence, electroluminescence, cathodoluminescence, mechanoluminescence, photoluminescence, radioluminescence, sonoluminescence, and thermoluminescence. The origin of each type of luminescence is self-explanatory from their nomenclature. However, bioluminescence and chemiluminescence essentially originate from chemical reactions, but in the case of bioluminescence, it is a biological reaction, often, seen in a living organism. However, in electroluminescence, the origin is mostly the radiative recombination of electrons and holes. Hence, it is mostly observed in semiconductor materials. On the other hand, cathodoluminescence occurs when an electron beam interacts with a luminescent material such as a phosphor. The origin of mechanoluminescence is the broken bonds of a crystal due to sudden mechanical actuation such as crushing and rubbing. Photoluminescence takes place when an electron reaches higher or lower energy levels by absorbing or radiating photons. Radioluminescence takes place when specific materials get exposed to ionizing radiations like α, β, and γ. While sonoluminescence and thermoluminescence involve acoustic and thermal excitation, respectively.

7.2.5 Fluorescence

Fluorescence [7] is an optical phenomenon where the emission of light from a material takes place after that material absorbs light or other electromagnetic radiation. This is also a kind of luminescence, but in this case, it does not involve any other energy for the emission of light, rather the absorbed optical energy itself undergoes emission. Mostly, the emitted optical wavelength is higher than the absorbed radiation, i.e. higher energy gets absorbed and lower energy comes out of the material. Interestingly, in many cases, the absorbed radiation is in the ultraviolet (UV) range and hence invisible to the human eye, but the emission is in the visible range. Thus, those materials work only under UV radiation. The fluorescent materials stop glowing almost immediately after the incident or absorbed radiation turns off. Figure 7.3 shows the origin of absorption, fluorescence, phosphorescence, and other emissions with respect to the energy states present in a material. Moreover, different types of scattering are illustrated in

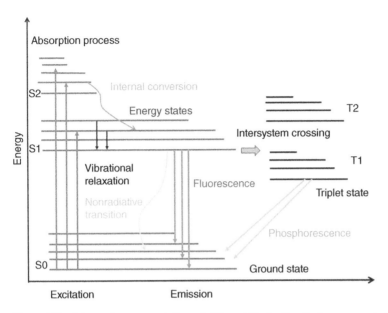

Figure 7.3 Schematic representation of different kinds of optical processes in terms of energy bands.

Figure 7.4a, where the scattering in terms of excitation wavelength (λ_{ex}) has been discussed. Moreover, the difference in fluorescence with Rayleigh and Raman (Stokes and anti-Stokes) scattering (discussed in Section 5.4.5) is described in connection with energy states in Figure 7.4b, c shows the typical spectrum for Raman and fluorescence for a given excitation light source.

7.2.6 Circular Dichroism

Circular dichroism [8, 9] is an optical phenomenon and is associated with polarized light absorption. The polarization of light depends on the associated field of optical radiation. The optical signal is an electromagnetic transverse wave that has two components, namely electric field (E) and magnetic field (H), oscillating perpendicularly to each other and the direction of propagation. In linearly polarized light, the electric field of the wave oscillates in one plane, and for circularly polarized light, the electric field vector oscillates circularly around the direction of propagation. Figure 7.5 shows the schematic representation of the linear and circularly polarized lights. In right circularly polarized (RCP) light, if the propagation is toward the observer, the electric field vector rotates clockwise, whereas for left circularly polarized (LCP) light, the electric vector rotates counterclockwise. When a

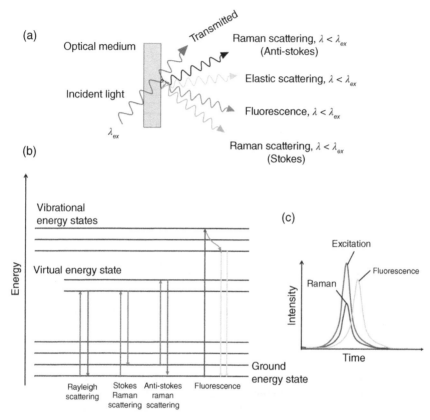

Figure 7.4 Schematic representation of the different scattering processes (a) and the schematic illustration in terms of energy bands (b) and intensity spectrum (c).

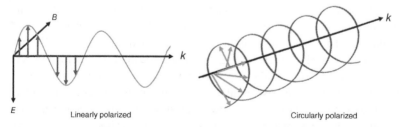

Figure 7.5 Schematic representation of linear and circularly polarized lights.

circularly polarized light passes through an optical medium, depending on the nature of the material, the speed of RCP (C_R) and LCP (C_L) differs, i.e. $C_R \neq C_L$, moreover, their wavelength ($\lambda_R \neq \lambda_L$) and the absorption also differ, i.e. molar excitation coefficients of RCL (ε_R) and LCP (ε_L) are different, i.e. $\varepsilon_R \neq \varepsilon_L$. The difference between the absorption of RCP and LCP is the circular dichroism.

In this case, differential absorption of two types of circularly polarized light (left-hand circular [LHC] or right-hand circular [RHC]) represents two different spin-angular momenta of the photon. It is used to understand the properties of optically active chiral materials. The LHC and RHC polarized lights behave differently when it passes through an optically active chiral material. The wavelength and absorption quantity varies for both LHC and RHC. In this technique, the delta absorbance, $\Delta A = (A_L - A_R)$, is measured. However, the difference between the molar excitation coefficient, $\Delta \varepsilon = (\varepsilon_L - \varepsilon_R)$, is termed as the molar circular dichroism. Here, the subscripts L and R correspond to LHC polarized light and RHC polarized light, respectively. Circular dichroism is widely used to detect different biomolecules. It is used in near-infrared to near ultraviolet spectrum. It has an application in the investigation of protein structure, charge transfer, and electronic structure of a material.

7.2.7 Z-Scan Technique

Z-scan is a technique that is used mostly in nonlinear optics. This technique is mostly used to measure nonlinear photonic parameters such as nonlinear refractive index, nonlinear susceptibility, and nonlinear absorption coefficient. This is a method to measure Kerr nonlinearity or nonlinear index n_2. In this case, a material is moved through the focus of a laser beam and the beam radius is measured behind the material. The two major Z-scan techniques that are utilized for the measurements are the closed aperture and open aperture Z-scan techniques. In case of a closed aperture technique, an aperture is placed to block a fraction of light before it reaches the detector; however, in an open aperture case, all the light is allowed to reach the detector. Another technique that is also used for high-sensitive measurements is known as eclipsing z-scan, where the light is blocked in the central region.

7.2.8 Förster Resonance Energy Transfer

A technique for measuring the separation between two chromophores known as a donor–acceptor pair is fluorescence resonance energy transfer (FRET). FRET has a limitation in that it only works when there are less than 10 nm between the donor and acceptor pairs. Contrarily, FRET, a phenomena that is strongly reliant on distance, has gained popularity as a technique for determining the dynamic activities of biological molecules at the nanoscale.

The acronym FRET stands for Förster (fluorescence) resonance energy transfer. In a nonradiative process known as the Förster energy transfer, an excited donor transfers energy – not an electron – to an acceptor group. This technique can be utilized to investigate biological structures since it is highly distance dependent. One common usage is simply measuring the distance between two points of interest on a big molecule, frequently a biological macromolecule, by adding sufficient donor–acceptor groups to the large one. If there is no conformational change during this procedure, the distance between the donor and acceptor groups of a big molecule with just one donor and one acceptor group may be easily computed. Additionally, the dynamical activities between two locations on this macromolecule, such as protein interactions, may be evaluated if the molecule experiences a large conformational alteration. These days many different applications employ this method, such as single-molecule research, molecular motors, biosensors, and DNA mechanical movements. As a result of its intrinsic simplicity, FRET is frequently referred to as the "spectroscopic ruler."

The theoretical analysis benefited greatly from Theodor Förster's contributions. According to Kasha's rule, a photon excites a donor group (D), which then relaxes to the lowest excited singlet state, S1. The energy released when the electron returns to the ground state (S0) may also excite the donor group if the acceptor group is nearby. This nonradiative phenomena is referred to as "resonance." The stimulated acceptor emits a photon and goes back to the ground state if none of the other quenching states are present.

The resonance process is related to the Coulombic interaction between electrons. Given that wave function overlap is required for the Dexter energy transfer, which involves an electron exchange, the relative distance of Coulombic contact between the donor–acceptor pair may be higher.

The FRET efficiency (E), which measures the number of energy transfer events that take place for every donor excitation event, is the quantum yield of the energy transfer transition. The source of the FRET efficiency is

$$E = \frac{k_{\mathrm{ET}}}{k_{\mathrm{F}} + k_{\mathrm{ET}} + \sum k_{\mathrm{i}}}$$

where k_{ET} is the rate of FRET, k_{F} is the rate of radiative relaxation, k_{i} is the nonradiative relaxation rates. With a 50% transfer efficiency, the FRET efficiency may be linked to the donor–acceptor distance, where r is the distance between donor and acceptor chromophores and R_0 is the characteristic distance (the Förster distance or Förster radius).

$$E = \frac{k_{ET}}{1 + (r/R_0)^6}$$

7.3 Materials for Optical Sensing

Optical sensors can be made using a variety of materials including liquid and semi-isolid materials, especially in spectroscopic detection techniques. However, in this chapter, highly used materials such as metal oxide, polymers, and carbon materials related to solid-state sensor fabrications are discussed.

7.3.1 Metal Oxide Materials

Metal oxides are extensively explored for several optoelectronic applications due to their suitability in solid-state device fabrication and optoelectronic properties. In this direction, nanostructures of metal oxides have contributed significantly. Nanostructures have a higher surface-to-volume ratio that provides more scope of optical absorption and in turn increases the efficiency of devices. Prior art has suggested one-dimensional (1D) metal oxides are the best candidate for exploring the novel optical phenomenon. The size and dimensionality dependence of the material is another aspect that has been explored significantly over the past few decades. Researchers have integrated these nanostructured materials with different geometries for photodetectors and optical switches. Most of the photodetector nanostructure geometries, especially the 1D structure, can be employed as a conductor whose conductions can be varied by charge transfer or as field-effect transistor whose channel properties can be varied by applying appropriate potential in gate terminal. Often, the materials are also functionalized to enhance the optical sensing properties. The most popular and widely used metal oxides for optical sensors or photodetectors are Cu_2O, Ga_2O_3, ZnO, SnO_2, CdO, CeO_2, Fe_2O_3, and In_2O_3. Most of the metal oxides are semiconducting and thus are very suitable for solid-state optical sensor fabrication. The wide band gap, growth on different substrates, composite preparation, and ease in nanostructure synthesis make it promising for optical sensing applications.

7.3.2 Polymer Materials

There is a significant advancement in polymer electronics in the past few decades. Polymer-based optoelectronic devices have attracted attention due to their ease of synthesis and integration. However, most of the polymers suffer from high-temperature processing, environmental effect, and shelf life. Among many other polymers, conductive and semiconductive polymers have proven their suitability for application in optoelectronics. In this direction, π-conjugated polymers showed better responses, and a number of recent reports have shown their applicability in optoelectronic and optical sensing applications. These polymers are broadly classified into degenerate (interchangeable structures like polyacetylene) and nondegenerate (noninterchangeable structures like poly[para-phenylene]). The band structure and conductivity of both polymers change with doping. The charge transfer mechanism in semiconductive polymers is largely explained in terms of HOMO and LUMO bands, generation and recombination of polaron–excitation, and intermolecular charge hoping. Aside from these materials, only conductive polymers such as polyaniline, PEDOT:PSS, polyacrylic acid are also used for these optoelectronic sensor fabrications. Nanostructure-based polymer nanocomposites such as metal or metal oxide nanoparticle-based composites are also explored extensively to fabricate optical sensors. There are polymer-based sensors such as polymer composites based photodetectors and optical fiber-based sensors.

7.3.3 Carbon Materials

Carbon materials like carbon nanotube (CNT), graphene, reduced graphene oxide (RGO), graphene oxide (GO), carbon dots, and functionalized carbon materials are explored extensively for different optical and other sensors. The electronic and mechanical properties of graphene attracted a lot of attention among the research community. Moreover, its optical, optoelectronic, and electronic properties have opened up a large number of opportunities in the field of optoelectronic devices like solar cells. Many recent reports have suggested that the use of graphene in photovoltaic and photoconductive devices has improved performance significantly. In this direction, the use of reduced graphene oxide has also attracted significant attention. Composite materials such as polymer composite, metal nanostructure, metal oxide, and semiconductor composites have been explored significantly. On the other hand, CNT has emerged as another wonder material in solid-state sensor fabrication. Interestingly, CNT exhibits both metallic and semiconductive properties, and thus attracted a lot of attention in device and sensor fabrication. However, the separation of these two types of CNTs from a solution is still a hectic process. In recent times, researchers have also shown high-performance optoelectronic devices and sensors using carbon dots. The interaction of carbon materials with electromagnetic radiations is interesting and hence the optical absorptions, scattering, and photoluminescence of CNT and other carbon materials attracted significant attention for optical sensor development. Hence, there is a future for carbon materials to be used as a good alternative in the optoelectronic device and sensor industries.

7.4 Optical Techniques for Sensing

Optical techniques of detection have been explored largely in sensing applications due to the ease of use. In this regard, several types of solid-state sensors are also developed. But the traditional sensing techniques are still in use for several sensing applications, such as health care, environment, andchemical detection. The major sensing techniques include Raman scattering, absorption spectroscopy, luminescence spectroscopy, or fluorescence spectroscopy. The basic principles related to these spectroscopic techniques are already discussed in the previous sections. Although the sample preparation and complexity associated with the analysis are the major drawbacks of these spectroscopic techniques, these are still highly preferred because of their specificity, sensitivity, and selectivity in the detection process. Many solid-state sensors have been developed for point-of-care or affordable optical detection in the recent past, but rarely any of them could reach the limit that the mentioned techniques offer. Different nanomaterials have been chosen as an alternative for onsite detection devices due to the advantage of nanomaterials in terms of tunability chemical, mechanical, and electrical properties. Many physical and chemical properties have been explored to enhance the sensitivity, specificity, and selectivity of sensing arrangements. Here in this section, the major techniques and spectroscopic methods are discussed.

7.4.1 SPR-Based Detection

SPR [10, 11] is an optical phenomenon that occurs due to the interaction of electromagnetic waves such as light with a thin film of metal when the incidence angle of light is higher than the critical angle. Recent studies have shown that selective nanoparticles can exhibit this phenomenon that can be utilized to develop or enhance the performance of optical sensors and devices. Many solar cells use gold nanoparticles on the surface to get the benefit of SPR that increases the performance of the device. The incidence of light creates resonant oscillation of conduction electrons at the interface of negative and positive permittivity materials. This optical phenomenon has been significantly explored in biosensing applications. Figure 7.6a describes the SPR phenomenon schematically. Moreover, the oscillation of charge in an SPR-sensitive nanoparticle under electromagnetic influence is described schematically in Figure 7.6b.

The basic SPR experiment is immobilizing the receptor on the detecting surface of a flow cell, then saturating the receptors with an analyte solution and monitoring the changes in the sensor signal as the molecules bind. Microfluidic devices known as flow cells have channels that are generally smaller than 1 mm in diameter. A thin metal sheet, usually made of gold or silver, serves as the sensing surface. It can be flat or have nanometer scale patterns like arrays of holes or line gratings in it. The metal sheet is exposed to light, and photons of light couple to free electrons in the metal to produce charge oscillations (plasmons) at the surface of metal that resonate at particular light wavelengths. The plasmons are sensitive to variations in the interfacial refractive index because they are confined to the metal surface. The refractive index changes and the plasmon resonance shifts

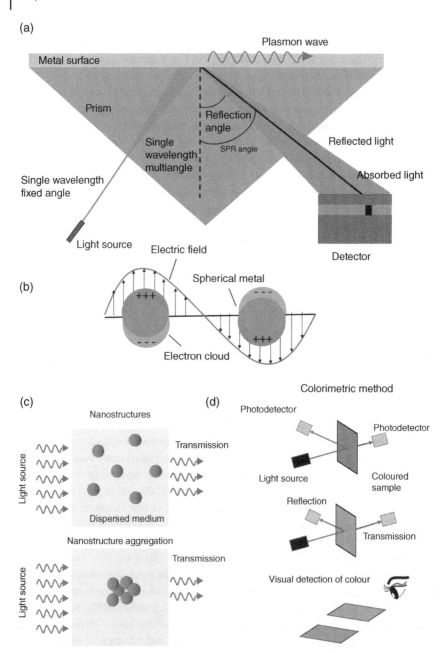

Figure 7.6 Schematic representation of the SPR phenomenon (a) and the oscillation of charges due to electromagnetic radiation (b). Images (c) and (d) schematically show the detection due to nanoparticle aggregation and colorimetric technique, respectively.

to larger reflection angles when the analyte binds to surface receptors. The quantity of bound analyte affects the shift's magnitude.

SPR has developed into a powerful optical detection technique for analyzing the interactions between immobilized ligands and analytes in solution. When fitting and figuring out the kinetic rate constants, the analyte concentration is very important. Several analyte concentrations must interact with the ligand in order to construct a whole kinetic profile for the interactions and determine binding constants. It is advised to use at least six concentrations, with eight or more being advantageous for reliable data and curve fitting.

7.4.2 Nanostructure Aggregation-Mediated Detection

Nanostructures such as nanoparticles, nanowires, and nanoribbons have shown an extensive involvement in the optical sensing developments. In this direction, the aggregation of nanostructures [12] has been explored to develop optical nanosensors. Experiments have proven that the aggregation of nanostructures shows a shift in the optical absorption frequency. Thus, similar to the varying size of the nanoparticles, aggregation of nanostructures has also been studied in different optical sensing applications. The change in the optical color sometimes is visually detectable. For example, the aggregation of Au nanoparticles in a solution turns the color of the solution from wine red to blue. The change in color can be attributed to optical absorption properties.

In many applications, controlling nanocrystal size is often an important but difficult goal to achieve. In this context, recent research shows that there are some number of atoms that create a stable cluster and hence the formation of the cluster can be tuned. In case of nanoparticles, the optimum number of atoms that form a stable configuration while maintaining a specific structure is known as structural magic numbers. Structural magic numbers correspond to minimum volume and maximum density configuration. The magic number of a nanoparticle can be calculated if the crystal structure is known. In case of an approximately spherical nanoparticle with FCC, the number of atoms, $N = (1/3)(10n^3 - 15n^2 + 11n - 3)$, and the number of atoms on the surface can be calculated using, $N_{surf} = 10n^2 - 20n + 12$, where n is the number of layers [13]. Table 7.1 shows number of total atoms and surface atoms for a nanoparticle with FCC crystal structure.

7.4.3 Micro/Nanofiber-Based Detection

Similar to other nanostructures, micro- and nanofibers [14] have been used extensively for the development of optical sensing. Long aspect ratio polymer nanofibers have been used to develop solid-state optical sensor development. Many optical components such as loop resonator, knot resonators, and coil resonators have been developed for different sensing applications. Most of the fiber fabrications

Table 7.1 Number of total atoms and surface atoms for a nanoparticle with FCC crystal structure.

Shell number (n)	Considering FCC closed pack structure		
	Diameter^a	Total atom (N)	On surface (N_{surf})
1	$1d$	1	10
2	$3d$	13	12
3	$5d$	55	42
4	$7d$	157	92
5	$9d$	309	162
6	$11d$	561	252
7	$13d$	923	362
8	$15d$	1415	492
9	$17d$	2057	642
10	$19d$	2869	812

^{a} The diameter d (is in nm for FCC) for Au 0.288, Ar 0.376, and Al 0.286.
Source: Poole and Owens [13]/John Wiley & Sons.

are done using electrospinning methods, and the developed polymer fibers provide benefits in terms of performance in many cases. Aside from different nanofiber materials, optical fibers in micron scale also been developed for different sensing applications. Hence, micro/nanofibers are an efficient technique of detection in optical sensing applications. Moreover, aggregation of nanostructures produces a shift in optical absorption and transmission that further can be detected for optical sensing. This method has a large application for water quality monitoring. Figure 7.6c shows the effect of nanoparticle aggregation schematically.

The micro/nanofibers are typically produced using electrospinning techniques. The fiber-based sensors are mostly the photodetectors and detect the optical signal pretty much the same way the other solid-state optical sensors do. However, for aggregation-based sensing, the nanofibers also show a change in optical properties upon aggregation similar to the nanoparticle.

7.4.4 Colorimetric Sensing

Change in color is the most common phenomenon in optical detections [15]. Many devices have been developed in the past few decades based on observing the color change of the sensor. This technique basically can lead to a visual detection system. However, associated electronics can make it even more sensitive as the electronics system provides more resolution in terms of color change and makes it independent of human eye limitations that vary from person to person. In this case, photodetectors and light sources such as LED and laser. play an important role. Most of the sensors and associated electronics work based on transmission, reflection, and combined mode. The photodetector analyzes either the transmitted light through or reflected light from the sensor in order to identify the color change. This technique has attracted significant attention in the recent past. Figure 7.6d shows a schematic illustration of the possible colorimetric sensing.

In this case, mostly a sensing chemistry involving a change in color is chosen. For example, the pH variation can change the color of different dyes, such as methyl orange, or some specific reactions like starch–iodide compound change color after reacting with amylase. Such reactions can be targeted for colorimetric sensors. The sensor can be in solid medium, for example a thin film of the compound on a substrate like glass, cellulose, and PVC or in liquid medium such as the solution of that compound in a small cuvette. For the electronic setup, the sensor, i.e. the thin film on the substrate or the solution in a cuvette, is placed in between a light source and a photodetector for transmission mode, or the light source and photodetector are placed at 45° angle with respect to the sensor for reflection mode as illustrated in Figure 7.6d. The output signal from the photodetector is then analyzed for the sensing operation.

7.4.5 Spectroscopy Techniques Associated with Sensing

The spectroscopic technique [16] of optical detection is very selective, sensitive, and precise. Most of this detection takes place in dedicated spectrometers. These techniques quantify material interaction with optical and electromagnetic radiations. All these spectrometers contain several solid-state sensors for their operation and detect responses from solid samples. Liquid samples are also acceptable in those spectroscopy instruments with different sets of arrangements or setups. A brief discussion about these techniques would provide an idea regarding optical detection techniques.

Before we go into the discussion of spectroscopic techniques, it would be helpful if we learn about different optical scattering mechanisms. Figure 7.5a shows different conditions for scattered and excitation wavelengths for different optical scattering, whereas Figure 7.5b shows in terms of the energy band diagram. The energy band diagram explains the Rayleigh (elastic), Stokes (inelastic), and anti-Stokes (inelastic) scatterings. When a photon from the excited state collides with the molecule, it goes to a short-lived virtual state and releases energy with a similar frequency to the incident photon to come back to its lowest vibrational state and this causes the Rayleigh scattering where the energy of the photon is conserved.

When the emitted energy has a net increase in the energy, then it is known as Stokes scattering, and when the emitted energy has a net loss, then it is known as anti-Stokes scattering.

7.4.5.1 Raman Spectroscopy

This technique of detection is based on optical scattering. As mentioned in the previous section, Raman scattering originates from the inelastic scattering of a monochromatic electromagnetic wave in the near-infrared to near-ultraviolet (UV) range. In this case, mostly a laser having a wavelength in the visible range is employed for the analysis. Once the laser interacts with a material surface, it excites the electrons to the higher energy states before it comes back to different vibrational and rotational ground states. The major reason for the Raman spectrum is the Stokes and anti-Stokes scattering. This transition of electrons at different levels is specific to a material at a given wavelength. Thus, this technique is used for different detection purposes.

In general, there is a difference between the intensities of Stokes and anti-Stokes components and it depends on the initial states of the molecule and the distribution of the same follows Boltzmann equation, $N \propto e^{-(E/k_B T)}$, where k_B is the Boltzmann constant and T is the temperature. There are very few molecules that initially are at a higher energy state and hence the anti-Stokes scattering is very less intense. This uneven distribution is due to the energy level spacing. If the spacing between the energy levels is small, the difference between the Stokes and anti-Stokes scattering is of almost the same magnitude and vice versa. Moreover, an increased temperature also creates almost similar Stokes and anti-Stokes scattering.

7.4.5.2 Luminescence Spectroscopy

Luminescence has been explained in the previous section of this chapter. This occurs due to the transfer of electrons from excited to ground state. Luminescence spectroscopy can be of many types similar to the types of luminescence. The mostly used luminescence are fluorescence, phosphorescence, and chemiluminescence. In these cases, the intensity of the emitted light is measured with different wavelengths.

7.4.5.3 Absorption Spectroscopy

Absorption spectroscopy measures the extent of optical absorption or associated property such as the transmission of a material under test at different wavelengths. This technique is a widely used optical measurement technique. The commonly used instrument for this purpose is a UV-Vis spectrometer that typically allows the measurement in 200–800 nm of the optical spectrum. This method can also be used for the measurement of the optical band gap of semiconductive material.

7.5   Fabrication Technique of Optical Sensors

The fabrication of optical sensors such as photodiodes and photovoltaic devices are mostly dependent on conventional micro/nanofabrication techniques such as the growth of oxide, deposition of material, deposition of metal, and in case of Si photodiode doping is also an essential process. Most of these processes are already discussed in Chapter 2 of this book. However, the advent of nanostructures has made the fabrication process easier as a material with desired properties is available (e.g. a typical ZnO nanowire UV photodetector [17] fabrication process). In this process, a bunch of n-type ZnO nanowires was grown using the vapor–liquid–solid (VLS) mechanism on a p-type Si having a thin layer of Au film thermally deposited on the substrate. Thereafter, dielectric materials spin-on glass to provide insulation and mechanical support to the top electrode. A thin Al (10 nm) and Ti (200 nm) were deposited on the ZnO wire as a top electrode.

In another process [18–20], the ZnO nanowires (ZnONW) were synthesized in a high-temperature horizontal furnace on c-plane sapphire substrates using a chemical vapor transport (CVT) technique in a quartz tube. Gold (Au)-assisted VLS approach was followed to grow the nanowires. In this case, a thin layer of Au was deposited on sapphire substrate and annealed at inert (Ar) atmosphere to create random distribution of Au nanoparticles (AuNPs). Furthermore, the ZnO micropowder was employed as Zn source and the powder was mixed with graphite powder at 1 : 1 ratio to form ZnO-C. Then, the mixture and the sapphire substrate were loaded in horizontal furnace using a crucible in a quartz tube. The CVT process was carried out at a high temperature of 950 °C. At this high temperature, carbothermal reduction of ZnO powder takes place resulting in Zn and O_2 gas species. In this situation, both $Zn(g)$ and $O_2(g)$ go near the sapphire substrate using different Ar fluxes, and VLS growth of ZnO takes place. In this process, both gases get absorbed by the Au liquid nanodroplet and when the Au nanodroplets are highly saturated, the $ZnO(s)$ precipitates at the interface of Au nanodroplet and sapphire substrate. An experimental setup is shown in Figure 7.7.

Aside from the microfabrication technologies, several printing technologies, drop casting, and brush coating were also explored for the development of different photodetector and optical sensors. A few of them have been discussed here.

7.5.1   Solution Process

The solution process has attracted overwhelming response in terms of affordable sensor fabrication. In this type of fabrication, many of the materials required for fabricating a sensor are prepared using a solution process or sol–gel method where the nanostructure dimension can be controlled by tuning different parameters such as stirring, temperature, and reaction time. Few optical sensors fabricated using the solution process, in the prior art, have shown a nice response.

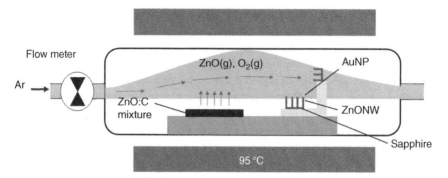

Figure 7.7 Schematic representation of ZnO nanowire fabrication process using chemical vapor transport (CVT) method.

7.5.2 Inkjet Printing

Many reports also have suggested an inkjet printing technique for the fabrication of different optical sensors. In this case, different flexible substrates are used for the sensor fabrication process. However, in a few cases, the inkjet printing was used to develop the electrodes only, whereas the active material was deposited using drop casting or some other techniques like a conventional deposition.

7.5.3 Screen Printing

Screen printing was similar to the inkjet printing, but allows to achieve less thickness of the deposited material. Many sensors also developed using the screen-printed techniques that allow a roll-to-roll fabrication of sensors. A variety of sensors were developed using screen-printed techniques previously [21–23]. A detailed discussion of different fabrication processes is discussed in Section 4.4 of Chapter 4.

7.6 Applications of Optical Sensing

The mentioned optical properties and techniques are employed extensively for the development of different sensors. In this section, a few of the applications will be discussed.

7.6.1 Environment Monitoring and Gas Sensing

As discussed earlier, optical sensors are mostly dependent on the reflection, scattering, absorbance, or similar optical phenomena while propagating through a medium. In the case of liquid medium, the SPR is an attractive option (Figure 7.8)

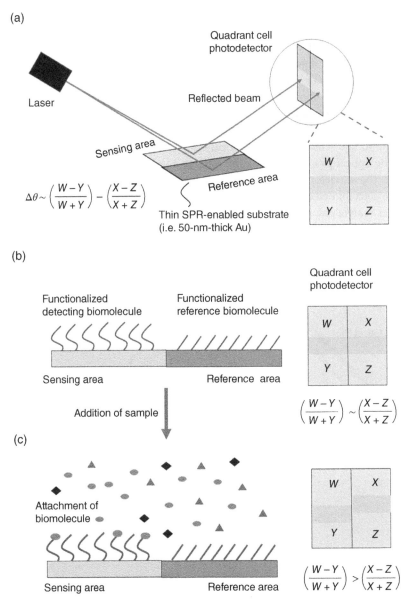

Figure 7.8 Schematic representation of the differential SPR setup. (a) shows the set-up for sensing schematically where a diode laser is focused with a lens on a gold film. The gold film is divided into two specified areas such as a sensing area and a reference area. The reflected beams from those areas are collected by a quadrant cell photodetector that simultaneously measures the SPR dips from the reference (X, Z) and sensing (W, Y) areas. (b) shows the schematic illustration of the system with sensing and reference areas having different functionalised biomolecules and the balanced photodetector with $(W - Y)/(W + Y) - (X - Z)/(X + Z) \sim 0$. (c) After the analyte interaction, the specific adsorption of metal ions on the sensing area shifts the SPR position so that $(W - Y)/(W + Y) > (X - Z)/(X + Z)$.

as described and discussed in the later section. However, for gas sensing, there are options like the use of optical fiber and photonic crystal. The sensor generally has an active area where the interaction of a photon with sensing material leads to an electronic signal. The same mechanism is also applicable in gas sensing. There are two commonly used optical sensors [24, 25] for gas detection: (i) optical fiber-based and (ii) photonic crystal-based sensors.

In case of fiber-optic sensors, the solid-state optical systems such as photodetectors and emitters such as LED or laser are used with optical fiber as the propagation medium. The active area of the optical fiber interacts with gas molecules that change the optical property such as the refractive index of the fiber surface. The change in the surface property in turn changes the intensity of the propagating light. The detector detects the intensity that can be correlated with the concentration of the analyte gas. Figure 7.9a shows the concept schematically. Many times,

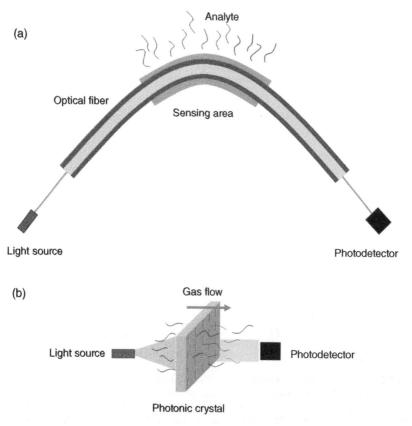

Figure 7.9 Schematic representation of different gas sensing techniques using (a) optical fiber and (b) photonic crystal.

fiber-optic sensors are used in arrays where individual fiber detects different analytes and hence can be used to detect multiple analytes at the same time. This type of sensor has very high selectivity, stability, and sensitivity. However, fiber-optic optical sensors have limitations in terms of miniaturization due to the size of the optical fiber.

The other type of sensor for gas detection is based on photonic crystal. Photonic crystals are refractive index-based materials. They constitute a periodic arrangement of dielectric materials having a different refractive index. These materials are expected to have very high sensitivity with the ability to detect nanosized chemical compounds and other physical parameters such as pressure, humidity, and temperature. They also provide design flexibility and can prevent electromagnetic interference from the electrical signals. This also helps in developing miniaturized optical integrated circuits.

Photonic crystal sensors consist of periodic arrangement of micro/nanostructure patterns of dielectric materials. This periodic arrangement of dielectric having different optical band gap allows only a certain wavelength to propagate through the material. The wavelength of the propagating light can be tuned by optimizing the material and pattern of the dielectric. Infrared photonic crystals are the one that are extensively used for gases, such as CO, CO_2, and CH_4 detection, as these gases have an absorption region in mid-infrared. The detection generally takes place due to a change in the effective refractive index in the photonic crystal because of the presence of gas or a change in the lattice distance of the periodic structure.

The sensing mechanism can be explained using Bragg's law, $m\lambda = 2nd \cdot \sin\theta$, where m and λ are the diffraction order and wavelength of light, respectively, n is the effective refractive index, d is the lattice distance between adjacent periodic pattern, and θ is the incident angle of light. In case of photonic crystal, all the parameters such as m, d, and θ in Bragg's law are constant aside from the wavelength incident light (λ) and effective refractive index (n). When a gas passes through the photonic crystal, it changes the effective refractive index of the system on the boundary of photonic crystal and the gas and each gas has a different refractive index and thus it shows the different response for different gases. Furthermore, some gas changes the lattice parameters as well, in that case the detection is based on a combined effect of lattice distance and effective refractive index. In general, the photonic crystal sensing layer is placed between a light source and a detector as shown in Figure 7.9b. The analytes interact with the periodic structure and changes the lattice parameters or refractive index as described earlier. Photonic crystals can reduce the dimension of the conventional gas sensors and can provide higher sensitivity. However, the use of external light sources and photodetector is still a problem for the miniaturization.

7.6.2 Health Monitoring

Optical sensor-based health monitoring techniques are very popular due to their structural simplicity and ease of implementation. In this direction, optical technique-based pulse oximetry is one of the common health monitoring devices in use. Pulse oximeters are a very essential tool for patients' health monitoring. It is useful in both intensive care and operation rooms in a hospital. The working principle of this kind of sensor is based on photoplethysmography technique that detects the oxygen saturation in arterial blood (SpO_2) noninvasively. This provides information regarding the percentage of oxygenated hemoglobin present in blood to monitor the cardiorespiratory health of a patient. The device consists of multiple solid-state optical devices such as LED and photodetector. The construction of the sensor is shown in Figure 7.9a. In general, the device consists of two LEDs having wavelength in red zone (~660 nm) and infrared zone (~940 nm). There is a solid-state detector such as photodiode on the opposite side of the finger and the detector detects the transmitted optical signal. The ratio of the absorption coefficients of these two transmitted lights gives the indication of oxygen content in the blood. The deoxygenated hemoglobin has an absorption peak close to 660 nm and oxygenated hemoglobin absorbs more light at 940 nm. The optical property of the oxygenated and deoxygenated hemoglobin helps in detection of the oxygen content.

Moreover, there are optical sensors for blood glucose monitoring and sweat analysis techniques. In most of the cases, the detection is based on fluorescence quenching and charge transfer that lead to the signal transduction. In many cases, the properties like pH are utilized to get some optical change in the sample after the interaction of analyte and the same is detected using the solid-state optical sensors and light source as explained in Figure 7.6d earlier.

7.6.3 Fingerprint Detection

Fingerprint detection has become one of the common techniques in personal verification systems. In every sector of professional places, fingerprint-based biometric system has arrived as the most efficient solution in attendance management. Among different methods of detection, optical sensors are one of the well-known techniques to detect the fingerprints. Solid-state optical sensors or devices are commonly used in this case. One of the techniques uses an array of solid-state optical sensors such as photodiodes and LEDs for the detection of fingerprints as described in Figure 7.10b. In this case, the photodetectors are integrated below the protective layer and the LEDs are integrated at the periphery of the device. A collimator layer below the protective coating helps in focusing the lights. The optical sensor in this case detects the reflected, scattered, and diffused light from the finger and processes it using an array of analog-to-digital converter (ADC) to

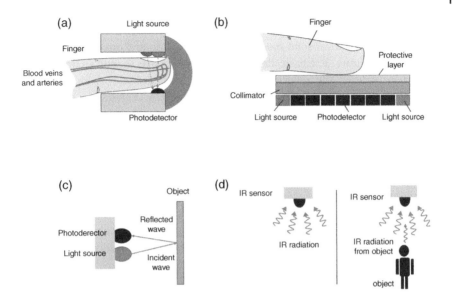

Figure 7.10 Schematic representation of different IR sensing techniques for (a) biomedical, (b) fingerprint detection, (c) distance measurement, and (d) motion sensing applications.

get the information. Similar techniques are used in many cases to develop fingerprint sensors.

7.6.4 Defense and Security

Optical techniques are used in variety of warfare equipment [26]. Laser sensing in this direction has attracted a lot of attention especially in defense and security applications that require counter measures. The application of laser sensing is frequently found in intelligence, surveillance, target acquisition, protection, positioning, communication, etc. The laser system is one of the multisensing systems. The basic principle of the sensing system for example for measuring distance is to send a narrow beam of laser and detect the reflected optical signal using solid-state optical sensors like photodiode.

Aside from laser sensing, there are a number of optical sensing systems as discussed earlier that are used to detect biological and nuclear threats and chemical warfare agents. Nerve agents are one of the specific chemical warfare agents that affect the central nervous system of a human being by blocking the activity of the enzyme acetylcholinesterase. There are molecular optical sensors that detect the presence of nerve agents and similar chemicals by measuring the change in fluorescence properties.

7.6.5 Motion Detection

There are several optical solid-state sensor-based motion and proximity detectors reported in literature. The basic idea of proximity sensing is to send an optical signal and to measure the reflected wave using a photodetector as illustrated in Figure 7.10c. Optical proximity sensing system generally consists of a solid-state light source mostly in the IR range and a solid-state optical photodetector-like photodiode that detects the reflected wave of the optical signal from the light source reflecting from an object in front. The response of the photodiode is then processed in order to get the information about the distance between the sensing system and object. Nearer the object, higher will be the response of the photodiode.

Similar to a proximity sensor, there are solid-state sensors for motion detection as illustrated in Figure 7.10d. In this case, passive infrared sensors are used to detect the infrared radiations from different objects. All the objects emit infrared radiation if their temperature is above absolute zero. The passive infrared sensors detect the changes in infrared radiation coming from the objects around them. Any motion changes the total amount of the infrared incident and thus changes the response such as the output voltage of the sensor that triggers the alarm. If any object such as a person or animal suddenly comes into the field of view of that sensor, the temperature of that specific place changes, which in turn changes the sensor response. The solid-state optical sensors are extensively used in this case.

The optical sensing in motion detection is now applied to many modern technologies like driverless cars and next-generation automotive. The mentioned technologies are modified based on the application and utilized in similar applications.

7.6.6 Water Quality Monitoring

In water bodies, heavy metals such as lead (Pb), mercury (Hg), and arsenic (As) are some of the major pollutants that cause deadly diseases like cancers. Heavy metals mostly enter the human body through water sources and cause serious health hazards due to prolonged exposure. These heavy metals cause harm to an aquatic living organism and destroy the ecosystem. It is very important to detect the level of these heavy metals before consumption.

The SPR of gold nanoparticles (AuNP) has been employed to detect heavy metals in water samples. The extent of SPR changes with the specific reaction of heavy metal ions with the functionalized AuNPs. Moreover, AuNPs' size can be varied to control the SPR properties. Solid-state sensors like photodiode have been implemented to detect metal ions in water. A high-resolution differential SPR sensor has been fabricated using photodiode to detect heavy metals by means of detecting the shift of SPR. Forzani et al. [27] reported that a quadrant cell photodetector can efficiently be used to sense the SPR. In this case, a gold film-assisted sensor surface has been modified into reference and sensing areas that are coated with specific peptides. The peptides provide a selective and accurate detection of metal ions like Cu^{2+} and Ni^{2+} in the range of ppt–ppb in drinking water. Figure 7.8a shows the schematic illustration of the setup to detect metal ions using photodiodes, and Figure 7.8b, c show the sensor geometry and response before and after analyte detection, respectively.

On the other hand, optical fibers are also been used for the detection of heavy metals. In this case, a molecule named calix[4]arene has been used to detect the copper and nickel metal ions. There are reports to analyze samples for heavy metal ions of lead, cadmium, copper, and zinc detection using differential pulse anodic stripping voltammetry (DPASV) and ions like nickel and cobalt using differential pulse adsorptive stripping voltammetry (DPASV). In these cases, dimethylglyoxime (DMG) is used as a complexing agent. In the case of optical fiber-based detection technique [28], calix[4]arene has been used to detect heavy metal ions. The optical absorption change was taken place as the detection parameter upon reacting with the complexing agent and ions due to the formation of the evanescent wave. Laser and photodiode were employed for the measurement. Figure 7.6a shows setup for sensing where the laser is passed through the optical fiber and the photodiode is placed to detect the intensity of the laser, whereas Figure 7.11b, c show the magnified image of the sensitive area of optical fiber and the detection mechanism before and after attachment of analyte, respectively.

7.6.7 e-Waste and Detection of Toxic Materials

Aside from heavy metals, there are a number of toxic compounds from e-waste or nuclear pollution in water such as cyanide and sulfite that also cause serious health hazards. Other than water, these toxic materials are also found in soil and food products. Hence, several detection systems in this direction have been reported based on optical sensing.

The SO_3^{2-} ion is highly used in food additives, preservatives, and fertilizers and hence, the concentration of SO_3^{2-} ion is highly prone to cross the safety limit in soil, water, and food products. Consumption of these food products can cause serious health damage. In this direction, a colorimetric sensor has been designed [29] to detect and monitor the SO_3^{2-} ions in water. As discussed earlier, the colorimetric sensors work based on the change in colors that can be detected visually or electronically. In this case, the oxidizing property of Co_3O_4 nanoparticles (NPs) was used to oxidize 2,2′-azino-*bis*(3-ethylbenzothiazoline-6-sulfonic acid) diammonium salt (ABTS) and 3,3′,5,5′-tetramethylbenzidine (TMB) to form colored

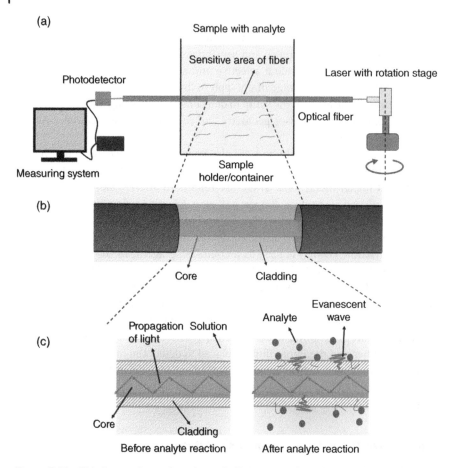

Figure 7.11 This image shows the schematic illustration of colorimetric detection using an optical fiber. Image (a) shows the detection setup, (b) shows the magnified sensitive part of the optical fiber, and (c) shows the detection mechanism schematically.

substance. This reaction was used to detect the sulfite ion that causes a change in the color of the abovementioned reaction. Aside from the colorimetry, fluorescence was also employed to detect the sulfite in food and living organisms. In this case, synthesized novel tetraphenyl imidazole compound 2-(4-(1,4,5-triphenyl-1H-imidazol-2-yl)ben-zylidene)malononitrile (TIBM) showed aggregation-induced emission enhancement (AIEE) upon reaction with sulfite.

Similar to sulfite, cyanide is another toxic compound that needs special attention as it can cause serious health hazards to living organisms including humans.

A recent report suggested a new simple oligothiophene-indenedione-based sensor that can detect cyanide [30] in aqueous media. This sensor provides both colorimetric and fluorometric detection of CN^-. The intramolecular charge transfers due to the reaction of CN^- and indenedione-vinyl group.

7.6.8 Detection of Microorganisms

Microorganisms play a vital role in the environment, food, and water monitoring system. Different microorganisms in water and food make it dangerous for human beings. There are several water-borne diseases like diarrhea and cholera that are caused due to the presence of a specific microorganism in the water. In this view, several viruses or bacteria detection systems [31] have been developed employing the optical techniques discussed previously such as SPR, photonic crystals, and waveguide. Many of the solid-state-based techniques alone are not feasible for the detection of microorganisms as the application demands disposable approach and hence colorimetric detection has attracted a lot of attention. Not only for health care, but for food, water, and environmental safety as well, the disposable technique is highly desirable. Recent research showed the detection of *Escherichia coli* bacteria [32] using calorimetric techniques where aptamer-modified nanoscale pentacosadiynoic acid (PCDA) vesicles were employed for the detection. The reaction between the bacteria and the aptamer-PCDA shows a change in color from blue to red, and this change can be detected using light source and photodetector setup as discussed in the previous section.

7.7 Prospects and Limitations

Optical sensing has merit to be employed as an affordable sensing system. This allows both conventional micro/nanofabrication as well as chemical and printing technologies. Thus, the optical system has attracted attention for a variety of sensing applications. However, due to involvement of both light sources and solid-state optical sensors, most of the sensing systems become bulky and hence it is a challenge to miniaturize the device. Moreover, few types of sensors such as optical fiber-based devices have very less scope for miniaturization due to the size of the fiber itself. However, the optical detection system provides affordable and high-performance solution to many real-life problems.

Colorimetric procedures give an advantage of visual detection that already has captured the market in terms of products like pregnancy test kits. However, precise detection demands electronic arrangements. In this direction, significant progress has been made but still, an optimum system with wireless capabilities is desirable. For this, a low-power system would be extremely helpful to deploy the same in remote areas. The solid-sate optical sensors have a huge scope in health care and biomedical applications as this approach provides simple, affordable, and disposable sensing arrangements keeping the electronics intact. This approach also offers very specific and sensitive detection systems based on spectroscopic techniques, though these are very expensive. Hence, development of affordable, hand-held spectrophotometers for sensing is still an open field of research.

Aside from this, different other optical techniques such as SPR, z-scan, and dichroism evolved as accurate and efficient optical characterization techniques. However, many of them are complex or suffer from commercial viability to common people. Further research is needed to make it affordable.

List of Abbreviations

ABTS	2,2'-Azino-*Bis*(3-Ethylbenzothiazoline-6-sulfonic Acid) Diammonium Salt
ADC	Analog-to-digital Converter
AIEE	Aggregation-induced Emission Enhancement
AuNPs	Au Nanoparticles
CMOS	Complementary Metal Oxide Semiconductor
CNT	Carbon Nanotube
CVT	Chemical Vapor Transport
FRET	Förster (Fluorescence) Resonance Energy Transfer
GO	Graphene Oxide
LCP	Left Circularly Polarized
LDR	Light-dependent Resistor
LED	Light-emitting Diode
LHC	Left-hand Circular
PCDA	Pentacosadiynoic Acid
RCP	Right Circularly Polarized
RGO	Reduced Graphene Oxide
RHC	Right-hand Circular
SPR	Surface Plasmon Resonance
TMB	3,3',5,5'-Tetramethylbenzidine
UV	Ultraviolet
VLS	Vapor–Liquid–Solid
ZnONW	ZnO nanowires

References

1 Hu, C., Muller-Karger, F.E., and Zepp, R.G. (2002). Absorbance, absorption coefficient, and apparent quantum yield: a comment on common ambiguity in the use of these optical concepts. *Limnology and Oceanography* 47 (4): 1261–1267.

2 McGill, T.C., Kurtin, S.L., and Shifrin, G.A. (1970). Optical reflection studies of damage in ion implanted silicon. *Journal of Applied Physics* 41 (1): 246–251.

3 Hehl, K. and Wesch, W. (1980). Calculation of optical reflection and transmission coefficients of a multi-layer system. *Physica Status Solidi (A)* 58 (1): 181–188.

4 Hovenier, M.I.M.J.W. (2000). Light scattering by nonspherical particles: theory, measurements, and applications. *Measurement Science and Technology* 11 (12): 1827–1827.

5 Brown, W. (1993). *Dynamic Light Scattering: The Method and Some Applications.* Clarendon Press.

6 Yukihara, E.G. and McKeever, S.W.S. (2011). *Optically Stimulated Luminescence: Fundamentals and Applications.* Wiley.

7 Jameson, D.M. (2014). *Introduction to Fluorescence.* Taylor & Francis.

8 Micsonai, A., Wien, F., Kernya, L. et al. (2015). Accurate secondary structure prediction and fold recognition for circular dichroism spectroscopy. *Proceedings of the National Academy of Sciences* 112 (24): E3095.

9 Klös, G., Miola, M., and Sutherland, D.S. (2019). Increased refractive index sensitivity by circular dichroism sensing through reduced substrate effect. *The Journal of Physical Chemistry C* 123 (12): 7347–7355.

10 Nguyen, H.H., Park, J., Kang, S., and Kim, M. (2015). Surface plasmon resonance: a versatile technique for biosensor applications. *Sensors* 15 (5).

11 Schasfoort, R.B.M. (2017). *Handbook of Surface Plasmon Resonance*, 2e. Royal Society of Chemistry.

12 Jiang, Z., Le, N.D.B., Gupta, A., and Rotello, V.M. (2015). Cell surface-based sensing with metallic nanoparticles. *Chemical Society Reviews*, 10.1039/C4CS00387J 44 (13): 4264–4274.

13 Poole, C.P. and Owens, F.J. (2003). *Introduction to Nanotechnology.* Wiley.

14 Li, K., Zhou, W., and Zeng, S. (2018). Optical micro/nanofiber-based localized surface plasmon resonance biosensors: fiber diameter dependence. *Sensors* 18 (10): 3295.

15 Yoon, J., Chae, S.K., and Kim, J.-M. (2007). Colorimetric sensors for volatile organic compounds (VOCs) based on conjugated polymer-embedded electrospun fibers. *Journal of the American Chemical Society* 129 (11): 3038–3039.

16 Gauglitz, G. and Vo-Dinh, T. (2006). *Handbook of Spectroscopy.* Wiley.

17 Luo, L., Zhang, Y., Mao, S.S., and Lin, L. (2006). Fabrication and characterization of ZnO nanowires based UV photodiodes. *Sensors and Actuators A: Physical* 127 (2): 201–206.

18 Núñez, C.G., Vilouras, A., Navaraj, W.T. et al. (2018). ZnO nanowires-based flexible UV photodetector system for wearable dosimetry. *IEEE Sensors Journal* 18 (19): 7881–7888.

19 Franco, F.F., Manjakkal, L., Shakthivel, D., and Dahiya, R. (2019). ZnO based screen printed aqueous ammonia sensor for water quality monitoring. *IEEE Sensors* 2019: 1–4.

20 Liu, F., Kumaresan, Y., Shakthivel, D. et al. (2019). Large-area, fast responding flexible UV photodetector realized by a Facile method. *IEEE Sensors* 2019: 1–4.

21 Manjakkal, L., Shakthivel, D., and Dahiya, R. (2018). Flexible printed reference electrodes for electrochemical applications. *Advanced Materials Technologies* 3 (12): 1800252.

22 Manjakkal, L., Sakthivel, B., Gopalakrishnan, N., and Dahiya, R. (2018). Printed flexible electrochemical pH sensors based on CuO nanorods. *Sensors and Actuators B: Chemical* 263: 50–58.

23 Manjakkal, L., Vilouras, A., and Dahiya, R. (2018). Screen printed thick film reference electrodes for electrochemical sensing. *IEEE Sensors Journal* 18 (19): 7779–7785.

24 Singh, M., Raghuwanshi, S.K., and Prakash, O. (2019). Ultra-sensitive fiber optic gas sensor using graphene oxide coated long period gratings. *IEEE Photonics Technology Letters* 31 (17): 1473–1476.

25 Anamoradi, A. and Fasihi, K. (2019). A highly sensitive optofluidic-gas sensor using two dimensional photonic crystals. *Superlattices and Microstructures* 125: 302–309.

26 Burnworth, M., Rowan, S.J., and Weder, C. (2007). Fluorescent sensors for the detection of chemical warfare agents. *Chemistry – A European Journal* 13 (28): 7828–7836.

27 Forzani, E.S., Zhang, H., Chen, W., and Tao, N. (2005). Detection of heavy metal ions in drinking water using a high-resolution differential surface plasmon resonance sensor. *Environmental Science & Technology* 39 (5): 1257–1262.

28 Benounis, M., Jaffrezic-Renault, N., Halouani, H. et al. (2006). Detection of heavy metals by an optical fiber sensor with a sensitive cladding including a new chromogenic calix[4]arene molecule. *Materials Science and Engineering: C* 26 (2): 364–368.

29 Qin, W., Su, L., Yang, C. et al. (2014). Colorimetric detection of sulfite in foods by a TMB–O_2–Co_3O_4 nanoparticles detection system. *Journal of Agricultural and Food Chemistry* 62 (25): 5827–5834.

30 Guo, Z., Hu, T., Sun, T. et al. (2019). A colorimetric and fluorometric oligothiophene-indenedione-based sensor for rapid and highly sensitive detection of cyanide in real samples and bioimaging in living cells. *Dyes and Pigments* 163: 667–674.

31 Rajapaksha, P., Elbourne, A., Gangadoo, S. et al. (2019). A review of methods for the detection of pathogenic microorganisms. *Analyst* 144 (2): 396–411.

32 Wu, W., Zhang, J., Zheng, M. et al. (2012). An aptamer-based biosensor for colorimetric detection of *Escherichia coli* O157:H7 (in eng). *PloS One* 7 (11): e48999–e48999.

8

Magnetic Sensors

8.1 Introduction

Magnetism, a very ancient discovery [1], was first discussed in ancient Indian, Chinese, and Greek documents somewhere between the sixth and third century BC. Despite the mentioned fact, magnetism has been explored by scientists and engineers even today in order to make electronic devices, sensors, magnetic equipment, etc. A number of inventions and discoveries have been made harnessing the properties of magnetism. With time, the phenomenon was explored in smaller sized particles that gave rise to micro- and nanomagnetism [2]. It was observed that almost all physical properties change or few of the properties become intense in smaller dimensions and hence nanomagnetism has attracted the attention of the large scientific community for the past few decades. Many of the electronic properties and electron transport such as electron-spin dependency become controllable in the nanomagnetic regime. This provides an open scope and a possibility to develop a large scale of futuristic devices. The prior art has shown the capability of utilizing nanomagnetism in multidimensional applications such as magnetic memory or storage devices, magnetic recording heads, magnetic resonance imaging, biomedicine, and drug delivery. Aside from these, nanomagnetism is presently used in spin-dependent devices such as spin-logic, spin-torque oscillators, and magnonic crystals.

The magnetic storage phenomenon is nothing but the sensing of external magnetic fields and hence a brief discussion on the consequence of nanomagnetism in data storage technology would be helpful. Magnetism was utilized to develop storage more than a hundred years ago and the same was in use for storing audio, video, and data as nonvolatile memory since its discovery. Magnetic tape technology was significantly used for storage purposes before IBM first demonstrated a magnetic hard disk drive in the year of 1956. However, due to limited storage capacity, the magnetic disk was not very popular. The development of

Solid-State Sensors, First Edition. Ambarish Paul, Mitradip Bhattacharjee, and Ravinder Dahiya.
© 2024 The Institute of Electrical and Electronics Engineers, Inc.
Published 2024 by John Wiley & Sons, Inc.

magnetoresistance and nanomagnetism has altered conventional storage technology, and in the last few years, storage space for data has risen by a factor of more than 20 million. The storage density has also been increased and hence higher storage capacity was enabled in a small area. Although storage devices are a significant development and outcome of nanomagnetism, several other solid-state devices are also developed for sensing and other electronic applications. Nanomagnetism has also enabled spin-dependent devices like metal oxide semiconductor field effect transistor (MOSFET) which provides better performance. However, a lot of these gadgets still need technological adjustments to be utilized in practice, and a lot of research is being done in this area right now. Hall effect magnetoresistance has a wide application in different sensing applications. Different types of magnetoresistance have been employed to develop high-precision sensors and devices.

Researchers have been interested in magnetic sensors for quite some time. Several magnetic phenomena have been utilized for sensing applications. Moreover, magnetic phenomena, like magnetic induction, Hall effect, and magnetoresistance, are used to create sensors for different applications such as biosensing, storage devices, and current sensing. People have used the magnetic properties of materials such as diamagnetism and paramagnetism to fabricate different sensors. In this chapter, different material properties, magnetic phenomena, fabrication techniques, and a few applications will be discussed.

8.2 Materials' Magnetic Properties

Various magnetic materials have different sorts of magnetic characteristics [3]. For that reason, different sensing applications require different magnetic materials. Among others, paramagnetism is highly desirable in most magnetic sensors. In this section, the properties are explained briefly. Figure 8.1 shows the alignment of magnetization in different types of materials.

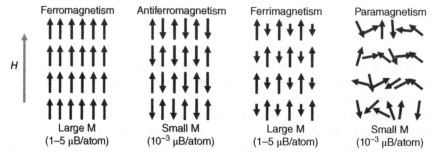

Figure 8.1 The magnetization alignment inside different kinds of materials and the typical magnetization values.

8.2.1 Diamagnetism

Due to a substance's diamagnetism, which is a magnetic property, the magnetic field applies the repelling force on that material. A material often experiences repulsion when the magnetic field is applied because it creates a magnetic field inside it that is normally in the opposite direction. Diamagnetism is a quantum mechanical phenomenon, which can be found in all materials. But, in a few cases, this phenomenon is outplayed by other associated processes such as the creation of magnetic dipole. Due to the way, it interacts with the field of magnetism, diamagnetism is the sole common occurrence among diamagnetic materials.

According to traditional theory, a magnetic moment (μ) is created when an atom is in a closed orbit or an electron is rotating around its axis. The moment can be defined as, $\mu = -(2/m_e)I$, where m_e is electron rest mass and I is the angular momentum. The I can be the spin angular momentum, orbital angular momentum, or total angular momentum of an electron because this is a classical theory. However, the electron's magnetic moment will suffer a torque under an exterior magnetic field, $dI/dt = (\mu \times H) = -(2/m_e)(I \times H)$.

I will process around H at a frequency indicated by the curl of I and H in the equation above, $\omega_L = eH/2m_e$, this frequency is known as Larmor frequency. This phenomenon causes a magnetic moment, $\mu_i = -(e^2/6m_e^2)H\bar{r}^2$, where $\bar{r}^2$ is the mean square distance of the electron from the nucleus. For an atom having z electrons, the susceptibility per unit volume can be given as,

$$\chi = -\frac{Ne^2}{6m_e^2}\sum_{i=1}^{z}\bar{r}^2$$

where N is the volumetric atom density. All materials have a constant, temperature-independent negative diamagnetic susceptibility. χ is 10^{-6} cm^3/mol in order of magnitude.

8.2.2 Paramagnetism

Paramagnetism, in contrast to diamagnetism, is a magnetic property due to which a material experiences weak attraction under a magnetic field. This characteristic aids in inducing an internal magnetic field in a substance, directed in the path similar to the exterior magnetic field.

A permanent magnetic moment μ is connected to paramagnetic materials. A magnetic moment μ with total angular momentum J (where $\mu = g\mu_B J$, g is the Lande g factor) has $J(J+1)$ distinct values in that direction of external magnetic flux density B. The component of the μ along the direction of B can be defined as, $\mu = m_J g\mu_B$, where μ_B is the Bohr magneton and m_J is the azimuthal quantum number that takes values from $J, J-1, J-2, ..., -J$. The system's energy is defined as, $E = -\mu \cdot B = -m_J g \mu_B B$.

At room temperature, i.e. $m_J g \mu_B J H/kT \ll 1$ (where H is magnetic field strength, T is temperature), the magnetization of the system is determined using Boltzmann distribution and the magnetization can be written as,

$$M = \frac{Ng^2 J(J+1)\mu_B^2 H}{3kT}$$

Hence, the paramagnetic susceptibility can be defined as,

$$\chi_p = \frac{M}{H} = \frac{Ng^2 J(J+1)\mu_B^2}{3kT} = \frac{N\mu^2}{3kT} = \frac{C}{T}$$

where Curie constant, $C = \left(Ng^2 J(J+1)\mu_B^2\right)/3k$ and this is recognized as the Curie law of paramagnetism. Hence, from the above equation, it is evident that the paramagnetic susceptibility is independent of the biased magnetic field and inversely related to temperature. For one Bohr magneton's magnetic moment (μ_B), $\chi \sim 1/30T$, hence, $\chi \sim 10^{-4}$ cm^3/mol at $T = 300$ K.

8.2.3 Ferromagnetism and Antiferromagnetism

Ferromagnetism is the strongest responsive magnetic property of a material. Ferromagnetic materials are often used to make permanent magnets. Ferromagnetic materials experience a strong attractive force under the magnetic field. Iron, cobalt, and nickel are well-known ferromagnetic materials.

Antiferromagnetism is similar to ferromagnetism. But the electron spins are arranged in the material in an orderly fashion. The pattern is temperature dependent and at high temperatures, the pattern vanishes.

Ferromagnetic materials have spontaneous magnetization which is the manifestation of the strong interaction between atomic dipoles. Weiss' definition of the typical molecular field effect on a magnetic dipole is, $H_m = H + \gamma M$, where M denotes the magnetization, H denotes the applied magnetic field, and γ denotes the Weiss constant. If one substitutes this equation into the paramagnetic susceptibility, $\chi_p = M/H = C/T$, it becomes,

$$\chi_p = \frac{M}{H_m} = \frac{M}{H + \gamma M} = \frac{C}{T}$$

Further, this equation gives,

$$\chi_f = \frac{C}{(T - C\gamma)} = \frac{C}{(T - T_c)}$$

This equation is known as the Curie–Weiss law and T_c is Curie temperature. Below T_c, ferromagnetic material shows spontaneous magnetization.

Weiss omitted any discussion of the field of molecules. The explanation of the molecular field was given by Heisenberg by providing quantum mechanical exchange interaction. The Heitler–London model for hydrogen molecules served as the foundation for Heisenberg's theory, the wave function of two electrons can be defined as, $\psi(r_1, s_1)$ and $\psi(r_2, s_2)$, where r_i and s_i are the spatial and spin coordinates, respectively. When two electrons get closer, the two wave functions overlap and create a combined wave function which is antisymmetric as per Pauli's exclusion law and the combined wave function becomes, $\psi = \phi(r)\Gamma(s)$, where $\phi(r)$ is the function related to spatial coordinate and $\Gamma(s)$ is a function for spin coordinate only. Hence, the antisymmetric wave function can be defined as, $\psi_{anti} = \phi_{sym}(r_1, r_2)\Gamma_{anti}(s_1, s_2)$ or $\psi_{anti} = \phi_{anti}(r_1, r_2)\Gamma_{sym}(s_1, s_2)$. More explicitly, the singlet ψ_S and triplet ψ_T states of the combined wave function can be represented as,

$$\psi_S = A[\phi_a(r_1)\phi_b(r_2) + \phi_a(r_2)\phi_b(r_1)]\left[\Gamma_\alpha(s_1)\Gamma_\beta(s_2) - \Gamma_\alpha(s_2)\Gamma_\beta(s_1)\right]$$

$$\psi_T = B[\phi_a(r_1)\phi_b(r_2) - \phi_a(r_2)\phi_b(r_1)]\begin{bmatrix} \Gamma_\alpha(s_1)\Gamma_\alpha(s_2) \\ \Gamma_\alpha(s_1)\Gamma_\beta(s_2) + \Gamma_\alpha(s_2)\Gamma_\beta(s_1) \\ \Gamma_\beta(s_1)\Gamma_\beta(s_2) \end{bmatrix}$$

In the singlet state, the entire spin quantum number $S = 0$ as the spins are in the antiparallel state, whereas the entire spin quantum number $S = 1$ for triplet state, and therefore there are $(2S + 1)$ degenerate states.

For a hydrogen molecule with two nuclei, let us assume a and b, the Hamiltonian H_{12} may be stated as,

$$H_{12} = \frac{e^2}{r_{ab}} + \frac{e^2}{r_{12}} - \frac{e^2}{r_{1b}} - \frac{e^2}{r_{2a}}$$

where the separation between two nuclei, the distance between electrons, and the distance between nuclei and electrons are denoted by the letters r_{ab}, r_{12}, r_{1b}, and r_{2a}, respectively. The energy of the singlet and triplet states are expressed as,

$$E_S = A^2(K_{12} + J_{12})$$

$$E_T = B^2(K_{12} - J_{12})$$

where J_{12} is the exchange integral and K_{12} is the average Coulomb interaction energy J_{12} must be positive for spins to align parallel.

In this case, the direct interaction of two overlapping wave functions is taken into consideration and thus it describes the direct exchange. In some ferromagnetic materials, a direct exchange might not represent a meaningful interaction though. When 4f electrons are strongly localized in a material and do not participate in bonding, the indirect exchange must be crucial.

In almost all cases, magnetic properties are important to be considered in designing a magnetic sensor. Therefore, Table 8.1 provides the magnetic characteristics of several soft magnetic materials.

Table 8.1 Magnetic properties of selected soft magnetic materials.

Material	Composition	Maximum permeability	Coercivity (A/m)	Saturation flux density (T)
Cobalt	$Co_{99.8}$	250	800	1.79
Permendur	$Fe_{50}Co_{50}$	5000	160	2.45
Iron	$Fe_{99.8}$	5000	80	2.15
Nickel	$Ni_{99.8}$	600	60	0.61
Silicon–Iron[a]	$Fe_{96}Si_4$	7000	40	1.97
Hiperco	$Fe_{64}Co_{35}Cr_{0.5}$	10,000	80	2.42
Supermendur	$Fe_{49}Co_{49}V_2$	60,000	16	2.40
Ferroxcube 3F3[b]	Mn-Zn-Ferrite	1800	15	0.50
Manifer 230[c]	Ni-Zn-Ferrite	150	8	0.35
Ferroxplana[d]	$Fe_{12}Ba_2Mg_2O_{22}$	7	6	0.15
Hipernik	$Fe_{50}Ni_{50}$	70,000	4	1.60
78 Permalloy	$Fe_{22}Ni_{78}$	100,000	4	1.08
Sendust	$Fe_{85}Si_{10}Al_5$	120,000	4	1
Amorphous[e]	$Fe_{80}Si_{20}$	300,000	3.2	1.52
Mumetal 3	$Fe_{17}Ni_{76}Cu_5Cr_2$	100,000	0.8	0.90
Amorphous[e]	$Fe_{4.7}Co_{70.3}Si_{15}B_{10}$	700,000	0.48	0.71
Amorphous[e]	$Fe_{62}Ni_{16}Si_8B_{14}$	2,000,000	0.48	0.55
Nanocrystalline	$Fe_{73.5}Si_{13.5}B_9Nb_3Cu$	100,000	0.4	1.30
Superalloy	$Fe_{16}Ni_{79}Mo_5$	1,000,000	0.16	0.79

[a] Nonoriented.
[b] At 100 kHz.
[c] At 100 MHz.
[d] At 1000 MHz.
[e] Annealed.
Source: Adapted from Ripka [2].

8.3 Nanomagnetism

Magnetism studied in submicron range particles gives rise to nanomagnetism [2–4]. The magnetic properties are very different in the nano or submicron scale compared to bulk. The properties vary even at the interfaces of materials. In the nanoscale, these properties are very important. The major properties that influence mostly the magnetic nanomaterials are magnetic anisotropy, interlayer exchange coupling, and exchange bias coupling.

8.3.1 Magnetic Anisotropy

Magnetic anisotropy is the relationship between a crystal's energy and the orientation of the magnetic field regarding the crystalline axes. In an ultrathin film, the magnetic anisotropy [4] plays an important role due to magnetic spin–orbit interaction at the interfaces. Thus, perpendicular magnetic anisotropy (PMA) which is the change of orientation from in-plane to perpendicular to plane is very important.

8.3.2 Interlayer Exchange Coupling

For a multilayer construction comprising nonmagnetic and magnetic materials, interlayer exchange coupling [5] is a significant nanomagnetic phenomenon. Between binary ferromagnetic layers that are spaced apart using a slim nonmagnetic layer, this phenomenon will be seen. The giant magnetoresistance (GMR) effect in a magnetic multilayer structure is based on this. This phenomenon can be observed between two ferromagnetic layers separated by a thin nonmagnetic layer. This is the basis of the GMR effect present in the magnetic multilayer structure. The spin-polarized electron present in the ferromagnetic layer experiences a magnetic coupling. However, the interlayer coupling is highly dependent on the Fermi surface electron arrangement in the nonmagnetic layer. It is experimentally found that the exchange strengths show damped oscillations due to the increase in thickness.

8.3.3 Exchange Bias

Exchange bias [6] is a magnetic phenomenon where the hysteresis loop of ferromagnetic and antiferromagnetic materials shifts along the field axis after cooling to a temperature.

8.3.4 Spin-Polarized Transport

Spin-polarized transport [7] has been reported long back in 1936. Mott realized in a study that at very low-temperature magnon, a quasi-particle and a collective excitation of the spin structure of electron in a crystal, when scattering becomes insignificant, the majority and minority spin with parallel and antiparallel magnetization in a ferromagnet, do not blend in the process of scattering. Thus, the conductivity is formed as the total of two separate, unequal spin components, resulting in spin-polarized current in the material. This type also goes by the name "two current." This model has been employed extensively to explain the different magnetoresistive phenomenon.

One of the first experimental studies on the spin-polarized transport of carriers used the tunneling current. Two junctions between nonmagnetic (N) and ferromagnetic (M) metal N/F/N create IV characteristics which could be modified by an applied exterior magnetic field. It was also detected that an unpolarized current becomes spin-polarized when it passes across a ferromagnetic semiconductor. Furthermore, a series of experiments showed that the tunneling current in an F/insulator (I)/F junction remains polarized even outside the F material. The F layers' magnetization affects the tunneling current and thus the resistance or conductance of the entire F/I/F system changes with the applied external magnetic field or change in magnetization in F layers. Figure 8.2 describes this schematically and the change in resistance gives rise to tunneling magnetoresistance (TMR) which can be expressed as,

$$\text{TMR} = \frac{\Delta R}{R_{\uparrow\uparrow}} = \frac{R_{\uparrow\downarrow} - R_{\uparrow\uparrow}}{R_{\uparrow\uparrow}}$$

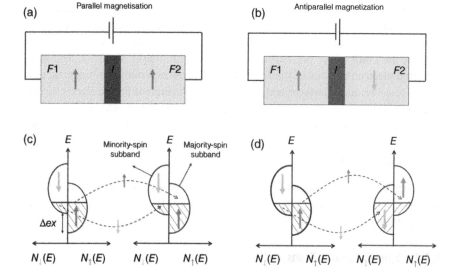

Figure 8.2 The tunneling junctions between two ferromagnetic materials in (a) parallel and (b) antiparallel magnetization. The corresponding (c) up-spin and (d) down-spin electron density in the material.

where R denotes resistance; ↑↑ and ↑↓ show the parallel and antiparallel magnetization in the F layers, respectively. Figure 8.2 shows the tunneling junctionsbetween two ferromagnetic materials in parallel (a) and antiparallel (b) magnetization. Images (c) and (d) show the corresponding up-spin and down-spin electron density in the material and Δex is the exchange spin splitting.

8.4 Magnetic Sensing Techniques

The magnetic sensors [8] generally utilize the generated magnetic field from an external source or vary a magnetic field in the system to initiate the sensing mechanism. In most cases, one or more of the parameters such as permanent magnets, electromagnets, magnetic beads, and magnetic particles are employed for sensing purposes. Two prevalent mechanisms of solid-state magnetic sensors are Hall effect and magnetoresistance.

8.4.1 Hall Effect Sensors

This effect [9, 10] was discovered by E. H. Hall in the year 1879. After his name, the effect is called the Hall effect. As per the effect, whenever a current carrying semiconductor or conductor block is kept under a perpendicularly aligned magnetic field regarding the electric current as illustrated in Figure 8.3, then the magnetic field applies a transverse force on the electric charges flowing through the medium and thus the charges start accumulating at one side of the conductor or semiconductor and thus a voltage is formed between two sides of the material block. The

Figure 8.3 The Hall effect experimental setup.

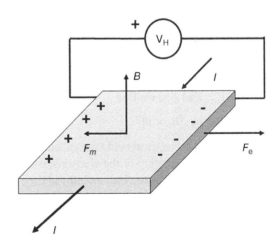

Hall effect is the name given to this phenomenon, and Hall voltage is the resultant voltage.

The fundamentals of the Hall effect are as follows. If a lengthy strip with current flowing through it is exposed to a dc magnetic field, the Hall effect manifests in its most basic and traditional form. The Lorentz force affected the carriers present in the strip.

$$F = e \cdot E + e \cdot (v \times B)$$

Here, e stands for electrical charge of a carrier, E for local electrical field, v for charge carrier's velocity, and B for the magnetic flux density, which we assume to be parallel to the strip plane.

Suppose the material of the strip is an N-type highly extrinsic semiconductor. We ignore the fact that there are holes. An external electrical field called Ee is applied in the x direction along the length of the strip. That external field is mostly responsible for the electrical field E. The electrical field outside affects how the electrons behave. The electrons move down the strip with the average drift velocity in response to the external electrical field, $v_{dn} = \mu_n \cdot E_e$, μ_n being the drift mobility of electrons. The corresponding current density is provided by, $J_n = q \cdot n \cdot \mu_n \cdot E_e$, where q denotes the fundamental charge.

The carrier velocity v is due to the thermal agitation and drift. Let us neglect for a moment the thermal motion. The magnetic component of the Lorentz force is then provided by,

$$F_{mn} = q \cdot \mu_n \cdot [E_e \times B]$$

That force pushes the electrons toward the strip's upper edge. Due to this, the concentration of electron at the strip's top and lower edges, respectively, rises and falls. An electrical field emerges between the strip edges as a result of those space charges. This electrical field acts on the electrons by a force,

$$F_{en} = -q \cdot E_H$$

That force tends to decrease the excess charges at the strip's edges. As a steady state, the two transverse forces F_{mn} and F_{en} balance. By equating both the equations of F_{mn} and F_{en}, we find

$$E_H = \mu_n \cdot [E_e \times B]$$

The transverse electrical field E_H is called the Hall electric field. Without neglecting the thermal agitation of the electrons, we get, $E_H = -\mu_{Hn} \cdot [E_e \times B]$.

Here, $\mu_n \cdot H_n$ denotes the Hall mobility of the electrons. The Hall mobility differs a little from the drift mobility: It is given by, $\mu_{Hn} = r_H \cdot \mu_n$, where r_H is the Hall scattering factor. This numerical parameter captures the impact of the carriers' thermal motion and their scattering on the Hall effect. Another useful expression for the Hall electrical field is obtained when the external electrical field expressed before is expressed by the current density, $E_H = -R_H \cdot [J \times B]$.

Here, R_H denotes the Hall coefficient, in this case given by, $R_H = \dfrac{1}{q \cdot n}$, where n denotes the density of free electrons. Here, again we use the appropriate sign because we neglected the thermal agitation of charge carriers. Without neglecting their thermal agitation instead of we obtain, $R_H = r_H/(q \cdot n)$.

The development of a measurably high voltage between the strip's edges is the hallmark of the Hall effect. It is known as the Hall voltage. Let us choose two points on the opposite corners of the strip in Figure 8.3 so that their potential difference is zero when $B = 0$. The Hall voltage is then provided by, $V_{Hall} = \int_{S_1}^{S_2} E_H ds$.

In this particular case, we find that $V_{Hall} = \mu_n \cdot H_n \cdot E_e \cdot B \cdot w$, where w denotes the width of the strip.

Once more, the strip's current density may be measured by, $J = \dfrac{I}{t \cdot w}$.

Here, I denote the current in the strip and t is the thickness of the strip. So, the source of Hall voltage is, $V_{Hall} = (R_H/t) \cdot I \cdot B$.

The above equations describe the basics of Hall effect. However, considering correction factors and other practical factors, the magnitude of Hall voltage increases with the increase in applied magnetic field B. Additionally, the current and voltage biasing, respectively, can be provided via,

$$V_{Hall} = G\frac{w}{l}\mu_H V_{Bias}B = S_{V_V}V_{Bias}B \tag{8.1}$$

$$V_{Hall} = G\frac{r_H}{n \cdot e \cdot t}I_{Bias}B = S_{V_I}I_{Bias}B \tag{8.2}$$

Here, G = geometrical correction factor, μ_H = Hall mobility of majority carriers, r_H = the Hall scattering factor, V_{Bias} = total bias voltage, n = carrier concentration, e = electron charge, I_{Bias} = total bias current, t = thickness of the n-well implantation, B = external magnetic field, S_{V_V} = voltage-related voltage-mode sensitivity, and S_{V_I} = current-related voltage-mode sensitivity.

The voltage-related voltage-mode sensitivity (S_{V_V}), which is influenced using the geometry and Hall mobility, is significantly reliant on temperature. Hence, S_{V_V} will change with temperature as well. On the other hand, the carrier concentration (n) has an inverse relationship with the current-related voltage-mode sensitivity (S_{V_I}). This term is constant for most of the operating temperature and also for a plate doping density in the range of 10^{15} and 10^{17} cm^{-3}.

Selective material for a Hall sensor can increase the device performance and hence extremely significant in selecting a perfect material for the fabrication. According to Eq. (8.2), materials having increased mobility and decreased

Table 8.2 Materials for making a Hall plate along with their desired properties at room temperature [11].

Material	E_g (eV)	μ_n (cm^{-2} V^{-1} s^{-1})	n (cm^{-3})	R_H (cm^3/C)*
Si	1.12	1500	2.5×10^{15}	2.5×10^3
InSb	0.17	80,000	9×10^{16}	70
InAs	0.36	33,000	5×10^{16}	125
GaAs	1.42	85,000	1.45×10^{15}	2.1×10^3

*RH is computed for a certain doping level.

conductivity will work well as Hall effect sensor materials. Thus, metals which have increased conductivity and low mobility are not a good option for Hall sensors. The sensors are generally fabricated using n-type semiconductors since the majority carrier, i.e. electron, has higher mobility than holes. The suitable materials, in this case, are Si and III–V semiconductors like InSb, InAs, and GaAs. High mobility and significant conductivity are features of III–V semiconductors. Si is preferred to fabricate the Hall device despite having moderate mobility since it is compatible with IC technology.

Table 8.2 describes different parameters such as the energy band gap (E_g), the carrier mobility (μ_n), and Hall coefficient (R_H) at 300 K of various semiconductors suitable for the fabrication of Hall plates. The Hall coefficient (R_H) in this case is also calculated for a fixed doping density.

The voltage mode operation of the Hall sensor is highly popular. In this case, the output is measured in terms of voltage. The current is the sensor's output in the case of a current mode Hall sensor though. However, the current mode is very seldom used in practice. The offset voltage or output voltage when zero magnetic field is applied is one of the key issues with Hall devices. The current mode Hall sensor has a similar construction to already available gadgets. The same option of offsetting the offset brought on by mismatch exists with the present mode device. The way in which signals are taken out makes the most impact.

The voltage mode Hall sensor is illustrated in Figure 8.4a, where the bias current I_{Bias} flows from D to B arm of the plate. In this case, a symmetric structure is fabricated for current spinning, which reduces the offset of the device. Hall voltage develops across the orthogonal arm (AC) under an applied magnetic field (BZ) in the plane's direction. The Hall voltage, in this case, can be calculated from the Eq. (6.2).

Figure 8.4b shows the current mode, where the current has been injected laterally in two adjacent arms (A and B) and an applied magnetic field causes unbalanced output current from the other two adjacent arms (C and D). The difference

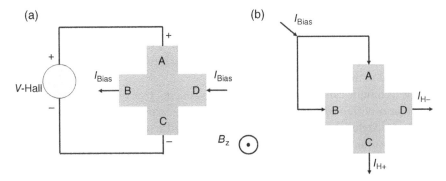

Figure 8.4 (a) Hall plate in voltage mode operation. (b) Hall plate in current mode operation.

between the output current, in this case, can be characterized as a corresponding current source of a Hall current I_{Hall}. The principal of the same can be found in previous reports [12], where the Hall current is defined as,

$$I_{Hall} = \mu_H \frac{w}{l} Bl_{Bias} \tag{8.3}$$

where I_{Bias} = total bias current, μ_H = Hall mobility of majority carriers, w = width-to-length ratio of the plate, and Bl = normal magnetic field.

To comprehend its use and application, this technique has undergone thorough study. The expression of output currents (I_{H+}, I_{H-}) can be given as,

$$I_{H^+} = \frac{I_{Bias}}{2} + \frac{I_{Hall}}{2} \tag{8.4}$$

$$I_{H^-} = \frac{I_{Bias}}{2} - \frac{I_{Hall}}{2} \tag{8.5}$$

In this case, the Hall current (I_{Hall}) is relational to the biasing current (I_{Bias}), external magnetic field (B_z), and magnetic resistance coefficient (β). The Hall plate current with a cross shape may be expressed as,

$$I_{Hall} = \mu_H \frac{\beta B_z I_{Bias}}{1 - (\beta B_z)^2} \tag{8.6}$$

where β is calculated as:

$$\beta = \frac{R_{(B_z)} - R_{(B_z = 0)}}{R_{(B_z)} R_{(B_z = 0)}} \tag{8.7}$$

where $R(B_z = 0)$ and $R(B_z)$ represent the Hall plate resistance in the absence and presence of an external magnetic field, respectively.

For a high mobility plate, the current mode Hall sensor offers superior resolution in frail magnetic fields. Additionally, the system may be made smaller by using fewer terminals thanks to the existing style of operation. The present mode is anticipated to receive greater attention in the near future due to the current technical trend of miniaturization.

This technique can be used to figure out the doping type of a semiconductor as the polarity of Hall voltage depends on the majority carrier present in the semiconductor. Hall effect has major applications in industries to manufacture low-power sensors, position detection, and contactless switching. These magnetic sensors are often high performance and cost-effective. These are also integrable with CMOS technologies. Although temperature has effects on the performance of different Hall sensors, temperature drift and offset management can provide optimal behavior and high sensitivity. Often sensor geometries are optimized in Hall sensor fabrications to get good results. However, reports have shown that the geometry change may create a significant offset variation. Several sensors for position detection, biomolecule detection, chemical detection, and gas detection have been reported using the Hall effect sensing mechanism.

8.4.2 Magnetoresistive Sensors

The phenomenon known as magnetoresistance [3, 7, 13, 14] describes how the material's resistance alters under an exterior magnetic field. This is a quantum mechanical phenomenon and is generally based on quantum mechanical effects like exchange coupling, spin-dependent scattering, exchange coupling as discussed in the previous section. The MR can be classified into several types such as tunnel magnetoresistance (TMR), anisotropic magnetoresistance (AMR), GMR, colossal magnetoresistance (CMR), and ordinary magnetoresistance (OMR).

8.4.2.1 Ordinary Magnetoresistance

The OMR refers to the change in resistance due to the different Fermi configurations and the variation in resistance is positive for parallel and antiparallel configurations. This is the basic form of magnetoresistance which is visible in normal nonmetals.

Magnetoresistance (MR) is commonly defined as $MR(B) = [\rho(B) - \rho(0)]/\rho(0)$, where the electrical resistivities in applied magnetic field B and the zero field, respectively, are represented by $\rho(B)$ and $\rho(0)$. Negative MR has lately been seen in magnetic materials and some topological materials, while metals, semiconductors, and semimetals frequently exhibit positive MR. The majority of nonmagnetic materials have a small MR effect, which is characterized by a low-field quadratic field dependence that saturates to a few percent for metals. Due to various causes, MR effects come in many distinct forms.

Numerous presumptions have been made.

a) The magnetic field would not have an impact on the Fermi surface; rigid band approximation is what is being used here.
b) The approximate relaxation periods.
c) No consideration is given to Berry phase or Berry curvature.

The conductivity tensor is measured under an applied magnetic field which is as follows, in accordance with Boltzmann transport theory and the assumptions indicated earlier:

$$\sigma_{ij}^{(n)}(B) = \frac{e^2}{4\pi^3} \int dk \tau_n v_i^n(k) v_j^n(k) \left(-\frac{\partial f}{\partial \epsilon} \right)_{\epsilon = \epsilon_n(k)}$$

where e = electron charge, n = band index, τ_n = relaxation time of nth band, likely to be self-governing on the wave vector k, f = Fermi–Dirac distribution, $v_i^n(k)$ = velocity well-defined using band energy gradient.

$$v_i^n(k) = \frac{1}{\hbar} \frac{\partial \epsilon_n(k))}{\partial \epsilon}$$

and $v_i^n(k)$ = weighted average velocity of the charge carrier over time.

$$\frac{dkn(t)}{dt} = -\frac{e}{\hbar} v_n(k_n(t)) \times B$$

with $kn(t) = k$. The trajectory $k_n(t)$ can be found, integrating the above equation. Consequently, $v_n(k)$ is measured as the weighted average velocities with the route $k_n(t)$.

$$v_n(k) = \int_{-\infty}^{0} \frac{dt}{\tau_n} e^{t/\tau_n} v_n(k_n(t))$$

Since $v_n(k)$ is perpendicular to the Lorentz force, charge carriers are unaffected by it. As a result, energy $\epsilon_n(k)$ is constant, while k varies over time, showing that k is accordingly lateral to the constant energy surface and that v is normal to it. k is also normal to the constant energy surface. Therefore, the k vector creates an orbit that creates a cross-section of the Fermi surface since k is orthogonal to B as well.

In the implementation, usually, we put $B\tau_n$ together. Reformulate as

$$v_n(k) = \int_{-\infty}^{0} \frac{dBt}{B\tau_n} e^{Bt/B\tau_n} v_n(kn(t))$$

$$\frac{\sigma_{ij}^{(n)}(B)}{\tau_n} = \frac{e^2}{4\pi^3} \int dk v_i^n(k) v_j^n(k) \left(-\frac{\partial f}{\partial \epsilon} \right)_{\epsilon = \epsilon_n(k)}$$

8.4.2.2 Anisotropic Magnetoresistance

In 1857, AMR was revealed by Lord Kelvin. This phenomenon causes due to the change in orientation of magnetization of transition metals with respect to the current passing through the material. This is also observed in different alloys. In this case, the spin–orbit coupling, and band splitting are the major reasons behind the phenomenon.

The anisotropic dispersion of conduction electrons from the band in this exchange split band with uncompensated spins is the foundation of the AMR effect. The quantum mechanical exchange energy is the difference between the two states' energies of the magnetic spin moment. Ferromagnetism and ferrimagnetism are caused by the electrons. The majority of the materials exhibit positive AMR coefficients, which shows that the parallel alignment of the spontaneous magnetization M_f and current density J is what causes the high resistivity state.

By splitting the issue into dual portions – (i) relationship of resistivity ρ with the course of M_f and (ii) connection between magnetization direction and applied field H – it is possible to simplify the explanation of the complicated behavior of a generic magnetoresistor.

Resistance and Magnetization:
The AMR is a two-dimensional issue in soft magnetic thin films with a single area state with the coordinate system and the thin, rectangular ferromagnetic film's dimensions (length l, breadth b, and thickness d). An applied field H_y causes M_f to spin from the uniaxial anisotropy's relaxed axis direction in the direction of its hard axis direction. The angle ($\theta = \varphi - \psi$) between M_f and J determines the resistivity. With

$$\rho(\theta) = \rho_0 + \left(\rho_p - \rho_0\right)\cos^2\theta = \rho_0 + \partial\rho_0\cos^2\theta \tag{8.8}$$

and $\rho = \rho_p$ for M_f parallel J and $\rho = \rho_0$ for M_f parallel J orthogonal, the magnetoresistive coefficient, expressed as the quotient $\partial\rho/\rho_0$, may amount to many percentage points. With the resistance

$$R(\theta) = \rho(\theta)\frac{l}{bd} = R + \partial R\cos^2\theta \tag{8.9}$$

the voltage in x direction is

$$U_x = I\frac{l}{bd}\left(\rho_0 + \partial\rho_0\cos^2\theta\right)$$

The tensor property of ρ governs yet another AMR-related phenomenon. Perpendicular to the current density's underlying electrical field, E_x, J_x is a field of electricity.

$$E_y = J_x\partial\rho\sin\theta\cos\theta$$

The phenomenon is referred to as the planer or unusual Hall effect because of its orientation. The planner Hall voltage

$$U_y = I \frac{\partial \rho}{d} \sin \theta \cos \theta$$

Magnetization and Applied Field:
It should be remembered that the flux density B is measured by all magnetic sensors. The torque operating on magnetic moments and the Lorentz force operating on electrical charges in motion result in a physical effect. When considering the following factors, H is typically utilized since B and the functional field H of sensor's outside has a general connection $(B = \mu_0 H)$.

If H is in y direction, then the angle

$$\theta = \arcsin \frac{H_y}{H_0}$$

Between M_f and J (for $-1 < H_y/H_0 < 1$) results in the resistance's dependency on the field.

$$R(H_y) = R_0 + \partial R \left[1 - \left(\frac{H_y}{H_0} \right)^2 \right]$$

8.4.2.3   Giant Magnetoresistance

GMR is a nanomagnetic phenomenon in magnetic multilayers separated by nonmagnetic layers. In 1988, Fert and Grünberg made this discovery. The resistance across the layer falls to a certain value and eventually saturates beyond a specific magnetic field intensity when a multilayer of ferromagnetic (FM)/nonmagnetic (NM)/ferromagnetic (FM) material is maintained in the rising magnetic field. The cause of the phenomena was determined to be the antiferromagnetic exchange coupling and scattered condition of spin-up and spin-down electrons. In this situation, the current-in-plane (CIP) and current-perpendicular to the plane (CPP) geometries are predominantly employed. In the geometry of CIP, the flow of current is with the multilayer and in CPP geometry the current flows across the layers. Later, studies revealed that the GMR can be found in uncoupled magnetic layers if the magnetizations of FM layers can be changed from parallel to antiparallel configuration. The magnetoresistance varies depending on the width of the sandwiched nonmagnetic conducting layer, which is typically quite thin.

Magnetoresistance values that were first measured in single crystalline, (100) Fe/Cr/Fe sandwiches and (100) oriented Fe/Cr multilayers were substantially greater (50% at low temperatures). Moreover, the behavior of the Fe layers themselves may be able to explain the higher magnetoresistance. These structures' Cr layer thicknesses match a value that was previously discovered to provide an anti-ferromagnetic interaction between the Fe layers. These original, independent observations, of what is now known as GMR, provided much impetus to the development of spin electronics, because of their immediate implementation in magnetic recording head technology. The magnitude of GMR is calculated either as $\delta R/R = R_{AP} - R_P/R_{AP}$ or $\delta R/R = (R_{AP} - R_P)/R_P$, the latter giving a much larger value, where, $R_{AP} > R_P$, are the resistances observed with the Fe layers aligned antiparallel (initially) and parallel (on the application of a magnetic field), respectively. Subsequently, other ferromagnetic/nonmagnetic metallic multilayer structures have also shown GMR, most notably Co/Cu, which at first had some of the highest values. Moreover, these polycrystalline Co/Cu multilayers produced by sputtering showed an oscillation (period $\sim$9 Å) of the saturation magneto-resistance with Cu spacer thickness; large GMR values were observed for thicknesses corresponding to antiferromagnetically coupled layers, whereas for thicknesses initially showing ferromagnetic coupling, the external field had little effect on comparative placement of Co layers, which displayed weak GMR. Interestingly, even for very large Cu spacer thicknesses, with very weak interlayer coupling, GMR is detected, provided the magnetic layers are broken into spheres. Factually, antiferromagnetically coupled multilayers are not at all a precondition for GMR, and similar effects were reported in inhomogeneous magnetic systems, such as granular alloys, containing a uniform dispersion of magnetic particles in a nonmagnetic matrix.

A simple model to explain GMR is the resistor network, shown in Figure 8.1. Here, according to the two-spin-channel model of electrical conduction, we assign two different resistivities, $\rho\uparrow$ and $\rho\downarrow$, for the spin$\uparrow$ and spin$\downarrow$ channels, respectively, in the ferromagnet. In the nonmagnetic spacer layer, the resistivity, ρN, is identical for both spin channels. The coefficient $\alpha = \rho\downarrow/\rho\uparrow$, where $\alpha > 1$, may be used to describe the spin asymmetry of the resistivities of the two channels. The spin electrons are therefore lightly scattered in the parallel configurations while being substantially dispersed in both layers. The current is shorted by the spin$\uparrow$ channel, giving $R_P = (2\rho\uparrow\rho\downarrow)/(\rho\uparrow + \rho\downarrow)$. In the antiparallel configuration, each of the spin channels is either strongly or weakly scattered in the layers, and the resistance is averaged, with $R_{AP} = (\rho\uparrow + \rho\downarrow)/2$. Thus, GMR ratio is provided as

$$\text{GMR} = \frac{\delta R}{R} = \frac{R_{AP} - R_P}{R_P} = \frac{(\rho\uparrow + \rho\downarrow)^2}{4\rho\uparrow\rho\downarrow}$$

8.4.2.4 Tunnel Magnetoresistance

TMR is a magnetic phenomenon depending on the spin-dependent tunneling current through a tunnel junction comprising magnetic material, insulator, and magnetic material layer. Spin-polarized tunneling of electrons, in this case, is dependent on the magnetization of materials. In this case, the insulating layer thickness should be very small in order to facilitate tunneling through the barrier. This type of magnetoresistance can show up to 6000% resistance change under the influence of magnetic fields.

Magnetic tunnel junctions (MTJs), formed of two ferromagnets parted using a slim insulator, are where TMR occurs. Electrons can tunnel between two ferromagnets if the insulating barrier is small (naturally a few nanometers). Due to the fact that this process is prohibited in classical physics, the tunnel magnetoresistance is solely a quantum mechanical phenomenon.

Thin film technique is used in the production of magnetic tunnel junctions. Molecular beam epitaxy, pulsed laser deposition, and electron beam physical vapor deposition are also used on a laboratory size, while film deposition is being done on an industrial scale. Photolithography is used to prepare the connections.

The impact amplitude or relative resistance changes are defined as TMR: $=$ $(R_{ap} - R_p)/R_p$ where R_{ap} and R_p are electrical resistances in the antiparallel and parallel states, respectively.

The ferromagnetic electrodes' spin polarizations can also be used to explain the TMR phenomenon. At the Fermi energy, the spin-dependent density of states $\mathcal{D}$ (DOS) is used to compute the spin polarization P: $P = (\mathcal{D}(E_F) - \mathcal{D}(E_F))/(\mathcal{D}(E_F) + \mathcal{D}(E_F))$.

Spin-up electrons are oriented parallel, whereas spin-down electrons are aligned antiparallel to the exterior magnetic field. Two ferromagnets, P_1 and P_2, have polarizations of spin which provide information on the relative resistance change:

$$TMR = \frac{2P_1P_2}{1 - P_1P_2}$$

The junction experiences equal rates of electron tunneling in both directions in the absence of any voltage. At a bias voltage U, electrons tunnel predominantly to the positive electrode. The current may be modeled as two currents on the presumption that spin is preserved throughout tunneling. Two partial currents, one for spin-up electrons and the other for spin-down electrons, split the overall current. Depending on the magnetic states of the connections, these change.

A known antiparallel state can be reached in one of two methods. Using ferromagnets having various coercivities (utilization of various materials or thickness of film) is the first way. Second, exchange bias refers to the connection between an antiferromagnet and one of the ferromagnets. The uncoupled electrode's magnetization is still "free" in this instance.

If P_1 and P_2 or if both electrodes have 100% spin polarization, equal 1, the TMR becomes infinite. The magnetic tunnel junction behaves like a switch in this case, switching magnetically between infinite and low resistance. Ferromagnetic half-metals are the materials that are taken into account for this. The spin polarization of their conduction electrons is complete. Although this behavior is theoretically expected for a variety of materials, there has been some quiet disagreement on its experimental proof. Bowen et al.'s findings of 99.6% spin polarization at the edge between $La_{0.7}Sr_{0.3}MnO_3$ and $SrTiO_3$ essentially equate to evidence of the characteristic obtained experimentally if one just takes into account the electrons that participate in transport.

An increase in both temperature and bias voltage causes a reduction in TMR. Both can theoretically be explained by magnon excitations, interactions with magnons, and tunneling with regard to localized states brought on by oxygen vacancies.

8.4.2.5 Colossal Magnetoresistance

The CMR occurs at a range of temperatures due to the phase transition from ferromagnetic to paramagnetic. This phenomenon is usually predominant in perovskite materials. Generally, a double exchange of electron spin orientation and a proton is expected in this scenario.

The ability of some materials, primarily perovskite oxides based on manganese, to drastically change their electrical resistance due to the magnetic field is called CMR. Materials having CMR may exhibit resistance changes of orders of magnitude, whereas conventional materials' magnetoresistance only permits resistance swings of up to 5%. The large series of ferromagnetic and AF spin configurations of the system may be explained by a hypothesis of magnetic exchange via covalent bonds. According to the "double" exchange hypothesis, there is a struggle for control of the magnetic order between "double" exchange, which prefers order of ferromagnetic spin, and super-exchange, which promotes an AF spin configuration. The characteristic of the charge carrier and the properties of transport must still be understood in order to fully realize the magnetic structure on the basis of spin–spin interaction under Jahn–Teller lattice distortion and "double" exchange.

8.5 Fabrication and Characterization Technologies

Magnetic sensors are developed using both conventional and nonconventional processes. Conventional micro/nanofabrication techniques are used for the progress of a variety of magnetic sensors. The details of fabrication are discussed in Chapter 2 of this book. However, a brief process flow and a few new fabrication techniques are discussed here.

8.5.1 Conventional Fabrication

The major fabrication technology for the magnetoresistive sensor is the deposition of metal layers. Sputtering, thermal deposition, and e-beam evaporation are extensively used for the magnetic metal layer depositions. The deposition of extremely thin layers of magnetic and nonmagnetic metal elements is one of the main difficulties in this situation, thus optimizing the deposition parameters is crucial. Aside from the micro/nanofabrication techniques, people have also used the solution process, drop casting, and printing technologies to fabricate magnetic sensors.

8.5.2 Solution Process

The solution process has also been employed for fabricating granular magnetoresistive sensors. This type of fabrication involves synthesizing magnetic or nonmagnetic metal nanoparticles of different sizes. Most of the cases, the metal salts are reduced to form different nanoparticles, however, controlling the oxidation of the metal nanoparticles is a challenging aspect of this process. But this process offers a synthesis of different-sized nanostructures by tuning different parameters such as stirring, temperature, and reaction time.

8.5.3 Printing Technologies

Although creating nanofilms of metal is not possible using the existing inkjet or screen-printed techniques [15, 16], different magneto-responsive inks have been employed for the development of magnetoresistance devices. Printed Hall sensors are also reported in this direction. The advent of the super inkjet printer has provided higher resolution of up to 1 μm, which can help in developing different types of magnetoelectronic sensors.

8.6 Magnetic Sensor Applications

Magnetic sensors find huge applications from electronics to biosensing. The advent of GMR has made the biosensing platform more sensitive. Moreover, the magnetoresistive phenomenon is also used in read heads for magnetic drives and current sensing. Herein, a few of these applications are discussed in brief.

8.6.1 Biosensors

Different magnetoresistive sensors [17, 18] are employed for the development of a variety of biosensors. Magnetic microbeads are used extensively for immobilizing biomolecules for the detection of biomarkers. Different GMR sensors have been reported in this direction that is capable of high-precision detection. For example, DNA concentration was detected by functionalizing the streptavidin-conjugated magnetic beads to react with the DNA on the GMR sensor. Multiple GMR devices are reported to detect bioanalytes to make point-of-care testing devices.

Several sensors are also developed based on TMR. Hall effect sensor also showed promising results to be used as a sensitive biosensor. Several magnetic nanostructure materials are used to make biosensors in this regard.

In recent times, the magnetic sensors are also integrated with systems such as mechanical, microfluidic, and microelectromechanical devices [19–21]. In microfluidic applications [22], the sensors are used to detect magnetically labeled cells and quantify the count of cancer cells in a flowing system. Systems are also developed to detect multiple biomarkers in smartphones using GMR-based sensors. Figure 8.5 shows the scheme of detecting biomolecules using magnetic sensors like the Hall sensor (a) and GMR sensor (b). In both the cases, magnetic beads are functionalized with specific biomolecules that react with the analyte and attach with the sensor surface. As a result, there is a magnetic field generated close to the sensor, and the magnetic field from the bead is used to detect the object.

8.6.2 Magnetic Storage and Read Heads

GMR has been explored significantly for developing read heads and storage devices [23–25]. The GMR devices are incorporated in commercial hard disk drives. The CPP geometry of GMR is now being explored for developing high-performance magnetic read-head technology. The AMR and GMR read heads were commercially introduced by IBM for the first time in 1990 and 1997. The binary counts such as 0 and 1 are detected in terms of change in magnetizations by GMR read heads. There is a huge demand from the industry side to develop high-density memory devices to miniaturize the memory components. Aside from GMR, the TMR and AMR are also explored significantly to develop magnetic storage devices.

8.6.3 Current Sensing

Magnetic sensors are very sensitive to current sensing [26, 27]. Figure 8.6a shows the schematic illustration of the current detection mechanism. A magnetic sensor may sense the magnetic field created by a current flowing through a wire around it. Several sensors are present in the market which work on Hall effect, capable of detecting current in a system. Unlike conventional current sensors, AMR sensors are capable of detecting current in contactless mode. AMR sensors are also used as analog multipliers due to their resolution and bandwidth which in turn provides higher sensitivity. However, many of the AMR sensors suffer from input signal leakage, nonlinearity, and heating. Recent reports have also shown the current

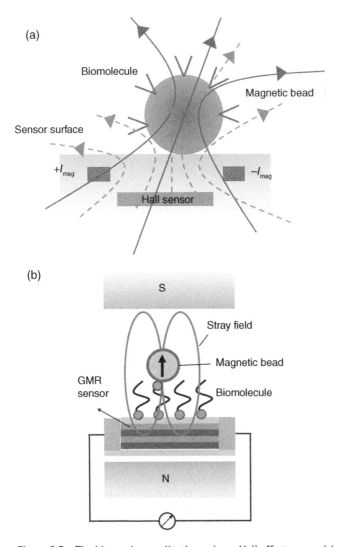

Figure 8.5 The biosensing application using a Hall effect sensor (a) and a GMR sensor (b).

sensor based on TMR as a built-in chip. Further development has also indicated the capability of the GMR sensor to be used as an on-chip current sensor. GMR sensor provides high sensitivity, linearity, and ease of tuning. GMR eddy current sensor has been employed for the detection of cracks. The schematic representation of the eddy current sensor, which detects changes in magnetic fields caused by eddy currents, is shown in Figure 8.6b. Recently, CMOS technology compatible GMR sensor was developed for current sensing in integrated circuits.

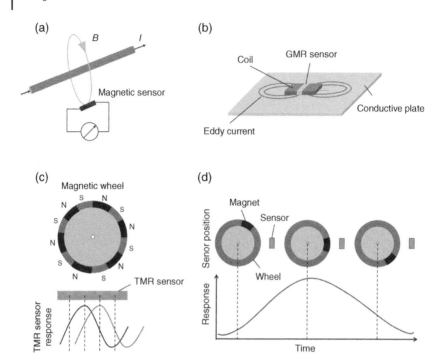

Figure 8.6 Different sensing mechanisms in brief such as current sensing (a), eddy current sensing (b), angle sensing (c), and position sensing (d).

8.6.4 Position and Angle Sensors

Magnetic sensors are also used for position and angle detection [28, 29]. Figure 8.6c, d show the schematic illustration of the position sensing using magnetic sensors. In this case, either a magnet or a sensor is connected to the wheel which interacts with one of its fixed counterparts to generate signals. The delay and phase of the consecutive signals are then used to determine the position and angle. AMR sensors are being used extensively by the automotive industries for the angle and position sensing for the past few decades. However, the advent of GMR technology helped in manufacturing high-precision angle and position detection. These sensors can withstand high temperature and pressure. The sensors were also placed between a rotating wheel and a permanent magnet in antilock braking systems (ABS). Spin valve was used for the contactless detection of position. These sensors have two layers – one is a pinned rigid magnetic layer and other one is a lenient magnetic layer. These sensors' responses are largely influenced by the applied external magnetic field. In this direction, TMR-based sensors were developed for angle displacement. Time pulse count of phase delay of TMR and a reference sensor output has been exploited to measure the angular displacement.

8.7 Prospects and Limitations

There is a great deal of potential for the currently available sensors to be employed in different areas of applications, including magnetic storage, current sensors, and different environmental sensing applications. However, the sensitivity and complex fabrication limit these sensors to be used as a disposable one. Magnetoresistive sensors have shown potential in the ultrasensitive detection of biomolecules and use in magnetic storage devices. However, the sensors require more attention in terms of the fabrication process. The existing fabrication techniques are not suitable for disposable sensor fabrication as it will not be economically viable and hence, the sensors may find difficulties in terms of commercialization in biosensing or environmental applications. Moreover, the sensors are highly sensitive toward external magnetic fields and hence proper packaging and magnetic shielding are desired. Since the measurements might be impacted due to the existence of an outer magnetic field, isolation is now another difficult task.

List of Abbreviations

AMR	Anisotropic Magnetoresistance
CIP	Current-in-Plane
CMR	Colossal Magnetoresistance
CPP	Current-Perpendicular to the Plane
FM	Ferromagnetic
GMR	Giant Magnetoresistance
MOSFET	Metal Oxide Semiconductor Field Effect Transistor
NM	Nonmagnetic
OMR	Ordinary Magnetoresistance
PMA	Perpendicular Magnetic Anisotropy
TMR	Tunnel Magnetoresistance

References

1 Meyer, H.W. (1971). *A History of Electricity and Magnetism*. MIT Press.
2 Ripka, P. (2001). *Magnetic Sensors and Magnetometers*. Artech House.
3 Krishnan, K.M. (2016). *Fundamentals and Applications of Magnetic Materials*. Oxford University Press.
4 den Broeder, F.J.A., Hoving, W., and Bloemen, P.J.H. (1991). Magnetic anisotropy of multilayers. *Journal of Magnetism and Magnetic Materials* 93: 562–570.
5 Stiles, M.D. (1999). Interlayer exchange coupling. *Journal of Magnetism and Magnetic Materials* 200 (1): 322–337.

6 Nogués, J. and Schuller, I.K. (1999). Exchange bias. *Journal of Magnetism and Magnetic Materials* 192 (2): 203–232.

7 Tsymbal, E.Y. and Zutic, I. (2016). *Handbook of Spin Transport and Magnetism.* CRC Press.

8 Asfour, A. (2017). *Magnetic Sensors: Development Trends and Applications.* IntechOpen.

9 Paun, M.-A., Sallese, J.-M., and Kayal, M. (2013). Hall effect sensors design, integration and behavior analysis. *Journal of Sensor and Actuator Networks* 2 (1): 85–97.

10 Popović, R.S. (1989). Hall-effect devices. *Sensors and Actuators* 17 (1): 39–53.

11 Kvitkovic, J. (1997). Hall generators. *CERN Accelerator School on Measurement and Alignment of Accelerator and Detector Magnets*, Anacapri, Italy (11–17 April 1997), vol. 98-05.

12 Sze, S.M. and Lee, M.K. (2013). *Semiconductor Devices, Physics and Technology.* Wiley.

13 Tumanski, S. (2001). *Thin Film Magnetoresistive Sensors.* CRC Press.

14 Ennen, I., Kappe, D., Rempel, T. et al. (2016). Giant magnetoresistance: basic concepts, microstructure, magnetic interactions and applications. *Sensors (Basel, Switzerland)* 16 (6): 904.

15 Manjakkal, L., Sakthivel, B., Gopalakrishnan, N., and Dahiya, R. (2018). Printed flexible electrochemical pH sensors based on CuO nanorods. *Sensors and Actuators B: Chemical* 263: 50–58.

16 Manjakkal, L., Shakthivel, D., and Dahiya, R. (2018). Flexible printed reference electrodes for electrochemical applications. *Advanced Materials Technologies* 3 (12): 1800252.

17 Krishna, V.D., Wu, K., Perez, A.M., and Wang, J.-P. (2016). Giant magnetoresistance-based biosensor for detection of influenza A virus (in english). *Frontiers in Microbiology, Methods* 7 (400): 1–8.

18 Wang, S.X. and Li, G. (2008). Advances in giant magnetoresistance biosensors with magnetic nanoparticle tags: review and outlook. *IEEE Transactions on Magnetics* 44 (7): 1687–1702.

19 Pekas, N., Porter, M.D., Tondra, M. et al. (2004). Giant magnetoresistance monitoring of magnetic picodroplets in an integrated microfluidic system. *Applied Physics Letters* 85 (20): 4783–4785.

20 Trontelj, J. (1999). "Optimization of integrated magnetic sensor by mixed signal processing. *IMTC/99. Proceedings of the 16th IEEE Instrumentation and Measurement Technology Conference (Cat. No.99CH36309)*, Venice, Italy (24–26 May 1999), vol. 1, pp. 299–302.

21 Herrera-May, L.A., Aguilera-Cortés, A.L., García-Ramírez, J.P., and Manjarrez, E. (2009). Resonant magnetic field sensors based on MEMS technology. *Sensors* 9 (10): 7785–7813.

22 Xianyu, Y., Wang, Q., and Chen, Y. (2018). Magnetic particles-enabled biosensors for point-of-care testing. *TrAC Trends in Analytical Chemistry* 106: 213–224.

23 Comstock, R.L. (2002). Review modern magnetic materials in data storage. *Journal of Materials Science: Materials in Electronics* 13 (9): 509–523.

24 Irshad, M.I., Ahmad, F., and Mohamed, N.M. (2012). A review on nanowires as an alternative high density magnetic storage media. *AIP Conference Proceedings*, Kuala Lumpur, Malaysia (26 September 2012), vol. 1482, no. 1, pp. 625–632.

25 Stevens, L.D. (1981). The evolution of magnetic storage. *IBM Journal of Research and Development* 25 (5): 663–676.

26 Dogaru, T. and Smith, S.T. (2001). Giant magnetoresistance-based eddy-current sensor. *IEEE Transactions on Magnetics* 37 (5): 3831–3838.

27 Ouyang, Y., He, J., Hu, J., and Wang, S.X. (2012). A current sensor based on the giant magnetoresistance effect: design and potential smart grid applications. *Sensors* 12 (11): 15520–15541.

28 Burger, F., Besse, P.A., and Popovic, R.S. (1998). New fully integrated 3-D silicon Hall sensor for precise angular-position measurements. *Sensors and Actuators A: Physical* 67 (1): 72–76.

29 Mapps, D.J. (1997). Magnetoresistive sensors. *Sensors and Actuators A: Physical* 59 (1): 9–19.

9

Interface Circuits

9.1 Introduction

The advent of semiconductor evolution, cheaper mass-produced integrated circuits, and Internet connectivity has opened a lot of opportunities for many application-based smart systems [1]. One of the primary targets in modern research is to interface computers and intelligent computing facilities with the outside world for monitoring the environment, health, agriculture, etc. [2]. For this purpose, detecting and quantifying various physical, chemical, biological, or any other parameters are necessary and sensors play an essential role in that. Hence, to connect those sensors with computing facilities, it is essential to have a suitable and proper interface circuit [3]. All the sensors need suitable circuits to operate for a specific application [4, 5]. For accurate performance and safety of the device, the sensor often requires a customized circuitry based on its application [6]. For this reason, it is equally essential that the sensor engineer is aware of the possible circuitry and user interface options. Furthermore, an idea of the computing and processing unit can help the sensor designer and the user to extract valuable information from measured parameters [7]. This chapter is focused to discuss various interfacing circuits, different electronic components for amplification, noise and its reduction, exciting sources, batteries, various energy-harvesting devices, and various data transmission systems used for sensor operations.

9.1.1 Functions of Interface

Interfacing of sensors and readout circuits is most important for accessing the physical quantity sensed by the sensor. An input/output (I/O) interface is the part of a circuit or a circuit itself that is used to connect an I/O device to a system such as computer bus [8]. On one side of the interface, address data and control of a bus signal are available and on the other side, a data path with associated controls is

Solid-State Sensors, First Edition. Ambarish Paul, Mitradip Bhattacharjee, and Ravinder Dahiya.
© 2024 The Institute of Electrical and Electronics Engineers, Inc.
Published 2024 by John Wiley & Sons, Inc.

provided to convey data between the interface and the I/O device. This is referred to as a port and can be either parallel or serial in nature. A parallel port transfers data in the form of several bits, often 8 or 16, to or from the device simultaneously [9]. A serial port sends and receives data bit by bit. For both formats, bus communication is the same; parallel to serial conversion, and vice versa, occurs within the interface circuit. The I/O interface performs the following functions [10]:

- Provides at least one word (or one byte, in the case of byte-oriented devices) data storage buffer
- Contains status flags to detect whether the buffer is empty (for output) or complete (for input)
- Contains address-decoding circuitry to determine when the CPU is addressing it
- Generates the proper timing signals for bus control system
- Executes any format conversions required to transport data between the bus and the I/O device, for example, in the case of a serial port, parallel–serial conversion.

9.1.2 Types of Sensor Interfacing Circuits

There are typically two kinds of highly preferred sensor interface circuits: namely, resistive and voltage-producing sensor circuits. Resistive sensors are the ones that act as a variable resistor due to the influence of stimuli or input [11]. These sensors typically have a base resistance or resistance in the absence of input, and minimum and maximum resistance values that match the input stimuli's extremes. These values can be found in the data sheet of the sensor under consideration or during fabrication. Otherwise, user can measure those values. These values are important for the designing of the interface circuit. Similarly, there are sensors that provide a variable voltage as output. The voltage output varies with the input. The range of voltage, i.e. maximum to minimum, can be found in the datasheet of the sensor. In this case, the lowest or minimum value is more crucial in developing the interface circuit. However, the maximum value is also important.

Before connecting a sensor with a computing system, the signal must be delivered to an interfacing device of some sort and then it goes to a further computing system [12]. In this scenario, it is vital to notice that the signals from the majority of the sensors are analog. Hence, the interfacing circuit begins with an analog front end to which various analog components are attached. Interfacing circuits for each type of sensor or the same sensor for various applications may differ as the desired parameters to be extracted are different. Furthermore, almost all cases, the signal that comes out of a sensor needs conditioning before it goes for further processing. The conditioning by the interface circuit includes signal filtering and amplifying. Almost all interfacing circuits operate and digitize a voltage signal in 0–5 V range.

The signal conditioning part in the interface circuit takes care of the same. These operations are usually done stepwise as (i) converting resistance into voltage, (ii) dividing the voltage into desired values, (iii) amplifying the voltage to a desired value, and (v) shifting the voltage wherever necessary. The mentioned processes can be summarized or realized using the following circuit diagram.

The output voltage (V_{out}) obtained from the Figure 9.1a is

$$V_{out} = \frac{V_{in} \times R_2}{R_1 + R_2} \tag{9.1}$$

From the equation above, the ratio between output and input voltage can be given by,

$$\frac{V_{out}}{V_{in}} = \frac{R_2}{R_1 + R_2} \tag{9.2}$$

The output voltage (V_{out}) obtained from the Figure 9.1b is

$$V_{out} = V_{in} \times \left(1 + \frac{R_2}{R_1}\right) \tag{9.3}$$

Shifting Voltages: As discussed earlier, sensors that produce a voltage output that is symmetrical around 0 V are also available. For instance, if a sensor has a voltage range of $-X$ to $+X$ volt, and analog-to-digital converters (ADCs) require a positive voltage, then these voltages must be shifted upward so that they can work from 0 to

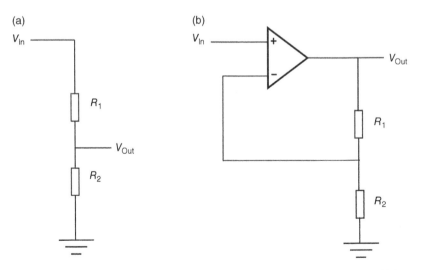

Figure 9.1 (a) Voltage divider circuit for resistance to voltage conversion. (b) Voltage amplifier circuit to amplify the voltage.

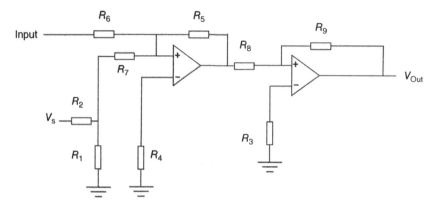

Figure 9.2 Circuit for shifting of voltage for sensor circuit design.

$2X$ volts. The voltage shifting circuit is more complicated than the preceding circuits. It adds a set voltage to the input signal using a twin op-amp. The values of two resistors R_1 and R_2 in a voltage divider determine the voltage to be applied. The circuit for shifting of voltage for sensor circuit design is shown in Figure 9.2.

It is vital to keep in mind that each conditioning step introduces some signal contaminants. This indicates that the signal will get less pure as it passes through additional phases. Furthermore, signals that undergo significant alterations will have more contaminants than signals that undergo just minor changes.

9.1.3 Battery

This section deals with a very important component of readout circuit that is battery. A battery is a device that provides supply voltage to the circuit and made up of two or more cells. For example, an electric vehicle's driving system comprises two batteries, a motor, and two converters [13]. Since these components are interconnected, they generate ripple signals. The proper interface between the batteries and the converters can extend the life of the system. In designing an interface circuit, it is usually assumed that the battery is a simplified voltage source.

But, to make the interface design more efficient, the internal parameters of the battery should also be taken into account. Three distinct models for modeling battery properties are discussed. Figure 9.3a shows the schematic representation of the ideal battery. The batteries considered in general are lead–acid ones. In this ideal model, the battery is represented as an analogous circuit by a single voltage source. The model is fairly basic because the internal parameters are omitted. The model, however, does not reflect the battery's intrinsic features. Internal resistance is taken into account in this linear model. Figure 9.3b depicts a linear model of

Figure 9.3 Models for electric battery. (a) Ideal, (b) linear, and (c) Thevenin models.

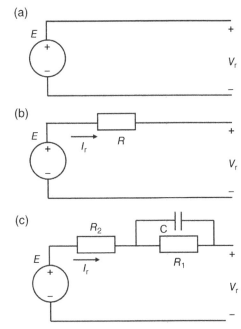

battery with internal resistance. The voltage and internal resistance depend on the state of discharge and other parameters and are represented as follows:

$$E = E_0 - k \cdot f \tag{9.4}$$

$$R = R_0 - K_R \cdot f \tag{9.5}$$

where E_0 = fully charged no load voltage, f = state of discharge, R_0 = internal resistance for fully charged condition, k, K_R = constants associated with experiments.

The values of different electrical parameters, i.e. no-load voltage (E), internal resistance (R, R_1), and parallel combination of C and R_2, can be obtained from this Thevenin model, as shown in Figure 9.3c. Electrically this model gives accurate result than the linear model to simulate or to characterize the behavior of a battery system.

9.1.4 Battery Characteristics in System Analysis

To analyze the battery model composed of capacitors, resistors, and other components, two methodologies are utilized [14]. One methodology is to study the system by analyzing its waveform using a computer program, while the other method is to simulate the system.

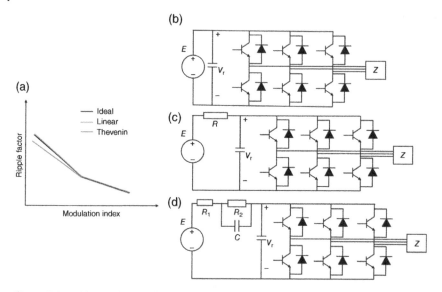

Figure 9.4 (a) Ripple factor of battery output current connected to a three-phase inverter. Inverter system with three phases and condenser filters. Using the (b) ideal, (c) linear, and (d) Thevenin battery models.

Various models of batteries are used for the analysis. The ideal battery model is then used to study the system's characteristics in Figure 9.4a. It is a graph between ripple factor and modulation index. Figure 9.4b, d indicates the different models of inverter system with three phases and condenser filters. Figure 9.5 represents ripple factor w.r.t. modulation index of the battery output current related to condenser filters for (i) single-phase and (ii) three-phase inverter.

In general, the transfer function (switching) of an inverter, H, can be calculated using the Fourier series which is given below:

$$H(\omega t) = \sum_{n=1}^{\infty} h_n \sin n(\omega t) \tag{9.6}$$

Load current I_0, output voltage V_0, battery current I_i are represented by:

$$(\omega t) = V_i \cdot H(\omega t) \tag{9.7}$$

$$I_0(\omega t) = V_0(\omega t) \cdot Z^{-1}(\omega) \tag{9.8}$$

$$I_i(\omega t) = I_0(\omega t) \cdot H(\omega t) \tag{9.9}$$

where V_i is the inverter input voltage and Z is the load impedance.

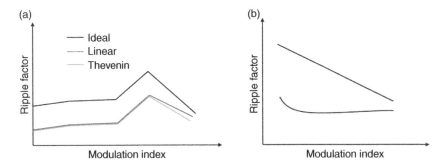

Figure 9.5 Ripple factor of battery output current with condenser filters. (a) Single-phase inverter and (b) three-phase inverter.

The output current of the battery is given by [From (9.3) to (9.6)]:

$$I_i(\omega t) = I_0(\omega t) \cdot H(\omega t)$$

$$= \left[E \cdot \sum_{n=1}^{\infty} \frac{h_n}{|Z_n|} \cdot \sin(n\omega t - \phi_n) \right] \cdot \left[\sum_{n=1}^{\infty} h_k \cdot \sin(k\omega t) \right]$$

$$= \frac{E}{2} \sum_{k=1}^{\infty} \sum_{n=1}^{\infty} \frac{h_n \cdot h_k}{|Z_n|} \{ [\cos(k-n)\omega t + \phi_n] - [\cos(k+n)\omega t - \phi_n] \}$$

$$(9.10)$$

where $Z_n = \sqrt{R^2 + (n\omega L)^2}$, R = load resistance, and L = load inductance.

Because the inverter input voltage V_i is the same as the DC voltage E in the ideal model, the exact representation of the battery output current is produced. However, an assumption must be made in the linear model.

$$V_i(\omega t) = V_{dc} + \sum_{k=1}^{\infty} C_k \cos(k\omega t + \theta_k). \tag{9.11}$$

Using the equations from (9.3) to (9.6) once more, the output current can be calculated as follows:

$$I_i(\omega t) = I_0(\omega t) \cdot H(\omega t) \tag{9.12}$$

Then V_i is the inverter input voltage,

$$V_i(\omega t) = E - I_i(\omega t) \cdot R. \tag{9.13}$$

which means that both values of V_i are the same.

Thevenin model analysis: The output current and V_i of the battery can be determined using the linear model. Obtaining exact values, on the other hand, is

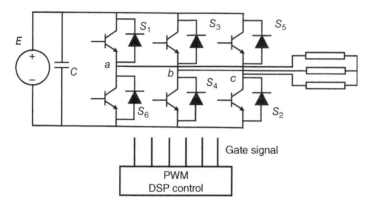

Figure 9.6 Circuit diagram for three-phase inverter.

extremely difficult and time-consuming. Thevenin model is used to assess the battery's internal properties. But the linear model can be used for a simple and easy representation and modeling. It demonstrates that with a battery-powered supply, the needed size of the filter can be substantially smaller than the others like a bridged type AC power supply or DC source. A bridge-type three-phase inverter circuit has been shown in Figure 9.6.

Input Impedance: Input impedance of the interface circuit is an important parameter for interfacing of sensor and readout circuit. It depicts the loading effect of the interface circuit on the sensor, which is stated as:

$$Z = V/I \tag{9.14}$$

The input impedance of the interface circuit can be modeled as a parallel connection of an input resistance (R) and an input capacitance (C). Then, the complex input impedance can be given by:

$$Z = \frac{R}{1 + j\omega RC} \tag{9.15}$$

If the supply frequency is very low, the capacitive reactance can be negligible and can be treated as open circuit. So, the input impedance value becomes very close to only the value of the resistance:

$$Z \cong R. \tag{9.16}$$

9.1.5 Applications of an I/O Interface Device

This section deals with the practical implementation of interfacing the devices. Any CPU that wants to communicate with its I/O devices needs a surface [15].

The interface is used to interpret addresses generated by the CPU. As a result, a surface is used to interact with I/O devices, i.e. an interface is used to transport information between the CPU and the I/O devices, which is referred to as the I/O interface. I/O allows an interface to open any file without knowing anything about it, even if the file's essential basic information is unknown. It is also capable of adding extra devices to a computer system without interfering with the operating system. It can also be used to identify generic sorts of I/O devices so that distinctions between them can be abstracted. Each of the generic types is accessed via an interface, which is a standardized collection of functions.

9.1.6 Importance of Input Impedance

If there is a significant difference between input impedance of the interface circuit and the sensor's output impedance, we need to consider impedance matching. Consider a fully resistive sensor coupled to the input impedance of an interface circuit as shown in Figure 9.7. The input voltage to the interface circuit as a function of frequency (f) can be written as [16]:

$$V = \frac{E}{\sqrt{1 + \left(\dfrac{f}{f_c}\right)^2}} \tag{9.17}$$

$$f_c = (2\pi RC)^{-1} = \text{corner frequency} \tag{9.18}$$

Assuming 1% amplitude detection accuracy is required, the maximum stimulus frequency is given by:

$$f_{max} \approx 0.14 f_c, \; f_c \approx 7 f_{max} \tag{9.19}$$

From the above calculation it is clear that, to process the sensor signal, the corner frequency of the interface circuit must be sufficiently larger than the maximum frequency of the sensor signal. The corner frequency depends on the R and C values, i.e. input impedance of the interface circuit. Thus, the interface circuit with

Figure 9.7 Input impedance circuit.

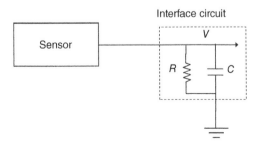

proper input impedance must be chosen such that its corner frequency is sufficiently large. The frequency bandwidth of op-amps, which are the basic building blocks of any interface circuits, is often restricted. So, these considerations based on input impedance be taken seriously while choosing an appropriate interface circuit for the sensor in hand. Furthermore, depending on the sensor characteristics its output impedance must be taken into consideration while calculating the corner frequency. For instance, a capacitive sensor can be electrically modeled as a pure capacitor connected parallel to the adjacent input capacitance of the interface circuit. Hence, sensor performance and successful detection of any target parameter is highly influenced by the input impedance of the interface circuit. Also, the frequency response of the circuit decides the speed of the response. In this direction, a user of programmable operational amplifier can be a better option as the user can modify its bias current and, hence, the frequency response of the first-stage amplifier to have a flexible operation according to the sensor's demand. The speed of response in this case can be increased by increasing the bias current of the op-amp [17].

9.2 Amplifier Circuits

As discussed in the previous section, conditioning of the sensor signal is extremely important for accurate analysis and computing. In this direction, amplification of the analog sensor output plays an important role. For example, there are sensors that provide low-voltage signals as output and hence it is important to amplify the signal so that further processing can be performed. It is important to know the basics of the amplifiers for correctly designing the interface circuits. The amplifiers discussed in this section are ideal operational amplifiers, voltage followers, instrumentation amplifiers, and charge amplifiers.

9.2.1 Ideal Operational Amplifier (Op-amp)

The operational amplifier is a necessary component in the majority of interface circuits [18]. It is worth noting, however, that the fundamentals of op-amps may be found in any standard book on analog circuits. This section will go over the specifics of an operational amplifier. Figure 9.10 depicts the ideal operational amplifier's equivalent circuit model. This device contains two input terminals, i.e. one inverting and one noninverting, and one output terminal. A perfect op-amp has the following characteristics: infinite input impedance, zero output impedance, zero common-mode gain, and infinite open-loop gain [19]. The ideal amplifier is not used alone, instead this is a part of a circuit which also contains passive

Figure 9.8 Equivalent circuit model of an ideal operational amplifier.

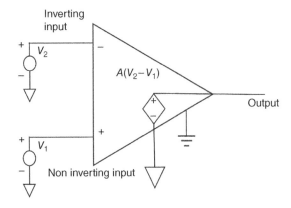

components in feedback and input and output sides. Figure 9.8 indicates the internal circuit components of op-amp.

9.2.2 Inverting and Noninverting Op-amps

In case of an ideal op-amp, if we are using a resistor as the passive component in the feedback loop, we get two configurations known as inverting and non-inverting op-amps [18]. The schematics of inverting and non-inverting op-amps are given in Figure 9.9.

Mathematical calculation of closed loop gains for inverting op-amps:

Consider the ideal behavior of an op-amp, in which the current entering the op-amp is zero and the potential difference between inverting terminal (V_1) and noninverting terminal (V_2) is zero due to virtual short assumption.

i.e. $V_1 - V_2 = 0$ (9.20)

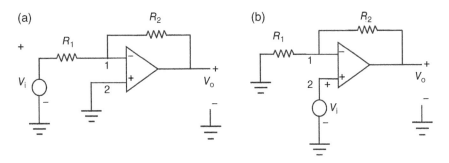

Figure 9.9 Schematic diagram of (a) inverting op-amp (b) noninverting op-amp.

Applying KCL at point 1, we get,

$$\frac{V_1 - V_i}{R_1} = \frac{V_0 - V_1}{R_2} \tag{9.21}$$

Therefore, the closed loop gain,

$$\frac{V_0}{V_i} = -\frac{R_2}{R_1} \tag{9.22}$$

Similarly, in case of noninverting op-amps, applying KCL at point 1 (inverting input point), we get,

$$\frac{V_i - 0}{R_1} = \frac{V_0 - V_i}{R_2} \text{ [as } V_1 - V_2 = 0] \tag{9.23}$$

Therefore, in the case of noninverting op-amps, the closed loop gain is,

$$\frac{V_0}{V_i} = 1 + \frac{R_2}{R_1} \tag{9.24}$$

9.2.3 Voltage Follower

If the noninverting terminal is connected to the output terminal, the configuration would have a unity voltage gain. This configuration has a high input impedance, allowing this circuit to operate as a buffer amplifier to connect a high impedance source to a low impedance load [19]. It is called unity gain amplifier, but commonly known as voltage follower [20]. The schematic of a general voltage follower amplifier is given in Figure 9.10.

In this case, also, we are considering the ideal behavior of the op-amps.

Voltage at inverting input point = Voltage at the noninverting input point
$$= V_i$$

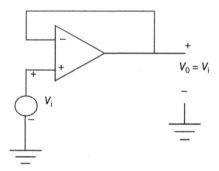

Figure 9.10 Schematic diagram of a voltage follower amplifier.

Since there is no passive component present in the feedback loop voltage,

Voltage at inverting input point = Voltage at the output point = V_0

From the above two equations, we get,

$$V_i = V_0 \tag{9.25}$$

9.2.4 Instrumentation Amplifier

Another useful amplifier in the field of readout circuit design is *instrumentation amplifier*. It is a type of differential amplifier, outfitted with input buffer amplifier [21]. This is normally used in measurement and test equipment because it eliminates the need of input impedance matching. The schematic diagram of a typical instrumentation amplifier is shown in Figure 9.11.

Inspecting Figure 9.4, we get that both the op-amps, A_1 and A_2, in the first stage have noninverting configuration, each with a gain of $(1 + R_2/R_1)$. The second stage of the amplifier is a difference amplifier. The input difference signal to the second stage is,

$$\left(1 + \frac{R_2}{R_1}\right)(V_{I1} - V_{I2}) = \left(1 + \frac{R_2}{R_1}\right)V_{Id} \, [V_{I1} - V_{I2} = V_{Id}] \tag{9.26}$$

Given the second stage's input, the amplifier's final output V_0 is,

$$V_0 = \frac{R_4}{R_3}\left(1 + \frac{R_2}{R_1}\right)V_{Id} \tag{9.27}$$

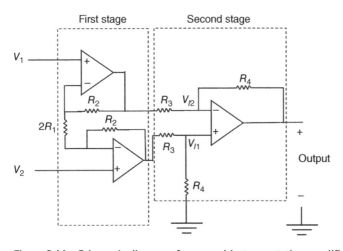

Figure 9.11 Schematic diagram of a general instrumentation amplifier.

Finally, the differential gain of the amplifier is given as,

$$A_v = \frac{R_4}{R_3}\left(1 + \frac{R_2}{R_1}\right) \tag{9.28}$$

9.2.5 Charge Amplifiers

In the feedback loop of an op-amp, a capacitor resistor parallel combination functions as an integrator. The charge amplifier is a type of electronic current integrator that generates a voltage proportional to the charge injected at the input side of the total integrated input current [22, 23]. It is commonly used at the output side of a sensor where charge output of the sensor (mainly piezoelectric sensors and photodiodes) needs to be converted to proportional voltage. Figure 9.12 denotes the schematic diagram of a charge amplifier along with a sensor at input side.

According to KCL,

$$q_{in} = -q_f \tag{9.29}$$

The feedback capacitor governs the amplification,

$$V_{out} = \frac{-q_f}{C_f} \tag{9.30}$$

9.2.6 Applications of Amplifiers

There are a wide variety of applications of op-amps in sensor circuits. In the case of general inverting and noninverting op-amps, the input voltage signal (say, pressure) is amplified with desired amplification factor by selecting the inverting and noninverting points. In the case of voltage follower, it acts as a buffer circuit and also plays an important role in impedance matching of the circuit. This voltage

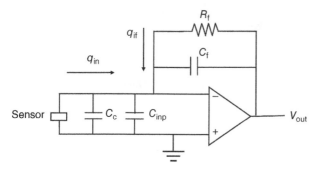

Figure 9.12 Schematic diagram of a charge amplifier along with a sensor at input side.

follower can be connected between high impedance source and low impedance load [24]. Instrumentation amplifier is commonly used in measurement and test equipment because it eliminates the need of input impedance matching. Charge amplifiers are commonly used at the output of a sensor which outputs charge [25]. These charge amplifiers produce output voltage proportional to induced input charge.

9.3 Excitation Circuits

Active sensors require external electricity to operate [26]. Power can be given in several forms such as constant current, constant voltage, or time changing current such as sinusoidal or pulsating current. The external power is referred to as the excitation signal. In many circumstances, the sensor's response and accuracy are directly dependent on the excitation signal. In this section, various excitation circuits are discussed with their respective circuit diagrams.

9.3.1 Current Generators

A current generator is an excitation circuit used to provide a constant current over a load regardless of external factors [27]. In terms of excitation, current generators are utilized as excitation to active sensors in a variety of applications. This is due to their useful property of supplying controlled amounts of current to the sensors. Ideally, these devices should generate an output current that follows the specified control signal provided by the user, without any error and should not be affected by the load. The key aspects of these devices are to have very large (as large as possible) output resistance and should provide the maximum required stable (i.e. unchanging) voltage across the load, i.e. voltage compliance. There are two types of current generators based on the direction of current flow:

1) Unipolar: flow in only one direction
2) Bipolar: flow in both directions

Both are based on op-amp operation and the circuits are given in Figure 9.13.

9.3.2 Voltage Reference

Another excitation circuit is the voltage reference circuit. The concept of this method is pretty much similar to that of current generators, just that now the voltage generated should remain constant/near to constant irrespective of factors such as power supply, temperature, and aging [28].

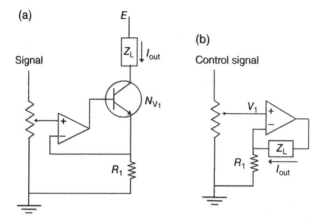

Figure 9.13 Op-amp-based (a) unipolar current generator and (b) bipolar current generators.

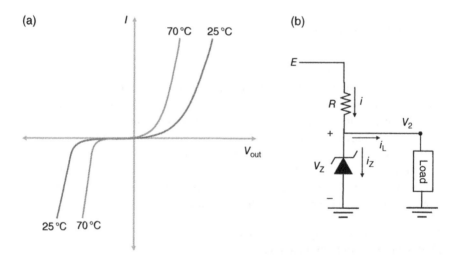

Figure 9.14 A circuit for voltage reference with its V-I characteristics. (a) I-V characteristics for different temperature and (b) Circuit diagram of a voltage reference circuit.

The Zener diode is connected in the reverse bias configuration and used for reference voltage. The circuit and the characteristics are depicted in Figure 9.14 for the reference. The reverse-biased Zener diode provides a constant voltage drop across the load, thus is chosen for the reference voltage. The current through the load is maintained according to the reference voltage V_Z. The *I–V* characteristics of the Zener diode with different temperature as shown in Figure 9.14a can be utilized to choose appropriate Zener diode for the desire value of V_Z at a given temperature range.

9.3.3 Oscillators

Oscillators are excitation devices used to produce periodic or variable electrical signals [29]. It consists of a gain stage and a positive feedback circuit wit embedded nonlinearity in the circuit. Due to positive feedback the oscillator is an unstable circuit by definition, but its timing characteristics must be steady or modulation of any predetermined function.

There exist three types of oscillators:

1) RC oscillators
2) LC oscillators
3) Crystal oscillators

Various multivibrators can be built with circuits containing logic gates such as NOR, NAND, and NOT gates. The circuits of square wave generators using logic gates and using comparator and operational amplifier are shown in Figure 9.15. When we mix these circuits with timers, then they are called relaxation oscillator circuits. To generate sine waves, we need to use LC networks as tuners of frequency and n-p-n transistors as amplifiers. The linear variable differential transformer is used for positioning the sensors and the sensor's transformer becomes a part of the oscillating circuit.

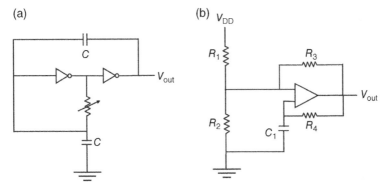

Figure 9.15 Square wave generators using (a) logic gates and (b) comparator and operational amplifier.

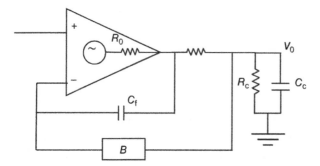

Figure 9.16 A driver circuit.

9.3.4 Drivers

Drivers are very similar to the voltage reference excitation circuits, but with very little dissimilarity that they must produce output voltages despite long-ranged loads and operating frequencies [30]. Figure 9.16 denotes a driver circuit. Based on the type of loads, the fabrication of drivers is dependent, such as when load is only of resistors, then a simple output stage with necessary current would suffice to be a driver. The real deal comes when the loads consist of capacitors and inductors, in which the design gets complex. Drivers are also called hard voltage sources.

To increase the tolerance of the driver stage toward the capacitance loads, we can isolate it by connecting it to a small resistor and capacitor.

9.4 Analog-to-Digital Converters

Another component in the field of readout circuit design for sensors is ADC. It is an important building block of the interfacing circuit as this one converts the analog signal from the sensor to digital output for further processing [31]. The digital output can be used for variety of subsequent processes such as alter, calculate, transfer, or store information as per requirement. ADCs are also very important components in signal processing and communication systems. In this section, basic concepts of ADC along with the working of various ADCs are provided below.

9.4.1 Basic Concepts of ADC

Analog signals are signals that contain a continuous stream of values. Almost all of the natural signals that are to be sensed by a sensor happen to be analog, i.e. continuous both in time and magnitude. A digital signal is defined as a sequence of

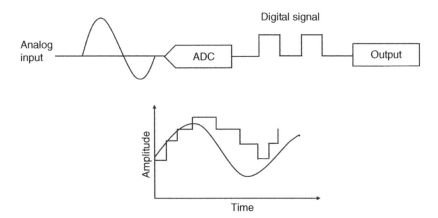

Figure 9.17 Typical block diagram of an ADC and a response associated with the same.

discrete values corresponding to discrete time series with appropriate sampling rate. ADCs are a sort of data converters that only transform signal from analog to digital. ADCs are frequently realized as serial or parallel converters [32]. Before being translated to the digital domain, an analog signal is converted to discrete form using a sample and hold circuit [33]. When a sample is compared to many quantization levels at the same time and converted to the digital domain in a single step, a parallel converter is utilized [34]. A serial converter is a system that compares sampled output to multiple quantization levels sequentially [35]. Figure 9.17 indicates the typical block diagram of an ADC and a response associated with the same. An ADC samples an analog waveform at regular intervals and assigns a virtual value to each pattern. The digital value is presented on the converter's output in binary code format. To calculate the value, divide the sampled analog input V by the source/reference voltage, then multiply by the number of digital codes.

9.4.2 V/F Converter

A voltage-to-frequency converter can be simply termed as an oscillator circuit whose frequency is modified by the control voltage. The VFC is monotonic and error free, with noise integration capabilities and low power consumption [36]. The VFC, which is small, inexpensive, and low powered, can be coupled to a subject and communicate with the counters via telemetry, as seen in Figure 9.18.

For a linear V/F converter, the relation can be given as:

$$f_{out} = GV_{in} \Rightarrow f_{out}/V_{in} = \text{Conversion factor}$$

Here, the frequency is directly proportional the provided input voltage.

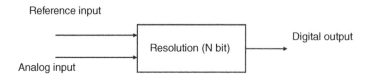

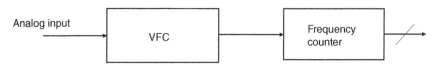

Figure 9.18 Schematic representation of voltage-to-frequency convertor.

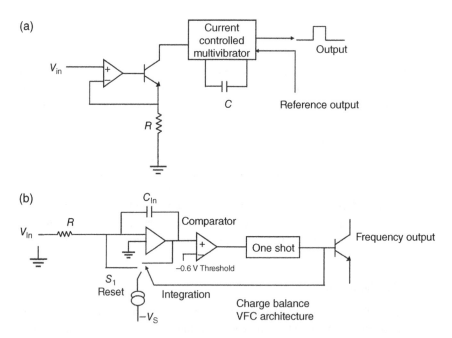

Figure 9.19 Architecture of VFCs (a) multivibrator type and (b) charge balance type.

There are two regularly used VFC architectures: (i) multivibrator type and (ii) charge balance VFC's [37]. Figure 9.19 denotes the two architectures.

In the multivibrator type V/F converter, the input voltage is connected to two voltage-to-current converters to have two current levels, i.e. i_a (charging) and i_b, respectively. The charging and discharging of a timing capacitor, i.e. each half cycle, is controlled by these currents i_a (charging) and i_b (discharging). By this process, the slope of charging and discharging of the capacitor are controlled and the desired output frequency can be obtained.

The charge–balance VFC employs a comparator, integrator, and precision charge source [38]. The integrator receives and charges the input. When the output of the integrator approaches the comparator threshold, it triggers the charge source, and in turn, the integrator generates a fixed charge. As the rate of the charge extraction is determined by its supply rate, the frequency of the charge source directly depends on the input to the integrator.

The multivibrator architecture has simple configuration and low power consumption, however, the charge balance architecture provides better high-frequency noise rejection facility.

9.4.3 Dual-Slope Converter

An indirect-type ADC does not convert analog signal into digital directly [39]. It turns the analog input into a linear function of time (or frequency) before producing the digital (binary) output in general. An example of an indirect type ADC is the dual-slope ADC. A dual-slope ADC creates an identical digital output for each analog input using the two (dual)-slope technique [40]. Figure 9.20 denotes the dual-slope ADC circuit.

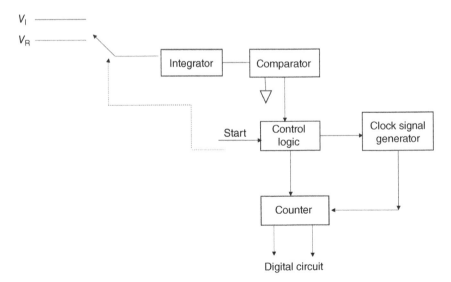

Figure 9.20 Dual-slope-type ADCs.

Here, the integrator generates two independent ramp voltages, V_A (analog input voltage) and $-V_{ref}$ (reference value). Therefore, it is named as a dual-slope *A–D* converter. When converting analog data to digital (binary) data, the dual-slope ADC is used in applications where precision is more important than speed. The major benefits of a dual-slope ADC are: averaging reduces noise on the input voltage, because they perform equally on the up-slope and down-slope, the capacitor and conversion clock settings have no influence on conversion accuracy. Linearity is good, allowing for extraordinarily high resolution of measurements. The biggest downside of dual-slope ADCs is their sluggish conversion rate, which is frequently in the range of 10 samples per second. In circumstances where this is not a concern, such as monitoring temperature transducers, a dual-slope ADC is a great solution.

9.4.4 Successive Approximation Converter

The principle of working of sequential approximation ADCs are analog-to-digital converters that undertake a binary search through all potential quantization levels before settling on a digital output for each conversion. The ADC of choice for low-cost medium to high-resolution applications is the successive approximation ADC [41]. Successive approximation ADCs offer sample speeds of up to 5 mega samples per second and resolutions varying from 8 to 18 bits (MSPs). It can also be designed in a small form factor with low power consumption, which is why this type of ADC is used in portable battery-powered equipment. Figure 9.21 denotes the successive approximation register ADC.

The SAR ADC performs the following duties for each sample:

It samples and saves the analog signals. The SAR logic delivers a binary code to the DAC for each bit based on the current bit under evaluation and the previously

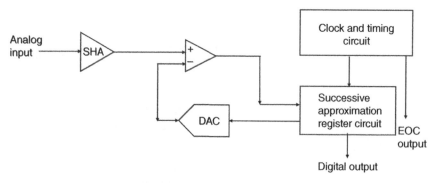

Figure 9.21 Successive approximation converter.

estimated bits. The comparator is used to determine the current bit's status. Once all bits have been approximated, the digital approximation is created at the end of the conversion (EOC).

The conversion time of SAC is fast, constant, and independent of the amplitude of the analog input signal V_A. The main disadvantages are that the circuit is intricate and the conversion time is longer than for a flash-type ADC.

9.4.5 Resolution Extension

The precision (resolution) of an ADC is its most important quality [42]. The higher the required precision, the more expensive the ADC. Higher ADC precision is attained by designing circuitry that quantizes the amplitude of the analog signal and converts it to a higher code-word digital signal. Practical ADCs have limited word lengths. Higher conversion accuracy is achieved by oversampling the low-resolution ADC integrated in a digital signal controller (DSC) and then processing the same with oversampled digital signal in software with the help of a digital filter and a decimator to successfully achieve a balance between system cost and accuracy. Oversampling is simply sampling a signal at a rate substantially faster than the Nyquist frequency. Although increasing the sample rate does not surely enhance ADC resolution, this approach more correctly monitors the input signal by better using the current ADC dynamic range by giving more samples. Oversampling increases the digital representation of the signal only up to the ADC's physical dynamic range limit (minimum step size), as should be obvious.

9.5 Noise in Sensors and Circuits

In this section, we are going to discuss about various types of noise present in the readout circuit that may be due to sensor or the circuit itself. This noise is responsible for obtaining false data from the sensor. So, it is useful to reduce or eliminate the noise present in the sensor and the circuit for best performance of the system and obtaining better results. Noises are of various types in nature. Those types are discussed here.

The general definition of noise is "unwanted disturbances of sound," but in electronics, it can be defined as the result of undesirable changes in voltage and current. In general, noise in an electronic system is classified according to its nature of origin. The following section will focus on the origin of different kinds of noise and different noise reduction techniques.

9.5.1 Inherent Noise

Noise is an extraneous signal that interferes with the necessary signal. It has no fixed frequency or amplitude and is highly unpredictable [43]. It can be classified into external noise and internal noise. External noise is generated from sources external to the system and cannot be analyzed quantitatively. Internal noise is generated within the system and is quantifiable; it is caused by the circuit elements like a resistor, transistor, and diode. Internal noise includes shot noise, partition noise, low frequency or flicker noise, high-frequency or transit-time noise, and thermal noise.

Thermal Agitation Noise (Johnson noise or electrical noise): Inside resistor, electrons are not stationary and is bouncing in some random direction [44]. Its velocity depends on the temperature. When we add all the random velocities of all of these electrons, it does not add up to zero instead it adds up to a net velocity. As these electrons have charge, there is a net motion of charge in the resistor, which implies there is a net current at any given moment, which sets up a voltage. In plain terms, variations in electron density cause variations in voltage across the resistor. This voltage fluctuation across the resistor causes thermal noise, also known as thermal agitation noise or Johnson noise.

The mean square value of noise voltage that is equivalent to noise power can be expressed mathematically as:

$$\overline{e_n^2} = 4kTR\Delta f \; (V^2/\text{Hz}) \tag{9.31}$$

Here, Boltzmann constant $k = 1.38 \times 10^{-23}$ J/K, T = temperature (K), R = resistance (Ω), and Δf = bandwidth (Hz). The root mean square value, i.e., $\overline{e_n}$ noise density per $\sqrt{\text{Hz}}$, more often used to describe Johnson noise.

Shot Noise (due to the random movement of electrons and holes): This type of noise is produced in the device where a potential barrier is present. For instance, noise level caused by the random arrival of the carrier in the PN junction of a transistor can be classified as shot noise [45]. Here, the current flow is not continuous; instead, the carrier moves in a random path. The noise is dependent on the current flow and not on temperature. The shot noise can be mathematically described as:

$$i_{sn} = 5.7 \times 10^{-4} \sqrt{I \Delta f} \tag{9.32}$$

Here, I = semiconductor junction current (picoamps); Δf = bandwidth (hertz).

Partition noise: Partition noise is generated when the current in a circuit is divided between two or more paths [46].

Flicker noise: This type of noise is generated when the charge carrier is randomly trapped and released between the interface of two materials [47]. This is also known as "1/f noise," as magnitude of this type of noise is higher at lower frequency. Its effect is more prominent below 100 Hz. At lower frequencies it may dominate over Johnson noise and shot noise.

Transit time noise: This type of noise is observed in devices such as semiconductors when the time period of the signal is compared with the transit time of a charge carrier while crossing the junction.

Electrical interference: The main reason for electrical interference is because of interference of electronic devices with the magnetic field present around the device. To remove this, the general approach is shielding and grounding [48, 49]. The elimination of ground loops takes up the majority of the effort in grounding and shielding. The details of ground loops are given in later.

Mechanical interference: It refers to the vibrations in the environment. Any sensor measurements are also sensitive to changes in atmosphere. The simple and easy way to eliminate such noise is to isolate the device from such vibrations. Using shock absorbents, temperature and humidity conditioning would be enough to remove these unwanted disturbances.

To represent the combined effect of all the noises, each of the noise voltage is squared and the mean is taken. Mathematically the root mean square value of all the noise voltages is often used to represent the combined noise effect from different sources. That is:

$$e_{noise} = \sqrt{e_1^2 + e_2^2 + e_3^2 + ... + e_n^2}$$

In below subsections, various types of noise reducing techniques or circuits are discussed.

9.5.2 Electric Shielding

Electric shielding helps in reducing electromagnetic interference [50]. Shielding is one of the means of reducing noise coupling. It helps in confining noise in a small region which prevents noise from getting into nearby critical circuits, i.e. it protects a specific region of space or any sensitive instrument from the influence of an external field produced by an electric charge. The shields can consist of metal boxes around the circuit region. The metal box acts as a Faraday cage under an electric field. Charge resulting from an external potential cannot exist on the interior of a closed conducting surface. Here, the cause of a noise can be due to an electric field or magnetic field. It is more difficult to shield noise due to the magnetic field.

The noise processes discussed earlier demonstrate that ground path inductance is critical when building a PCB. A multilayer board allows us to have solid ground planes, which reduces ground inductance dramatically.

9.5.3 Bypass Capacitor

It is another technique in reducing the noise from the circuit. An integrated circuit system powered by a direct current power supply is employed. If the input supply contains some ripple or noise, the performance of the integrated circuits (IC) suffers. If this noise is present, high-performance digital ICs such as microprocessors and FPGAs will have difficulty operating. The power supply rejection ratio is one technique to define an analog IC's sensitivity to power supply changes (PSRR). PSRR is the ratio of the change in output voltage to the change in power supply for an amplifier. As a result, high-frequency noise must be kept away from the chip. Capacitors can be used to reduce high-frequency noise.

The parasitic inductance and resistance of the PCB traces, wires, component pins, connection, and cabling can introduce noise to the signal. The influencing noise is in the form of alternating current. At DC voltage, a capacitor functions as an open circuit. When there is a very high frequency alternating current component in the signals passing through the circuit, capacitors operate as a short circuit and can be used to prevent noise from the subsequent circuit. It charges up to its maximum level when the digital IC is switching and providing some signal to load, then it needs a remarkable amount of current immediately, which is provided by the capacitors (the capacitors can act as a local charge reservoir to provide the circuits' charge during their transition state). A capacitor is connected parallel to the circuits where it will suppress or bypass signals. The bypass capacitors are also known as decoupling capacitors [51]. They act as frequency-dependent resistors [52]. Noise through the powerline is entirely unpredictable and changes in nature every time. These capacitors should be put as close to the IC as practicable. The closer the capacitor is to the load, the more noise is bypassed. A bypass capacitor is chosen based on different parameters like frequency range of operation, board space on PCB, and cost.

9.5.4 Magnetic Shielding

Magnetic shielding is an attempt to change the magnetic field [53]. Magnetic shielding is required when there are adjacent devices that could be susceptible to magnetic interference, such as cardiac pacemakers or other sensitive electronic or medical equipment. It is particularly important when low-frequency external magnetic fields are utilized nearby they need to be insulated from the MRI magnet's magnetic field or fringe field. The entire environment around the MRI magnet must be shielded from the fringe and magnetic fields in general. The widely used shielding device is Mu metal. Its composition of 80% nickel, 4.5% molybdenum, and balance iron makes it very permeable. This indicates that the material is very magnetically susceptible to an applied magnetic field.

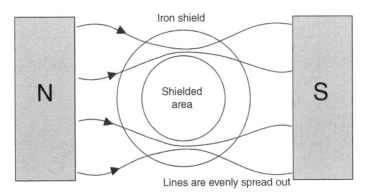

Figure 9.22   Magnetic shielding.

Figure 9.22 shows how a material is being shielded from magnetic waves. The magnetic field is attracted to the shielding material, which is one of the shield's important features [54]. The amount of material in the shield is also important, because the more material there is in the shield, the more magnetic field it can redirect.

9.5.5   Ground Planes

A ground plane on a printed circuit board (PCB) is a big piece of copper foil connected to the ground point of the circuit. It acts as the return path for current from a variety of components. The inductance of the signal return line is reduced by using a ground plane [55]. As a result, noise from transient ground currents is reduced. The signal always travels through the least impedance way. Consider infinite parallel thin traces as ground plane. The current will favor the traces with a lower impedance. We have talked earlier about return path and impedance; it is also important to understand how it reduces the noise. A simplified model for the return path inductance can be considered by putting an inductor in series with the ground of circuit layout. Figure 9.23 shows one example. The gate 1 output can be assumed as a logic high to low transition. As a result, the stored electric charge in C_{STRAY} will be discharged through the ground channel. The discharge will happen in a short amount of time (t) because of fast logic gates. The discharge current will pass through the inductance in the ground [56].

$$V = L\,(\Delta I / \Delta t) \tag{9.33}$$

Figure 9.23 denotes the above noise mechanisms. The inductance of the ground path is of paramount importance for PCB designing. It significantly reduces the ground inductance. The solid ground plane can be achieved with a multilayer board.

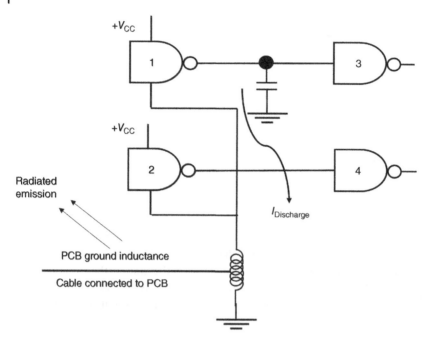

Figure 9.23 Circuit with ground plane.

9.5.6 Ground Loops and Ground Isolation

In general, a ground loop happens when two circuit points are expected to have the equal ground reference potential but have different ground potentials in reality. When enough current flows through the connection between the two ground points, a voltage drop occurs, resulting in two places with distinct potentials. Ground loops are a significant source of noise in electrical systems [27].

Common ground loops:

In an electrical equipment, hum and ground noise are signs of a ground loop caused by current through the ground or shield conductor of a wire [57]. In Figure 9.24, a signal cable S connects two electrical components, such as a line driver and receiver amplifiers (triangles). The chassis ground of each component is connected to the cable's ground or shield conductor. Signal V_1 between the signal and ground conductors of the wire (left) is applied by the driver amplifier in component 1. The signal and ground conductors are connected to a differential amplifier (right) at the destination and has an output voltage V_2. Subtracting the shield voltage from the signal voltage provide the signal input to the component that helps in eliminating the common-mode noise picked up on cable.

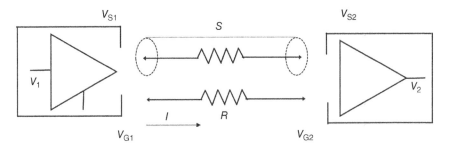

Figure 9.24 Schematic diagram of common ground loops.

$$V_2 = V_{S2} - V_{G2} \tag{9.34}$$

Isolation transformer:

An approach to reduce the ground noise and hum is the use of isolation transformer. Because it interrupts the DC connection between components while transmitting the differential signal over the line. No noise will be created even if one or both components are ungrounded (floating). Grounded shields are used between the two sets of windings in the better isolation transformers. In general, a transformer introduces some frequency response distortion. Figure 9.25 denotes the schematic diagram of isolation transformer.

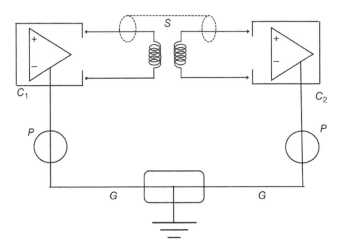

Figure 9.25 Schematic diagram of isolation transformer.

9.6 Batteries for Low-Power Sensors and Wireless Systems

Limited energy source is one of the challenges in wireless sensor systems [58]. The trade-off between energy and performance is one of the issues for electronic system design. We now understand that a wireless sensor network/system is made up of a base station and a number of wireless sensors (nodes) as shown in Figures 9.26 and 9.27 [59].

A wireless sensor system network must be compact in size, high in number, tether free, and low in cost.

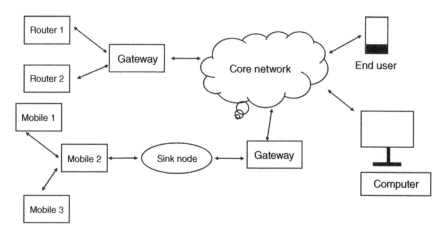

Figure 9.26 A node of wireless sensors network.

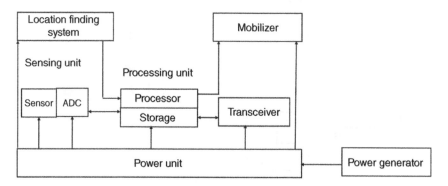

Figure 9.27 A complete wireless sensor system.

However, this is limited by energy, computation, and communication. A little battery means a small size. Low cost and energy consumption imply a low-power CPU, radio with a limited bandwidth and range. No maintenance or battery replacement is required with ad hoc deployment. One of the most essential concerns of WSN (wireless sensor networks) is scalability and reliability, which means that it should self-configure and be resistant to topology changes while maintaining connectivity and coverage.

The generic power profile of a node in WSN is [60]:

Here from Figure 9.28, the average power consumption can be calculated as:

$$P_{average} = DP_{active} + (1 - D)P_{sleep} \tag{9.35}$$

where D is duty cycle, i.e.

$$D = t_{on}/T. \tag{9.36}$$

So, lower the duty cycle, lower is the average power consumption.

Because these nodes are battery powered, each operation/data transfer pushes the node closer to death, as it will be difficult to change the batteries on a frequent basis in a remote deployment of such a network. We can see from the graph that lifetime of node is vital, and in order to preserve energy, we need a system that can: use sleep phase as much as feasible, only collect data if it is really necessary, make use of data fusion and compression, only transmit and receive if necessary, and less amount of receiving and sending of data.

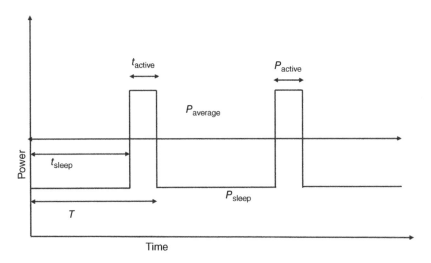

Figure 9.28 Power profile of a node in WSN.

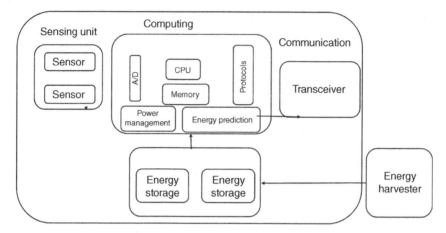

Figure 9.29 Block diagram of energy harvesting wireless node.

So, to check the issue, we need a proper or efficient battery and sensor system so as the power requirement is not much high and also the system provides good performance. For this, it is required to use power from batteries or harvest power from our environments natural resources and could store this energy along in power unit as shown in Figure 9.29 and use this power for our WSN network would have major advantages in above-discussed challenges.

An ideal battery for a WSN could be the one which has small size, high energy density, if rechargeable then should have high current charging and discharging, have a long-life cycle, low impedance, and the battery should be mechanically and electrically robust.

Currently low-power batteries for WSN are used in large number, but there has been a shift from this to energy harvesting and storing in WSNs.

9.6.1 Primary Cells

Primary cells are the batteries that are designed to be used only once, i.e. they cannot be recharged or used again [61]. The batteries have limited lifetime/power. An electrochemical cell transforms chemical energy to electrical energy. Based on the chemical composition of the cells, it can be further divided. Some of the most commonly used primary cells are: carbon–zinc, alkaline, and lithium coin cells [62, 63].

So, if we see the lifetime comparison between nickel metal hydride (NiMH), alkaline, and lithium coin cells, then we can see these graphs in Figure 9.30 for 9 V battery. It can be observed that lithium coin cell would be preferred as they offer a flat discharge cover and works for a longer time. In WSN, Li coin cell is a better option as it has small size and offer high energy density [64].

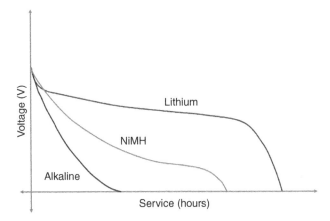

Figure 9.30 Graph plotted between voltage and service.

Now one of the issues with primary cell is also that the capacity of these batteries also changes with temperature, so we need to also keep this mind before deploying it. Cold temperatures cause the electrochemical reactions that take place within the battery to slow down and diminish ion mobility in the electrolyte, so cold temperature reduces battery voltage and runtime.

The main issue with primary cell is that it is limited and does not offer much usage and it increase the maintenance cost of our WSN [65]. Also, if we are not regular in changing batteries before they die, then we may lose critical information from the node as they would die out.

9.6.2 Secondary Cells

Secondary cells are ones that can be recharged and used repeatedly [66]. Now, the discharge and recharge occur via an electric current that is a reverse current, which aids in restoring the electrons to their original composition. Secondary cells, in general, have lower capacity and beginning voltage, a flat discharge curve, varied recharge life values, and a faster self-discharge rate. Flow batteries (e.g., where the cathode and anode are constructed of vanadium), lithium ion, lead acid batteries, and so on are popular secondary cells. Flow batteries are preferable in terms of life span, generally lasting for about 20 years (charge and discharge).

9.6.3 Energy Harvesting for WSN

The main problem with the batteries is that they have limited time span so a viable method is required for our sensor systems. Hence, from above we see that the primary and secondary cells are not a good or viable option if our deployment of ESN (energy harvesting WSN) is in remote areas. Thus, we need to use energy harvesting techniques which could harvest energy from available resources and used for the nodes. Now one of the most popular energy harvesting methods is by using solar-powered cells or solar panel. With addition to this, we require energy storage elements which could be supercapacitors, etc. Solar energy harvesting gives us directly DC voltage which can be used by the nodes. A typical block diagram of energy harvesting wireless node is shown in Figure 9.29 [67].

As a result of this energy waste, ideal listening, packet collision, and control packet overhead during transmitting, receiving, and listening may occur. The power management techniques to avoid these are:

- Duty cycling: using this, energy waste due to ideal listening is minimized [68].
- Data aggregation: used to minimize the communication overhead. This technique reduces energy consumption by eliminating redundant data.
- Data compression: using this, data are sent in compressed form, i.e. low-size data require less energy for transmitting.
- Data prediction: algorithms are employed in this case to forecast future data based on historical data and factors. In this, data from only that node are collected which deviates from predicted data.
- Dynamic scaling: the supply voltage and frequency of a WSN subsystem are modified based on instantaneous and expected workload. Other energy harvesting methods which can be used are: piezoelectricity (provide high voltage), radioenergy harvesting (enable wireless charging, thermal energy, wind energy, etc.).

Power generated on the node should now be greater than average power used on the node. Also, for proper functioning at the node at any given time, the energy stored in storage should be greater than the energy generated and energy consumed differ.

Also, from various state of art it is observed that energy harvesting of solar power provides a better solution than the batteries.

Now use of a single source for powering WSN is unreliable as these energy harvesting methods also depends on conditions, e.g. solar energy will vary according to the time in a day, season, and location. Radiofrequency energy is almost available everywhere, but it is useful in low-power applications as it has low-power density. So, we need a hybrid system depending on the application of use. The best solution for this could be to employ a hybrid system, which is a combination of an energy harvester and these batteries, where the battery delivers power when peak power is required, or energy harvesting is unavailable. When the energy harvesting device is operating, it supplies the standby power to the sensor system and charges the battery.

List of Abbreviations

ADC	Analog-to-digital Converter
ARW	Angle Random Walk
CPU	Central Processing Unit
EOC	End of the Conversion
I/O	Input/Output
NER	Noise Equivalent Rate
Op-amp	Operational Amplifier
PCB	Printed Circuit Board
PSRR	Power Supply Rejection Ratio
SAC	Successive Approximation Converter
VFC	Voltage-to-frequency Converter
WSN	Wireless Sensor Network

References

1 Kilby, J.S.C. (2001). Turning potential into realities: the invention of the integrated circuit (Nobel lecture). *ChemPhysChem* 2 (8–9): 482–489.

2 Singh, L., Baidya, D., and Bhattacharjee, M. (2022). Structurally modified PDMS-based capacitive pressure sensor. *2022 IEEE International Conference on Flexible and Printable Sensors and Systems (FLEPS)*, Vienna, Austria (10–13 July 2022). IEEE, pp. 1–4.

3 Jukan, A., Masip-Bruin, X., and Amla, N. (2017). Smart computing and sensing technologies for animal welfare: a systematic review. *ACM Computing Surveys (CSUR)* 50 (1): 1–27.

4 Bhattacharjee, M., Middya, S., Escobedo, P. et al. (2020). Microdroplet based disposable sensor patch for detection of α-amylase in human blood serum. *Biosensors and Bioelectronics* 165: 112333.

5 Baidya, D. and Bhattacharjee, M. (2022). Substrate optimization of flexible temperature sensor for wearable applications. *IEEE Sensors Journal*.

6 Gardner, J.W., Guha, P.K., Udrea, F., and Covington, J.A. (2010). CMOS interfacing for integrated gas sensors: a review. *IEEE Sensors Journal* 10 (12): 1833–1848.

7 Schwartz, D.E., Rivnay, J., Whiting, G.L. et al. (2016). Flexible hybrid electronic circuits and systems. *IEEE Journal on Emerging and Selected Topics in Circuits and Systems* 7 (1): 27–37.

8 Rosenfeld, A. (1988). Computer vision: basic principles. *Proceedings of the IEEE* 76 (8): 863–868.

9 Smith, S.W. (1997). *The Scientist and Engineer's Guide to Digital Signal Processing*. San Diego: California Technical Pub.

10 Li, J., Liao, W., Choudhary, A. et al. (2003). Parallel netCDF: a high-performance scientific I/O interface. *SC'03: Proceedings of the 2003 ACM/IEEE Conference on Supercomputing*, New York, NY, USA (15–21 November 2003), IEEE, p. 39.

11 Wilson, J.S. (2004). *Sensor Technology Handbook*. Elsevier.

12 Chong, C.-Y. and Kumar, S.P. (2003). Sensor networks: evolution, opportunities, and challenges. *Proceedings of the IEEE* 91 (8): 1247–1256.

13 Lin, Y.-S., Hu, K.-W., Yeh, T.-H., and Liaw, C.-M. (2015). An electric-vehicle IPMSM drive with interleaved front-end DC/DC converter. *IEEE Transactions on Vehicular Technology* 65 (6): 4493–4504.

14 Lukic, S.M., Cao, J., Bansal, R.C. et al. (2008). Energy storage systems for automotive applications. *IEEE Transactions on Industrial Electronics* 55 (6): 2258–2267.

15 Stalling, W. (2010). *Computer Organization and Architecture*, 8ee. Pearson.

16 Ramirez-Angulo, J., Carvajal, R., Torralba, A., Galan, J., Vega-Leal, A., and Tombs, J. (2002). The flipped voltage follower: a useful cell for low-voltage low-power circuit design. *2002 IEEE International Symposium on Circuits and Systems. Proceedings (Cat. No. 02CH37353)*, Arizona, USA (26–29 May 2002), vol. 3, IEEE, pp. III–III.

17 Wu, W.-C., Helms, W.J., Kuhn, J.A., and Byrkett, B.E. (1994). Digital-compatible high-performance operational amplifier with rail-to-rail input and output ranges. *IEEE Journal of Solid-State Circuits* 29 (1): 63–66.

18 Carter, B. and Mancini, R. (2017). *Op Amps for Everyone*. Newnes.

19 Safari, L., Barile, G., Ferri, G., and Stornelli, V. (2019). Traditional Op-Amp and new VCII: a comparison on analog circuits applications. *AEU-International Journal of Electronics and Communications* 110: 152845.

20 Erdi, G. (1979). A 300 V//spl mu/s monolithic voltage follower. *IEEE Journal of Solid-State Circuits* 14 (6): 1059–1065.

21 Balashov, E.V., Ivanov, N.V., and Korotkov, A.S. (2020). SOI instrumentation amplifier for high-temperature applications. *2020 IEEE East-West Design & Test Symposium (EWDTS)*, Virtual (4-7 September 2020). IEEE, pp. 1–4.

22 Karim, K.S., Nathan, A., and Rowlands, J.A. (2003). Amorphous silicon active pixel sensor readout circuit for digital imaging. *IEEE Transactions on Electron Devices* 50 (1): 200–208.

23 Vautier, B.J.G. and Moheimani, S. (2005). Charge driven piezoelectric actuators for structural vibration control: issues and implementation. *Smart Materials and Structures* 14 (4): 575.

24 Man, T.Y., Leung, K.N., Leung, C.Y. et al. (2008). Development of single-transistor-control LDO based on flipped voltage follower for SoC. *IEEE Transactions on Circuits and Systems I: Regular Papers* 55 (5): 1392–1401.

25 Fleming, A.J. and Moheimani, S. (2004). Improved current and charge amplifiers for driving piezoelectric loads, and issues in signal processing design for synthesis of shunt damping circuits. *Journal of Intelligent Material Systems and Structures* 15 (2): 77–92.

26 Wang, S., Lin, L., and Wang, Z.L. (2015). Triboelectric nanogenerators as self-powered active sensors. *Nano Energy* 11: 436–462.

27 Wang, J., Wu, Z., Pan, L. et al. (2019). Direct-current rotary-tubular triboelectric nanogenerators based on liquid-dielectrics contact for sustainable energy harvesting and chemical composition analysis. *ACS Nano* 13 (2): 2587–2598.

28 Dai, Y., Comer, D., Comer, D., and Petrie, C. (2004). Threshold voltage based CMOS voltage reference. *IEE Proceedings-Circuits, Devices and Systems* 151 (1): 58–62.

29 Auston, D.H. and Nuss, M.C. (1988). Electrooptical generation and detection of femtosecond electrical transients. *IEEE Journal of Quantum Electronics* 24 (2): 184–197.

30 Allen, P.E. and Holberg, D.R. (2011). *CMOS Analog Circuit Design*. Elsevier.

31 Murmann, B. (2006). Digitally assisted analog circuits. *IEEE Micro* 26 (2): 38–47.

32 Walden, R.H. (1999). Analog-to-digital converter survey and analysis. *IEEE Journal on Selected Areas in Communications* 17 (4): 539–550.

33 Dai, L. and Harjani, R. (2000). CMOS switched-op-amp-based sample-and-hold circuit. *IEEE Journal of Solid-State Circuits* 35 (1): 109–113.

34 Gersho, A. (1978). Principles of quantization. *IEEE Transactions on Circuits and Systems* 25 (7): 427–436.

35 Ishikawa, H., Nakahara, T., Sugiyama, H., and Takahashi, R. (2013). A parallel-to-serial converter based on a differentially-operated optically clocked transistor array. *IEICE Electronics Express* 10.20130709.

36 Gilbert, B. (1976). A versatile monolithic voltage-to-frequency converter. *IEEE Journal of Solid-State Circuits* 11 (6): 852–864.

37 Kester, W. and Bryant, J. (2009). voltage-to-frequency converters. *MT-028 Tutorial*.

38 Stork, M. (2002). New/spl sigma/-/spl delta/voltage to frequency converter. *9th International Conference on Electronics, Circuits and Systems*, Dubrovnik, Croatia (15–18 September 2002), vol. 2, IEEE, pp. 631–634.

39 Schuette, D.R. (2008). *A Mixed Analog and Digital Pixel Array Detector for Synchrotron x-Ray Imaging*. Cornell University.

40 Jafari, H., Soleymani, L., and Genov, R. (2012). 16-channel CMOS impedance spectroscopy DNA analyzer with dual-slope multiplying ADCs. *IEEE Transactions on Biomedical Circuits and Systems* 6 (5): 468–478.

41 Mortezapour, S. and Lee, E.K. (2000). A 1-V, 8-bit successive approximation ADC in standard CMOS process. *IEEE Journal of Solid-State Circuits* 35 (4): 642–646.

42 Abumurad, A. and Choi, K. (2012). Increasing the ADC precision with oversampling in a flash ADC. *2012 IEEE 11th International Conference on*

Solid-State and Integrated Circuit Technology, Xi'an, China (29 October–1 November 2012), IEEE, pp. 1–4.

43 Pal, S. and Mitra, M. (2012). Empirical mode decomposition based ECG enhancement and QRS detection. *Computers in Biology and Medicine* 42 (1): 83–92.

44 Mingesz, R., Vadai, G., and Gingl, Z. (2014). What kind of noise guarantees security for the Kirchhoff-law–Johnson-noise key exchange? *Fluctuation and Noise Letters* 13 (03): 1450021 https://doi.org/10.1142/S0219477514500217.

45 Friedrich, O., Halder, A., Boyle, A. et al. (2022). The PDF perspective on the tracer-matter connection: lagrangian bias and non-Poissonian shot noise. *Monthly Notices of the Royal Astronomical Society* 510 (4): 5069–5087.

46 Esho, I., Choi, A.Y., and Minnich, A.J. (2022). Theory of drain noise in high electron mobility transistors based on real-space transfer. *Journal of Applied Physics* 131 (8): 085111.

47 Shein-Lumbroso, O., Liu, J., Shastry, A. et al. (2022). Quantum flicker noise in atomic and molecular junctions. *Physical Review Letters* 128 (23): 237701.

48 Verma, S., Mili, M., Dhangar, M. et al. (2021). A review on efficient electromagnetic interference shielding materials by recycling waste – a trio of land to lab to land concept. *Environmental Science and Pollution Research* 28 (46): 64929–64950.

49 Zhu, S., Shi, R., Qu, M. et al. (2021). Simultaneously improved mechanical and electromagnetic interference shielding properties of carbon fiber fabrics/epoxy composites via interface engineering. *Composites Science and Technology* 207: 108696.

50 Jaroszewski, M., Thomas, S., and Rane, A.V. (2018). *Advanced Materials for Electromagnetic Shielding: Fundamentals, Properties, and Applications*. John Wiley & Sons.

51 Hosseini, M. and Bardin, J.C. (2021). A 1 mW 0.1–3 GHz cryogenic SiGe LNA with an average noise temperature of 4.6 K. *2021 IEEE MTT-S International Microwave Symposium (IMS)*, Atlanta, GA, USA (7–25 June 2021), IEEE, pp. 896–899.

52 Mishonov, T.M., Dimitrova, I.M., Serafimov, N.S. et al. (2021). Q-Factor of the resonators with frequency dependent negative resistor. *IEEE Transactions on Circuits and Systems II: Express Briefs* 69 (3): 694–698.

53 Karadakov, P.B. and VanVeller, B. (2021). Magnetic shielding paints an accurate and easy-to-visualize portrait of aromaticity. *Chemical Communications* 57 (75): 9504–9513.

54 Zhang, Z., Zhang, Y., Ming, W. et al. (2021). A review on magnetic field assisted electrical discharge machining. *Journal of Manufacturing Processes* 64: 694–722.

55 Moon, J.H., Fadda, D., Shin, D.H. et al. (2021). Boiling-driven, wickless, and orientation-independent thermal ground plane. *International Journal of Heat and Mass Transfer* 167: 120817.

56 Behi, H., Karimi, D., Jaguemont, J. et al. (2021). Novel thermal management methods to improve the performance of the Li-ion batteries in high discharge current applications. *Energy* 224: 120165.

57 Chen, L.H., Marek-Sadowska, M., and Brewer, F. (2003). Buffer delay change in the presence of power and ground noise. *IEEE Transactions on Very Large Scale Integration (VLSI) Systems* 11 (3): 461–473.

58 Azarpour, A., Suhaimi, S., Zahedi, G., and Bahadori, A. (2013). A review on the drawbacks of renewable energy as a promising energy source of the future. *Arabian Journal for Science and Engineering* 38 (2): 317–328.

59 Chae, M., Yoo, H., Kim, J., and Cho, M.-Y. (2012). Development of a wireless sensor network system for suspension bridge health monitoring. *Automation in Construction* 21: 237–252.

60 Ramadan, K.M., Oates, M.J., Molina-Martinez, J.M., and Ruiz-Canales, A. (2018). Design and implementation of a low cost photovoltaic soil moisture monitoring station for irrigation scheduling with different frequency domain analysis probe structures. *Computers and Electronics in Agriculture* 148: 148–159.

61 Hamm, A., Krott, N., Breibach, I. et al. (2002). Efficient transfection method for primary cells. *Tissue Engineering* 8 (2): 235–245.

62 Ovarfort, R. (1988). New electrochemical cell for pitting corrosion testing. *Corrosion Science* 28 (2): 135–140.

63 Gupta, S., Chang, C., Lai, C.-H., and Tai, N.-H. (2019). Hybrid composite mats composed of amorphous carbon, zinc oxide nanorods and nickel zinc ferrite for tunable electromagnetic interference shielding. *Composites Part B: Engineering* 164: 447–457.

64 Marks, T., Trussler, S., Smith, A. et al. (2010). A guide to Li-ion coin-cell electrode making for academic researchers. *Journal of the Electrochemical Society* 158 (1): A51.

65 Schlaermann, P., Toelle, B., Berger, H. et al. (2016). A novel human gastric primary cell culture system for modelling Helicobacter pylori infection in vitro. *Gut* 65 (2): 202–213.

66 Kamath, P.V., Dixit, M., Indira, L. et al. (1994). Stabilized α-Ni (OH) 2 as electrode material for alkaline secondary cells. *Journal of the Electrochemical Society* 141 (11): 2956.

67 Ulukus, S., Yener, A., Erkip, E. et al. (2015). Energy harvesting wireless communications: a review of recent advances. *IEEE Journal on Selected Areas in Communications* 33 (3): 360–381.

68 Carrano, R.C., Passos, D., Magalhaes, L.C., and Albuquerque, C.V. (2013). Survey and taxonomy of duty cycling mechanisms in wireless sensor networks. *IEEE Communication Surveys and Tutorials* 16 (1): 181–194.

Index

Solid-State Sensors, First Edition. Ambarish Paul, Mitradip Bhattacharjee, and Ravinder Dahiya.
© 2024 The Institute of Electrical and Electronics Engineers, Inc.
Published 2024 by John Wiley & Sons, Inc.

IEEE Press Series on Sensors

Series Editor: Vladimir Lumelsky, Professor Emeritus, Mechanical Engineering, University of Wisconsin-Madison

Sensing phenomena and sensing technology is perhaps the most common thread that connects just about all areas of technology, as well as technology with medical and biological sciences. Until the year 2000, IEEE had no journal or transactions or a society or council devoted to the topic of sensors. It is thus no surprise that the IEEE Sensors Journal launched by the newly-minted IEEE Sensors Council in 2000 (with this Series Editor as founding Editor-in-Chief) turned out to be so successful, both in quantity (from 460 to 10,000 pages a year in the span 2001–2016) and quality (today one of the very top in the field). The very existence of the Journal, its owner, IEEE Sensors Council, and its flagship IEEE SENSORS Conference, have stimulated research efforts in the sensing field around the world. The same philosophy that made this happen is brought to bear with the book series.

Magnetic Sensors for Biomedical Applications
Hadi Heidari and Vahid Nabaei

Smart Sensors for Environmental and Medical Applications
Hamida Hallil and Hadi Heidari

Whole-Angle MEMS Gyroscopes: Challenges, and Opportunities
Doruk Senkal and Andrei M. Shkel

Optical Fibre Sensors: Fundamentals for Development of Optimized Devices
Ignacio Del Villar and Ignacio R. Matias

Pedestrian Inertial Navigation with Self-Contained Aiding
Yusheng Wang and Andrei M. Shkel

Sensing Technologies for Real Time Monitoring of Water Quality
Libu Manjakkal, Leandro Lorenzelli, and Magnus Willander

Solid-State Sensors
Ambarish Paul, Mitradip Bhattacharjee, and Ravinder Dahiya

Printed and bound by CPI Group (UK) Ltd, Croydon, CR0 4YY